THE SMALL MATTER OF SUING CHEVRON

THE SMALL MATTER OF SUING CHEVRON

SUZANA SAWYER

DUKE UNIVERSITY PRESS — DURHAM AND LONDON — 2022

publication supported by a grant from
The Community Foundation for Greater New Haven
as part of the **Urban Haven Project**

Printed in the United States of America on acid-free paper ∞
Designed by Aimee C. Harrison / Project Editor: Lisl Hampton
Typeset in Portrait Text and Knockout by Copperline Books

Library of Congress Cataloging-in-Publication Data
Names: Sawyer, Suzana, [date] author.
Title: The small matter of suing Chevron / Suzana Sawyer.
Description: Durham : Duke University Press, 2022. | Includes
bibliographical references and index.
Identifiers: LCCN 2021036600 (print) | LCCN 2021036601 (ebook) |
ISBN 9781478015338 (hardcover) | ISBN 9781478017950 (paperback) |
ISBN 9781478022572 (ebook) | ISBN 9781478092612 (ebook other)
Subjects: LCSH: Chevron Corporation (2005)—Trials, litigation,
etc. | ChevronTexaco (Firm)—Trials, litigation, etc. | Peasants—
Ecuador—Social conditions. | Indigenous peoples—Ecuador—
Social conditions. | Environmental racism. | Petroleum industry
and trade—Environmental aspects—Ecuador. | Petroleum industry
and trade—Health aspects—Ecuador. | BISAC: SOCIAL SCIENCE /
Anthropology / Cultural & Social | HISTORY / Latin America /
South America
Classification: LCC HD1531.E2 S29 2022 (print) | LCC HD1531.E2 (ebook) |
DDC 305.5/63309—dc23/eng/20220114
LC record available at https://lccn.loc.gov/2021036600
LC ebook record available at https://lccn.loc.gov/2021036601

COVER ART: Photograph by the author.

This book is freely available in an open access edition thanks to
TOME (Toward an Open Monograph Ecosystem)—a collaboration
of the Association of American Universities, the Association of
University Presses, and the Association of Research Libraries—and
the generous support of the University of California, Davis. Learn
more at the TOME website, available at: openmonographs.org.

FOR ZOË...
my expanding pluriverse.

In nova fert animus mutatas dicere formas / corpora.
—P. OVIDUS NASO (8 AD), *Metamorphoses*

Of shapes transformde to bodies straunge, I purpose to entreate.
—Translation by ARTHUR GOLDING (1567)

Texaco's oil concession in the Ecuadorian
Amazon. Compiled from © 2021 Google Maps.

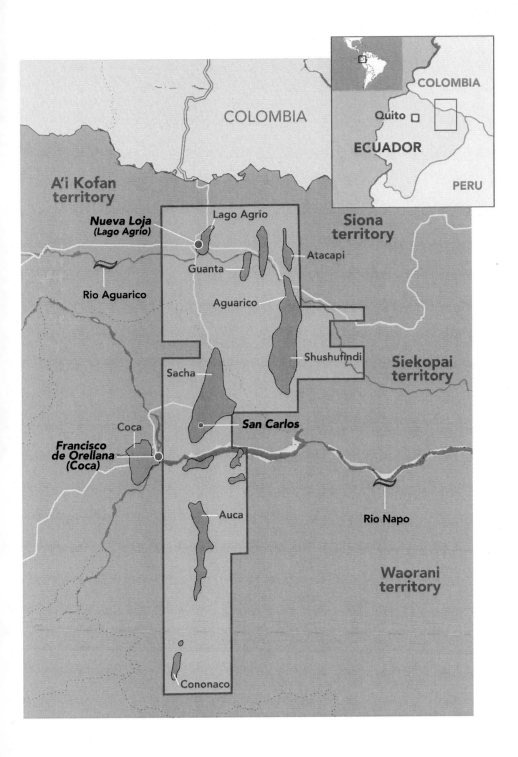

COLOMBIA

A'i Kofan
territory

Siona
territory

Nueva Loja
(Lago Agrio)

Lago Agrio

Atacapi

Guanta

Rio Aguarico

Aguarico

Shushufindi

Siekopai
territory

Sacha

San Carlos

Coca

Francisco
de Orellana
(Coca)

Auca

Rio Napo

Cononaco

Waorani
territory

COLOMBIA

Quito ☐

ECUADOR

PERU

CONTENTS

2001 — The US District Court for the Southern District of New York directs the lawsuit to be heard in Ecuador. (May)

2003 — The Lago Agrio plaintiffs sue Chevron in the Provincial Court of Justice of Sucumbíos. (May)

1993 — The Lago Agrio plaintiffs sue Texaco Inc. in the US District Court for the Southern District of New York. (November)

2001 — The Texaco and Chevron merger is finalized. (October)

2003 — The Ecuadorian litigation of the Lago Agrio plaintiffs' legal claim begins. (October)

TIME LINE

2011 — Chevron countersues the Lago Agrio plaintiffs in the US District Court for the Southern District of New York. (February)

2016 — The US Court of Appeals, Second Circuit, upholds the US District Court's 2014 ruling delegitimizing the Ecuador ruling. (August)

2009 — Chevron sues the Republic of Ecuador in the Permanent Court of Arbitration in The Hague. (September)

2013 — The trial of Chevron's countersuit commences in the US District Court for the Southern District of New York. (October)

2014 — The US District Court for the Southern District of New York finds the 2011 Ecuador judgment was procured through fraud. (March)

2011 — The Provincial Court of Justice of Sucumbíos finds Chevron liable for $9.5 billion in contamination cleanup costs. (February)

2018 — The tribunal of the Permanent Court of Arbitration in The Hague rules in favor of Chevron's bilateral investment treaty claim. (August)

ACKNOWLEDGMENTS

Researching and writing this book has been a passage, diverging and morphing over more than a decade. Along the way, many have generously helped shape this book—bringing their unique knowing to events surrounding this legal saga. Their stories, experiences, insights, and reflection sustained me during long episodes of excavating legal and technical archives in solitude.

My deepest gratitude goes to all those in Ecuador working on these court cases. The poise, acumen, and humor of the Ecuadorian legal team and the plaintiffs in this legal saga never cease to amaze. I am especially indebted to Pablo Fajardo, Julio Prieto, and Donald Moncayo. Each graciously indulged me over the years by spending hours discussing legal nuances, addressing my technical questions, and steering me through court documents and rainforest oil ruin. Emergildo Criollo, Mariana Jimenez, Toribio Piaguaje, William Lucitante, and Alex Lucitante generously shared their lived experiences, unease, and convictions in the face of excruciating odds. Over the decades, in various collectivities, these individuals have established two courageous organizations—the Unión de Afectados por Texaco and the Alianza Ceibo—that support environmental justice initiatives in the Ecuadorian Amazon. All proceeds from this book are directed to support them.

I am similarly indebted to the US side of the Ecuadorian plaintiffs' legal team. Conversations at different points with Cristóbal Bonifaz over the decades-long trajectory of this legal saga have made me all the wiser. I am especially grateful to the generosity of Steven Donziger and Aaron Marr Page. Both were unfailingly available to engage with my questions and to help track down obscure legal documents. The target of Chevron's unrelenting legal harassment, Steven has always shared with bigheartedness and humanity. It is my hope that this book (along with collaborative work with Lindsay Ofrias) reveals the corporation's legal campaign against Steven to be the unwarranted distraction it is and redirects scrutiny to Chevron.

A collective of Indigenous rights and environmental advocates, many with long-standing deep roots in the Ecuadorian Amazon, has been an unswerving source of support and assistance. Paulina Garzón's ethical clarity was an inspiration, beginning in 1993. The generosity and deeply committed work of Leila Salazar, Kevin Koenig, Paul Paz y Miño, and Atossa Soltani of Amazon Watch have provided a strong and sustained source of grounding since 2003. More recently, Mitch Anderson, Jenna Webb, Brian Parker, and Luke Weiss of Amazon Frontlines have opened ways of collaborative knowing for what might be. Last, the subtle practices of Lexie Groper of Amisacho Restauración engender healing worlds seen and unseen. The deep integrity and ethics that all of these individuals bring to their work in the northern Ecuadorian Amazon have been profoundly giving.

In the United States, I have been blessed by dear colleagues and friends. Marisol de la Cadena and Cristiana Giordano have been vast wells of warmth, inspiration, and provocation. Without their patient listening and challenging, this book would be otherwise. Kyu Hyun Kim, Alan Klima, Joe Dumit, Tim Choy, Tarek Elhaik, and Fatima Mojadedi have collectively and individually ever extended new ways of thinking and abiding. Similarly, a number of current and former graduate students have sustained me and my thought. I extend gratitude to Whitney Larratt-Smith, Kristi Onzik, Nathan Lane, Adam Kersch, Sofía Rivera, Jonathan Echeverri Zuluaga, Kristina Lyons, Jake Culbertson, Mariel Garcia Llorens, Duskin Drum, Stefanie Graeter, and Fabiana Li. I am especially indebted to Nathan Lane, whose attention to precision of thought helped bring this manuscript to production.

María Antonieta Guzmán-Gallegos needs a special thank you. Since my first fieldwork incursions in the Ecuadorian Amazon in 1992, María's grounded wisdom and insights of this region and beyond have been profoundly generative. Bill Durham also holds a special place. Bill's ongoing unwavering support since my decades-ago graduate years has been a rare, treasured gift. Other colleagues have also unsparingly supported this project, generating ideas to further nuance how to think about the controversies at the core of this book. Among them, I thank Andrew Barry, Doug Rogers, Paul Sabin, Sara Wylie, Matt Huber, Hannah Appel, Michael Watts, Arthur Mason, Andrea Ballestero, Dominic Boyer, John Sammito, Donna Goldstein, Lindsay Ofrias, Carol Greenhouse, Nancy Postero, John Andrew McNeish, Michael

Ziser, and Phil Kass for their generosity. Each has, in their own way, pushed me to hone my analytics and attend to more.

The embrace of deep sustaining friendships has been a continual source of replenishment: Marcus Grant (who passed too early from this life), Karina Morris, Guillermo Prado, Michael Moore, Eric Black, and Frauke Sandig have been rocks of stability, joy, and wonder. Jennifer Welwood has extended an embodied knowing that ever brings gifts. She and my sanghita sisters—Alejandra, Alexis, Annice, Amber, Amies, Charmaine, Karen, Janice, and Shanti—are tender community. Amrit Rai similarly has spun spirit.

Loving gratitude goes to my family near and far. Dorothy, Allan, John, Margaret, Leslie, Eric, Nancy, and Kirke have been breaths of love and laughter always. Helen, Mary, Anthony, Kate, Stefan, Florence, Lou, Jane, Steve, and Carolyn have recurrently extended delight and warmth. Josie and Kasya have made their zest for life infectious. Duncan has given in his own inimitable way in love and care. Zoë has embodied the density and vastness of love. I dedicate *The Small Matter of Suing Chevron* to her.

This book project has been generously supported by a fellowship from the American Council for Learned Societies, the University of California President's Faculty Research Fellowship in the Humanities, and the UC Davis Humanities Institute. Anitra Grisales was the first to read my manuscript-in-progress in its entirety and gave me the courage to push further. Lisl Hampton and Alejandra Mejía at Duke University Press meticulously attended to details to work their production magic. I am ever grateful for the fiercely gentle support of Gisela Fosado, editorial director at Duke University Press. Gisela's profound mind makes a difference that matters.

FRAUD

There is a video on the Chevron Corporation's website titled "The 'Legal Fraud of the Century' in 3 Minutes." It opens with an image of Steven Donziger, a longtime US advisory lawyer for Ecuadorian plaintiffs who, at the turn of the twenty-first century, had sued Chevron for contamination.[1] Seated with him are scientists who served as experts during the 2003–11 litigation against Chevron in Ecuador (figure 1).

Against gripping music, Steven's words sound: "Facts do not exist. Facts are created." One expert laughs. Across the screen in red letters emerges the word *FRAUD*. Next, *BRIBERY* is stamped on three still images—that of a scientific expert and of two former judges who had presided over the lawsuit. The photograph in the center is of Nicolás Zambrano, the Ecuadorian judge who found Chevron liable for $9 billion in February 2011 (figure 2).

FIGURE 2

Skillfully produced, over the next few minutes the video splices together a compelling cascade of "wrongdoing" in the Ecuadorian legal proceedings against Chevron. "Defending itself against false allegations" that the corporation was "responsible for alleged environmental and social harms in the Amazon region of Ecuador," Chevron countersued in the United States.[2] In 2014, the US District Court for the Southern District of New York "found [that] Donziger violated racketeering laws committing mail and wire fraud, money laundering, witness tampering, and obstruction of justice [in suing the corporation]. . . . Steven Donziger thought he was going to get rich by suing a big oil company" (figure 3). With a tinge of bravado, the final sentence

FIGURE 3

affirms: "But in the end, the US court's decision helped expose the fraud, the bribery, and most importantly, the *truth*."

In 2016, the US Court of Appeals for the Second Circuit upheld that "truth," and in 2018, under separate legal proceedings, the Permanent Court of Arbitration in The Hague did the same. Both judicial assemblies enacted a spectacular metamorphosis. The US court transformed a contamination lawsuit into a racketeering scheme, displacing attention onto a sole US lawyer. The Permanent Court of Arbitration transformed a contamination lawsuit into the breaching of a bilateral investment treaty and thus displaced liability onto the Republic of Ecuador. Codified in law, Chevron's fraud-of-the-century truth delegitimized a seven-plus-year litigation in Ecuador and a sovereign nation's judiciary. It made clear that neither should be taken seriously. *The Small Matter of Suing Chevron* intervenes to do precisely that.

Opening

Crude's Valence of Truths

Coalescence I

RENDERING TRUTHS

In 2004, the lawyers representing Ecuadorian *campesinos* (small farmers) and *indígenas* (Indigenous people) and the lawyers representing the Chevron Corporation followed the president of the Superior Court of Justice of Nueva Loja along a rivulet and down a precipice. Close behind were a court clerk, supporting counsel, technical experts, and local residents. The legal delegations stopped at the base of the ravine. Technical teams drove augers into and extracted samples from the swampy soils. And as the collective toured the site over the next day, lawyers expounded legal arguments on the presence or nonpresence of contamination, on the occurrence or nonoccurrence of cancer clusters, on the effectiveness or noneffectiveness of prior remediation, and on the existence or nonexistence of legal liability.

The legal entourage was in the midst of litigating a lawsuit against Chevron for Texaco's allegedly shoddy oil operations. Filed on behalf of thirty thousand Ecuadorian peasants and indígenas, the lawsuit alleged that Texaco (which had merged with Chevron in 2001) had used substandard technology to explore for and exploit hydrocarbons in Ecuador, and that this technology, in turn, systematically polluted the environment and endangered the health of local people. This was the first of over one hundred slated judicial inspections of former Texaco oil installations from which contamination allegedly seeped as a result of the company's operations between 1964 and 1990. In this, as with the other fifty-four judicial inspections that actually transpired during the litigation, legal teams disputed whether the crude oil visible in soils, embedded in sediments, and glistening on water was toxic. They disputed whether illness and poor health experienced by people living near oil facilities were the result of Texaco's activities. They disputed whether a remediation project undertaken by Texaco nearly a decade earlier was a sham. And they disputed whether a layered corporate subsidiary structure and previous state-corporate contracts shielded Chevron from liability.

In February 2011, after seven years of litigation, the president of the single-chamber court in Lago Agrio, a then rough and bustling Amazonian town, rendered a precedent-setting ruling. Judge Nicolás Zambrano found the Chevron Corporation responsible for polluting sizable tracts of the Ecuadorian Amazon and harming public health. He ordered that Chevron pay $8.646 billion in damages, monies to be used for "reparation measures": a sum soon increased to $9.5 billion to compensate the legal team.[1] Environmental justice movements around the world celebrated. The $9 billion fine made it the then largest ever to emerge from environmental litigation in history. And, for a while, the lawsuit was emblematic of the rapacious exploits of an arrogant oil company and the stalwart integrity of Indigenous, peasant, and green opposition.

On the eve of the 2011 Ecuadorian ruling, however, Chevron cried foul and filed a countersuit in a US district court. In March 2014, that court found that the Ecuadorian ruling had been procured through fraud, and the court placed an injunction on the ruling's enforceability in the United States.[2] Two years later, in August 2016, the US Court of Appeals for the Second Circuit upheld that ruling.[3] Concomitantly, in 2009, Chevron filed another legal claim with the Permanent Court of Arbitration (PCA) in The Hague against the Republic of Ecuador. In September 2018, the PCA's tribunal rendered its decision; in line with the US court, the tribunal determined that the republic had violated both justice and the US-Ecuador Bilateral Investment Treaty in upholding the Lago Agrio ruling. Irrespective of these actions by the US court and the investment tribunal, the 2011 Amazonian judgment against Chevron still stands—as it will into perpetuity.

Within the United States (and increasingly around the world), Chevron's legal-fraud worlding has succeeded in making "corruption" *the* optic through which to view the Ecuadorian litigation and judiciary. This worlding transmogrified an environmental contamination claim into a fraud and racketeering scheme (in the US counterlitigation), and then transmogrified it again into an international judicial and treaty violation (in the European counter-litigation). This successful deployment of "corruption" has had two primary effects. First, by branding the Ecuadorian lawsuit as fraudulent, it instructs that the Amazonian litigation need not be taken seriously. Second, by displacing questions of corporate wrongdoing, the "corruption" verdicts obscure the Amazonian litigation's far-reaching significance for transnational jurisprudence and environmental accountability. That the corruption optic seeks to foreclose further scrutiny is precisely why careful attention to how we reconcile challenging socioecological controversies—as well as make sense of formidable corporate adversaries—is called for. *The Small Matter of Suing Chevron* seeks to intervene toward that end.[4]

So how did it happen? How did an Ecuadorian judge assigned to a court-house in a city that started as a "jungle oil-camp," and still reeked, forgotten by twentieth-century petroempires, render a decision against Chevron, the second largest oil conglomerate in the United States and the fourth largest private petroleum company in the world? Rarely do complaints of contamination in marginalized places reach a court of law, let alone get litigated, much less prevail.[5] These litigations are not easy, often becoming spectacularly protracted. And how was it that after presiding over a seven-week bench trial, a US federal court judge delegitimized that decision three years later? A decision upheld on appeal by Ecuador's provincial Appeal Division, National Court of Justice, and Constitutional Court. Indígenas, campesinos, their lawyers, and their experts never imagined that corporate retaliation would condemn them to the underworld of mob extortion. And how was it possible—even after a decade of arbitration generated tomes of evidence supporting the Ecuadorian judgment—that an international tribunal ruled in concert with the US court's delegitimizing opinion and denounced the judgment internationally?

The Small Matter of Suing Chevron examines the processes that led to the precedent-setting Lago Agrio ruling and its tumultuous aftermath. Undoubtedly, the lawsuit in Ecuador—together with its New York pretrial hearings (from 1993 to 2002) and its countersuits in the same US federal court (from 2011 to 2016) and the PCA (from 2009 to 2018)—bears witness to daunting (at times debilitating) corporate opposition.[6] That Chevron has unleashed formidable legal defenses on both sides of the equator, as well as the Atlantic, is an understatement.

But the significance of the core litigation in Ecuador (2003–11) rests beyond the fact that, despite all, it found a multinational corporation guilty of negligence. The Ecuadorian litigation and the legal snarls that followed are momentous because they reflect how truths are legally scientifically made and also legally scientifically unmade. In Ecuador, the actual litigation from 2003 to 2011 intriguingly forged an inclusive, grounded, and experiential judicial assembly through which to address the scientific, technical, and legal controversies that too often mire contamination disputes. In the subsequent US forums, not only did the countersuit foreclose the possibility of such a judicial assembly, but also US legal procedures misattributed and demeaned the Ecuadorian judicial process and its radically different form. In good part, this was accomplished through Chevron's prodigious and fierce legal countercampaign, launched in the United States precisely when its prospects in Ecuador were not looking great. Yet it was also enacted through courtroom protocol and procedures. Held privately behind closed doors, Chevron's international arbitration further foreclosed possibilities of an inclusive judicial assembly. In late 2018, the PCA's tribunal not only embraced the ruling on the US federal court, but also significantly extended that opinion.

The Small Matter of Suing Chevron suggests that there is much to learn from these legal processes about crude's valence of truths—by which I mean the relational compositions through which truths are brought forth and consolidated. In the ruling of courts of law—in their *juris dictio*, that "language that speaks its performative authority into existence" and "simultaneously presupposes its power" (Richland 2013: 213)—legal truth is absolute in place and time. Upon considered deliberation, determining the finding of fact and application of legal principles, the language of law commands a singular legal truth.[7] A litigation may be complicated. A judge's decision may be complicated. But that judicial opinion is a coherent, rational, and inevitable legal truth based on the legal facts found and the legal principle applied. The verdict— the *verus dictum* (true saying)—is a "declaration or speaking of truth" (Constable 2010: 13).[8] In the juris dictio of courts of law, legal truth—and the facts determined legally veridical as a function of procedure and doctrine—is "found," not made. Through the work of "finding"—enabled by employing the "pragmatic warrants" substantiating legal authority—a court of law simultaneously settles which facts are "legally accepted" as true and renders *the* "authoritative account" to decisive effect (Mertz 2007: 67). This is the case, irrespective of the controversy surrounding a court's findings of fact or its role in authorizing the distribution of risks and harms (Beck 1992; Proctor 1995). There is, of course, a tension here. As Elizabeth Mertz and other legal scholars note, jurisprudence broadly recognizes that its truths are found as such by virtue of particular processes and procedures. And yet, in rendering a ruling and decreeing sanctioned penalties, judicial opinion often registers a conviction that implicitly attributes (more or less brazenly) "legal truth" to be real—or, at least, "highly authoritative as to epistemological certainty" (Mertz 2007: 67). Because the law invokes justice as fundamental to its being, *verus dictum* is normatively attributed to be *the* truth. That is, once loosened from its formal authority and exhausted of all processes of appeals, judicial opinion routinely lands as consequential truth in the world.

Exploring the Ecuadorian lawsuit against Chevron, its US countersuit, and its international arbitration, and in concert with the work of critical legal scholars (Eades 2008; Jacquemet 2009; Jain 2006; Jasanoff 1995, 2012), *The Small Matter of Suing Chevron* interrupts the notion that juris dictio "finds" the truth. Rather, the journey through these litigations follows how truth is *made* and also made complex. The paradox is that while the law cannot allow for the complexity of truths made, it partakes in enacting truth precisely as such. The task I have set for myself is, on the one hand, to show the complexity of the relationally contingent, sociomaterial compositions that produced legal truths over the course of these litigations. And, on the other, to signal the fraught tenor of law's complicity in the corporation's claim to being a sovereign moral actor.

The book's title, being plurivalent, seeks to signal this condition. For now, what's salient is the work of irony. Suing Chevron was clearly no small matter; it was a monumental feat that, in turn, generated monumental counterfeats. More trenchantly, that polite, if vaguely cheeky, rectifying formulation—"there is the small matter . . ."—tenders the proposition that the truths in these lawsuits were densely embedded in processes far exceeding their singularity or fixity, far exceeding the form of their authoritative rendering. The "small matter" seeks to rouse curiosity and materialize the contradictions inherent in liberal legality—that the objects of law and the subjects of law, far from being independent, autonomous, pregiven entities found with inherent facticity or will, all issue forth in practice as compositions saturated in relational contingencies.

The Small Matter of Suing Chevron proposes that the 2011 Ecuadorian judgment unfolded as it did in Ecuador not because, as Chevron claimed (and the US and international arbitration courts concurred), the Ecuadorian judiciary was corrupt. Nor did it find Chevron liable because, as the Lago Agrio plaintiffs (LAP) claimed, rightful science triumphed over evil, like David over Goliath. Rather, what I suggest was consequential in adjudicating this case were the procedures unique to Ecuador's civil law tradition, whereby geochemical, physiological, contractual, statutory, and experiential "facts" were both produced and subsequently argued as "evidence" to prove and disprove environmental, health, and contractual liability. Without doubt, stark disparities marked the lawsuit. And, clearly, missteps occurred. However, engaging in an analysis attuned to the complexity of the lawsuit, however, reveals that neither disparities, nor improprieties, nor scientific truth determined the legal outcome in Ecuador. Rather, the limitations and indeterminacies of science, the compromised quality of corporate contractual arrangements, the expanded modes of legal recognition, and the sociomateriality of "facts" and their making enacted a legal reality in Ecuador that led to this unparalleled and fiercely contested ruling.

COMPLEXITY'S AFFORDANCES

The Small Matter of Suing Chevron suggests that much has been lost in the US federal court and the PCA ruling that the Ecuadorian litigation was a "legal absurdity" (Chevron's phrase). Far from inconsequential, the nearly eight-year litigation against Chevron in Ecuador offers profound insights about truth and complexity. Countering Chevron's successful (in the United States and the world of commercial arbitration) "corruption" narrative, this book explores how (despite its flaws) the Ecuadorian litigation might serve as an instructive sociolegal assemblage for reckoning seemingly intractable contamination disputes.

The chapters that follow indirectly repudiate Chevron's successful legal counterclaim—and indeed the position of the US district court, the US court of appeals, and the PCA in The Hague—that portrayed the litigation in Ecuador as a sham and Ecuador's judicial system as corrupt. The text traces in detail how Ecuadorian court procedures gave form to a complex reality and made its slipperiness stable enough for judicial reasoning to distribute responsibility under the law. The "facts" generated from the court's unique judicial assembly of judicial inspections (fifty-four selected by the parties and nine by the court)—whether about hydrocarbon chemistry, disease etiology, business deals, laws of the Republic, local stories, or sensory perception—did not establish singular truths. Rather, the facts of chemistry hovered in the realm of the uncertain; those of disease in the realm of the indeterminant; those of contract in the realm of the dubious; those of law in the realm of the interpretive; those of testimony in the realm of the subjective; those of experience in the realm of the intuitive. They opened space for a reasoning that bowed toward precaution, among other things, to render a judicial decision.

The US court abided in and generated its own complexity, swayed by an imperious judicial hubris, the strictures of legal technique, and the litigating prowess of corporate lawyers. For one, it could not recognize the Ecuadorian proceedings as juris dictio. This was one of the foundations from which the US court determined that Chevron's accusations of subterfuge by LAP lawyers and Ecuadorian judges actually occurred.

Importantly, then, *The Small Matter of Suing Chevron*'s fundamental ethos is not to determine truths, or arbitrate truths, but to follow how truths were made. That is, the larger aim of this book is to understand the complexity by which the judicial rulings came to be made definitive. But what do I mean by complexity? A good jumping-off point might be a baseline understanding proposed by social theorists John Law and Annemarie Mol. Complexity exists, they suggest, when "things relate but don't add up," when "events occur but not within the processes of linear time," or when phenomena appear to "share a space but cannot be mapped in terms of a single" grid of intelligibility (2002: 1).[9] These sorts of predicaments infused both the litigation against Chevron and the corporation's countersuits. And this provokes reflection on the material, ethical, and legal conditions that juridical rulings and dispute parties sublimated in their determinations of truth. *The Small Matter of Suing Chevron* seeks to gather and recompose that which the law and those party to it often volatilized into the ether: the aberrant phenomena, the inconvenient practices, the dense repositories that made messy worlding processes.

So how were particular, highly contested legal truths derived? I explore this question by dissecting the competing facts that each side produced out of similar conditions. My analysis of the Ecuadorian litigation gives presence to a world in which absolutes rarely obtain, and yet decisions and actions are taken regardless. This is a world composed of complexity where, instead of certainty and fixity defining the ground from which to move, indeterminacy, unknowing, and ambiguity in part constitute the sphere of legal mastery. My analysis of Chevron's countersuit gives presence to a world in which reductive interpretation and discerned dissembling in part constitute the sphere of legal mastery. My analysis of the PCA arbitration gives presence to a world in which ambivalent legal technicalities, soldered together with performed misinterpretation, constitute the sphere of legal mastery. The court of each jurisdiction necessitated a distinctive mode of judicial mastery—techniques for assembling a unique jurisprudential grid of intelligibility—to render its legal truth. There was nothing straightforward about this, however. All knowing comes through method.

Holding that in mind and reworking Marilyn Strathern's words, it matters what method we use to think other methods with.[10] While researching and writing *The Small Matter of Suing Chevron*, chemistry recurrently surfaced as a problematic. As will soon be clear, chemistry influenced my thinking and informed my analytical point of entry. Leaning on chemistry, a discipline manifestly aware since the eighteenth century that its knowledge was methodologically mediated, underscored all the more the salience of method in my own knowing. A long line of philosophers of science suggests that what is known is contingent on the technical, procedural, and methodological circumstances of knowledge-making practice (Bachelard 1953; Canguilhem 1991; Rheinberger 2010). A number of science studies scholars extend this insight to propose not only that scientific methods describe their object of study but also that variable techniques of inscription produce it (Coopmans et al. 2014; Dumit 2004; Latour 1988; Latour and Woolgar 1987; Mol 2002; Vertesi 2015). The gnawing question: how can we seriously consider the proposition that method in the social sciences similarly describes and partakes in constituting its object of study (Latour 2005; Law 2004)? Chemistry inspired a method of delving into this legal trilogy—a legal saga whose meaning has been densely congealed through legal texts and public commentary—in a way that sustained an element of indeterminacy, openness, and surprise. It gave me alternative ways of inquiring and intervening, and for reflecting on how method manifests its object.

To begin, insights from chemistry shifted my analytical register away from a passion to denounce, away from a focus on overt power, away from a preoccupation with savvy charismatic characters, and away from the urgency to give voice to the forgotten. Others, scholars and journalists, have produced significant and moving work in these areas. Beyond the plethora of news articles on these legal claims and litigations, extended pieces of investigative reporting have appeared in *Vanity Fair* (Langewiesche 2007), the *New Yorker* (Keefe 2012), *Rolling Stone* (Zaitchik 2014), and the *Nation* (North 2015, 2021), each variously following two LAP lawyers, Steven Donziger and Pablo Farjardo. Two books present Chevron's case in a positive light (Barrett 2014; Goldhaber 2014a). A growing collection of anthropologists has analyzed the effects of Texaco's oil operations on Indigenous peoples (Cepek 2012, 2016, 2018; Krøijer 2017) and of the process of litigation on local populations, marginalization, and national identity (Fiske 2017, 2018; Ofrias 2017; Ofrias and Roecker 2019; Valdivia 2007). Law scholars have explored the cases' legal challenges (a limited list includes Alford 2012; Gomez 2013, 2015; Guamán 2019; Khatam 2017; Kimerling 1991b, 2006, 2013a, 2013b; Mella 2017), including an elegant case-study analysis (Aman and Greenhouse 2017). And a few documentaries detail the lawsuit, *Crude* (Berlinger 2009) being the most significant, with another currently in the making.[11] Ultimately, the legal team for the LAP will chronicle the events they lived. As may Chevron representatives as well.

Similarly, given that high emotions commonly infused discussion of this legal saga, thinking in parallel with chemistry directed me away from anticipated territories. Extending insights from chemistry compelled me to explore the lawsuit and the countersuits, and the discrete controversies within each, in ways other than through a lens of winners and losers, right and wrong, good and evil, exploited and exploiter, honorable humility and insatiable greed, noble redemption and shameless transgression. These labels, of course, have their place. But they easily flip and work for either side in the lawsuit and the countersuits, depending on one's moral persuasions and webbed positionality. Moving away from a good/evil analysis enabled me to think beyond the fact that what corporations do is simply lie. Not that they don't; lying was a recurrent practice. But what makes the oil industry, and Chevron in this case, so powerful is not that they lie about and falsify the real. It is that they generate entire worlds and those worlds enfold and recompose a plethora of entities and beings in coalescing "truths." Indulging in "the seductive clarity of denunciation" (Redfield 2005: 349) extends, for those of us in the petro-techno-zones of privilege, the illusion that we are not implicated in the very worlds the industry relationally elaborates. That is, it elides the dilemma that we are profoundly complicit in the very industry we condemn. The compulsion to denounce, rather than inspect, the relationships we sustain with and through crude oil is insufficient. Complicity invites discomfort and asks more of us—a tact, a discernment, a sensibility that eschews comforting binaries, hierarchies, and transcendence.

Having said this, one should not assume that condemnation, power, and the subaltern condition are absent from this book. Indeed, I take Chevron's relentless capacity in this legal saga as a sustained given. However—and here I recognize my debt to Michel Foucault (1980, 1995), Bruno Latour (2005; Latour and Woolgar 1987), and Gilles Deleuze and Félix Guattari (1987)—in *The Small Matter of Suing Chevron*, condemnation, power inequities, and the despair of contamination are effects of analysis, not the medium of analysis. Far from advocating agnosticism, the delve into complexity that I am encouraging in the chapters to come invites openness—becoming susceptible to what account-ability and respons-ability (cf. Haraway 2015) might entail.

In common speech, "valence" refers to the significance or emotional force that is generated, sustained, or repelled among entities in a particular context. It captures a realm of affective relationality and speaks to a bodied, intuitive, and prehensive capacity: that is, a capacity to discern a phenomenon, irrespective of cognition, such that it subsumes experience and transforms being (cf. Stengers 2011 on Whitehead). Valence registers an entity's relational proclivity toward being susceptible and responsive in rapport with others. Etymologically, "valence" comes from the Latin noun *valentia*, meaning "vigor, capacity, power," which derives from *valens*, the present participle of the verb *valere*, "to have strength, to be well." It resonates with a Deleuzian notion of affect, extending Spinoza's *affectus/affectio*: that state or mode of a body that slips along a passage of ever-growing and subsiding intensities as it simultaneously affects and is affected by another (Deleuze 1978). The concern being "the distribution" and "reciprocal influences" of affect across bodies (Jensen 2018: 32–33).

In chemistry, beginning in the mid-eighteenth century and extending for about one hundred years, chemists maintained that each known element had a fixed and specific valence. This valence was an affinity unit, or measure of rapport, numerically determined by the number of hydrogen atoms with which an atom of a given element could combine. Over the subsequent centuries, the chemical notion of valence became more nuanced. Today, valence can refer both to this simpler mechanistic definition (a fixed value) *and* to the combining power of chemical agencies more generally. Here the relative capacity of an element to connect, react, or meld—or to disavow or repel connection—is not static or stable but rather is ever-contingent on milieu. It reflects at its core a purely relational motive force. To speak of valence means to speak of a relationally constitutive reality in which entities are never singular or fixed but rather always emergences of collective composition. It is to hold the world and worlding as composed, in Marisol de la Cadena's words, of entities "with relations integrally implied" (2015: 32).

This book takes as its sphere of inquiry "crude's valence of truths." That is, leaning on chemistry and chemical philosophy, it delves into how competing truth facts at the core in this legal saga were, far from absolutes, emergences of collective composition: the often-arduous, agitated, viscously transformative combining effects of, with, and through crude oil. To be clear, I am not a chemist, nor do I claim chemical expertise. Rather, following the prodding of philosophers of chemistry, I seek to use chemistry—and, in particular, these scholars' writings on the historical practice of chemistry—as a muse capable of rousing novel discernment and leading me into the complexity of the amalgam between law, science, and crude. The world of chemistry is that of compositional entities. Chemistry offers a grammar for understanding as collective, for capturing the different modalities that constitute relational-being-ness, and for knowing that complex entities are never the sum of their parts. Chemical process and chemistry's insights give yet a denser imagination to the phrase "the small matter . . ."

But before I expand on configuring methods, let me provide some context.

GROUNDING SUBSTRATA

On May 7, 2003, forty-eight indígenas and campesinos filed a complaint in the Superior Court of Justice of Nueva Loja, soon to be renamed the Provincial Court of Justice of Sucumbíos, under the tort provisions of the Ecuadorian Civil Code (Articles 2214, 2215, 2229, and 2236) and the procedural authority of the 1999 Environmental Management Law (Ley de Gestión Ambiental).[12] Codified in 1861, the Civil Code provisions granted individuals the right to claim (either singly or via an *acción popular* [popular action]) that a tortfeasor remediate harm caused by negligent action. Thus, while the Civil Code long granted collectives the right to seek recompense from wrongs, the then-new Environmental Management Law elaborated on those rights and codified procedural rules.[13] It established the procedural regime by which individuals and groups affected by environmental degradation could pursue a legal claim on behalf of their communities with the intent of compelling remediation and recovering damages for environmental harm.

The lawsuit against Chevron was lodged on behalf of thirty thousand Indigenous and non-Indigenous inhabitants of the Ecuadorian Amazon. It alleged that Texaco had knowingly used substandard and obsolete technologies in its Amazonian oil operations between 1964 and 1990 and that these technologies systematically strewed industrial wastes throughout its vast oil concession, or area of operation. Over the course of Texaco's operations and into the unforeseen future, plaintiffs claimed, these industrial wastes threatened human and nonhuman well-being with death, disease, deprivation, and dislocation. And all to save a buck, the lawsuit alleged; implementing mid-twentieth-century, state-of-the-art technologies would have increased Texaco's per-barrel price of production and thus reduced profits. The company did not want either.

However, the life of the Lago Agrio legal claim predated its May 2003 filing in the Ecuadorian court. Indeed, the lawsuit was initially lodged against

Texaco ten years earlier in the District Court for the Southern District of New York, in November 1993. Greater detail of this history will soon follow, but for now suffice it to say that the lawsuit encountered a storied decade of pretrial hearings in the United States (between 1993 and 2002) as the case ricocheted back and forth, and back and forth again, between the US federal court and US court of appeals. In 2002, the federal courts sent the case to be litigated in Ecuador under specific conditions. Once in Ecuador, the case resided under the auspices of the Provincial Court of Justice of Sucumbíos, which held jurisdiction over the region in which Texaco's former oil operations resided. The actual litigation in Ecuador commenced six months later in October 2003. Over eight years (from May 2003 to March 2011), the case was overseen by the president of the court. Because the "presidency" was a rotating position and because two presidents recused themselves, in total six judges presided over the litigation, with two serving twice.[14]

As defined by the Environmental Management Law, the litigation process in Ecuador was divided into three distinct parts. The first day, October 21, 2003, hosted a "conciliatory hearing" aimed at finding a resolution between parties. When conciliation clearly was not in the cards—that first day, Chevron's counsel read its eighty-eight-page response in which the corporation contested the Ecuadorian court's competence and its jurisdictional authority, denied all alleged wrongdoing, and moved to dismiss the complaint—the case proceeded to the "evidentiary phase." This phase began with a week of court testimony and six days in which the parties outlined their requests for all present and future evidence they sought to prove their case. Parties requested documents, witness testimonies, and expert assessments, but most importantly they requested the onsite inspection of 122 allegedly contaminated oil-operation sites. These judicial inspections, and the extensive scientific labor associated with each, composed the bulk—five years—of the evidentiary phase. In 2010, the final judge, Judge Nicolás Zambrano, ended the evidentiary phase and embarked on the "judicial review and judgment" phase.

Ultimately spanning over two decades, three continents, and two legal systems, this legal saga is nearly overwhelming when one takes into account the Ecuadorian lawsuit against Chevron, Chevron's countersuit in the United States, the Chevron–Republic of Ecuador international arbitration in The Hague, and the multiplicity of derivative judicial proceedings. In 2012, Judge Gerard Lynch wrote, "The story of the conflict between Chevron and residents of the Lago Agrio region of the Ecuadorian Amazon must be among the most extensively told in the history of the American federal judiciary."[15] In terms of the volume of written pages and the size of case files, Lynch, a veteran judge, undoubtedly knows. Now multiply this. Both the case file in Lago Agrio and the one in The Hague similarly burgeoned beyond what the courts normally handled. While all three of these case files were related, they were far from identical; because they were different legal claims pursued within different legal traditions under different laws and procedures, each recognized and facilitated very different forms of garnering and submitting evidence. At its close, the case file in Ecuador alone was more than 230,000 pages—and much of the text on these pages was single-spaced.

Attending to this complexity is the focus of *The Small Matter of Suing Chevron*. Admittedly, my considered attention is partial. It does not engage every facet of this legal saga and it emerges through a particular method. As Strathern reminds us, "Ethnographic truths are similarly partial in being at once incomplete and committed" (2005: 39)—by which she means that there is always more data and that an analytic tack obliges distinct follow-through. Partiality, as I'll expand later, always entails particular connections. *Small Matter* explores specific dimensions that I feel are crucial to understanding crude's valence of truths. Given the authoritative weight of the US judicial systems and the ramifying consequences of its legal truths, the bulk of this book is an intervening rejoinder to the US court having delegitimized the 2011 Ecuadorian ruling. As such, the first five substantive chapters of this book descend into the density of the Ecuador litigation, suspending in the background puzzlement over Chevron allegations of Ecuador's judicial incompetence. Collectively these chapters suggest that the Ecuadorian litigation has much to teach about how to think: parts and wholes, sequences and compositions, individuals and mixts, precisions and veridictions, constrained and expansive relationalities. The final two chapters descend into the density of Chevron's countersuit and arbitration, respectively, and they signal how liberal legality can so brilliantly thrive on the more meager, isolate, and brittle terms of the pairs above. In doing so, these final chapters critique Chevron's legal-fraud worlding that both judicial bodies condoned and they surface, or distill, the compositional metamorphosis through which legal technique championed a reductive world.

The Ecuadorian litigation took place over seven-plus years and swirled, in my rendition, around three key controversies: (1) whether crude was toxic, (2) whether contamination had undermined human health, and (3) whether layers of contracts precluded corporate liability. The countersuits, each of different durations and intensities, coalesced, in my rendition, largely around misapplied chemistry, the contract form, and technicalities of law, variably enabling contamination concerns to percolate into their logic. This book's chapters (together with conceptual and empirical interstices) represent a latticework for thinking, in complexity, these dimensions and their excessive, unconsidered, and aberrant folds. Each chapter looks at practices that comprised and surrounded the litigation and its countersuits, focusing on how evidence and legal arguments came to take shape by virtue of litigation practices, judicial protocol, and legal philosophy. But the chapters don't seek to determine a truth around each key controversy. Rather, they seek to understand the larger scientific, judicial, and social debates within which and through which facts came to be fashioned and argued as legitimate legal truths. In my attempt to grapple with what was at stake, I have engaged in a plurality of research practices—transporting me to places that exceed the normative terrain of anthropology. I have examined and reexamined extensive court documents and legal doctrine. I have schooled myself in scientific debates around hydrocarbons, epidemiology, and environmental remediation. I have studied legal scholarship on corporate and contract law. And over the course of two decades, I have had in-depth conversations with lawyers, scientists, and indígenas and campesinos affected by Texaco's former oil operations in the Amazon.

Singularly striking when comparing the court documents from the Ecuadorian and US litigations is the recognition that different judicial traditions distinctively conditioned how facts could emerge and be argued as evidence. In Ecuador, controversies over the toxicity of crude, oil's effect on health, and a contract's capacity to dictate closure unfolded within the context of Ecuador's civil law tradition. Civil law is an "inquisitorial system" of law. Among other things, in Ecuador this meant that the court itself—along with opposing legal teams—was charged with investigating the issues at stake. As such, the bulk of the trial consisted of five years of onsite, official judicial inspections of oil-production sites. At each site, the judge, legal teams, scientific crews, local residents, and the press trekked through scrub forest to examine alleged contamination and its effects on human health. During each judicial inspection, technical teams retrieved water and soil samples for chemical analysis, local residents gave testimony about oil's incursions into their lives, legal teams advanced arguments to establish or absolve corporate liability, and the judge and his clerk viscerally experienced the sight, sound, smell, and feel of former Texaco operations. The judicial inspections thus served as the ground from which what would be argued as evidence—an array of sensory, geochemical, engineering, narrative, epidemiological, contractual, and statutory matter—emerged and was admitted to the court. The effect was that, in Ecuador, scientific controversy, far from curtailing judicial action (as often has occurred in US toxic torts), combined (generating unanticipated force) with sensory experience, oral testimony, and national statutes to form the basis for taking legal measures.

Chevron's countersuit, a seven-week bench trial litigated under the US common law system, focused on whether the Ecuadorian ruling had been procured through fraudulent actions. An "adversarial system" of law, common law litigation hinges significantly on the staging of a spectacle before a judge (and often jury) in which legal technologies and technicalities, lawyerly skill, witness preparation, and litigation financing, as well as judicial prerogative, can all shape legal proceedings and outcomes. At one level, the corporation's legal firepower outlitigated the Ecuadorians and their lawyers. With infinite economic resources, savvy corporate lawyers far outpaced their opposition and, splicing together improprieties garnered from the universe of their opponent's case documents (having severed attorney-client privileges), produced a near-airtight and convincing-enough narrative of partial truths. The impressive but often-overwhelmed LAP legal defense team, negligible witness preparation, English-to-Spanish translation problems, and restrictions on submitting evidence crippled the Ecuadorians—all giving greater plausibility to Chevron's convincing-enough narrative. The effect was that, in the United States, the immense force (think the thermodynamic energies of Chevron's two thousand counsel detailed on extensive discovery actions, unprecedented witness protection and preparation, and exquisite lawyering) needed to decompose the complexity of the Ecuadorian litigation and effectively recompose it through reductive, constrained, sequential elements had formed the basis of taking legal measures, despite never, not once, demonstrating substantive evidence of fraud.

The story, of course, does not end there; US courts do not have the final word. The US district court ruling is neither binding nor enforceable abroad. However, this is why the decision of the PCA in The Hague is so disturbing. And this is precisely why it is important to generate methods that interrupt the dominant trope for making sense of this legal saga. In his volume *Chemical Philosophy*, philosopher Manuel DeLanda remarks "what we consider real varies depending on whether we think of reality as that which we can correctly represent, or as that which we can affect and which can affect us" (2015: 186).[16] Although publicly committed to the former, Chevron's process was that of the latter. I, too, have espoused the latter. The difference: Chevron's method sought to reduce. Mine seeks to expand.

Coalescence II

CONFIGURING METHODS

When I first began exploring the case file of the Ecuadorian trial, I was perplexed by one of its primary questions: did the crude oil embedded and often visible in the landscape of Texaco's former oil concession contain dangerous elements? Indeed, I was befuddled by the amount of energy and concerted effort that scientists and lawyers had expended to demonstrate whether crude oil was, or was not, toxic. Wasn't contamination obvious and easy to prove? Quickly, it became apparent that my running assumption ("of course, crude oil is toxic") was naively inadequate for grasping the scientific and legal problematics at stake. My confusion only intensified as I delved further and looked at the actual data and analyses emerging from hundreds of soil and water samples collected during the trial's judicial inspections (which took place from January 2004 to March 2009). That confusion pushed me to learn about the actual chemistry of hydrocarbons and, more broadly, to read scholarship on chemical philosophy.

As noted, a significant focus of the judicial inspections was extracting material samples from the zone of Texaco's operations and analyzing their molecular content. But how was it possible for each party to make diametrically opposed claims about the reality of the material substances at the same judicial inspection sites when those claims were based on its chemistry? Of all the sciences, chemistry was, in my mind, a well-established, elemental science, hardly controversial. Of course, conflicting positions could arguably have resulted from one side or the other tampering with samples.[17] But especially in the early years of judicial inspections, the scientific results of laboratory assays detailing the molecular compounds found in soil and water samples were not dramatically different between parties. That is, the raw data that each legal party generated were not significantly dissimilar. This suggested that something else was to be learned from interrogating the systematic logic behind opposing arguments and that I needed to pry into the chemistry of crude oil.

The question of chemistry—and the chemistry of crude, to be specific—was not merely confined to the analysis of allegedly contaminated field samples. As I delved deeper into this legal saga, the question of chemistry did not disappear. Rather, it proved significant for understanding the configuration and trajectory of the lawsuits in general. As such, the problematic of chemistry emerges in every chapter. So, some words on configuring a method. Bear with me as I indulge for a moment in a bit of chemical philosophy. My intention is to invite you to consider some perhaps less-familiar substrates of thought.

CHEMICAL PHILOSOPHIES I

Since its alchemical beginnings, the practice of chemistry has long been concerned with the operations involved in reducing bodies into and reconstituting them from their constituent parts.[18] With the "chemical revolution" of the late eighteenth century, this pursuit came to be chemistry's defining project. Led by Antoine-Laurent Lavoisier, the popularly acclaimed father of chemistry, a world where all matter derived from four fundamental elements—air, fire, earth, and water—transformed into a world constituted by a plurality of simple substances in combination. Building on the work of many predecessors (for instance, that of Robert Boyle a century prior) and many contemporaries (Joseph Priestley and Henry Cavendish, among others), Lavoisier embarked on a series of meticulous and laborious experiments (from 1772 to 1794). By the late 1770s, he upended the reigning phlogistonist theory of combustion (i.e., the idea that entities burn because a component of fire [phlogiston] inheres within them). In 1777, he isolated "eminently respirable air" (Lavoisier 1790: 37) from metal acids and demonstrated that combustion derived from combining with this "air." And in 1783, he demonstrated that when "respirable air" combined with a second "inflammable air," they formed water. In short, Lavoisier and his team threw Aristotelean fundamental elements into question.

Shortly following these experimental demonstrations, in 1789, Lavoisier, together with colleagues (Louis-Bernard Guyton de Morveau, Claude Louis Berthollet, and Antoine François de Fourcroy), christened the nascent discipline with a new "chemical nomenclature." The new naming system captured the conceptual rigor and experimental protocols erupting from Lavoisier's laboratory. This scientific nomenclature determined new isolated substances by their competence—that is, what they were able to perform. For instance, "respirable air" became *oxygène* (from the Greek *oxys* [sharp, acid] and *-genes* [creation, formation]) because it was thought to be a constituent of acidification. And "inflammable air" became *hidrogène* (from the Greek *hydor* [water] and *geinomai* [to bring forth, engender]) because of its capacity to engender water.[19] Similarly, the new nomenclature deemed that a compound be called by the sequence of its component elements. Thus, the combination "water" resulted from the correct ratio of hidrogène and oxygène combining, resulting in the chemical equation: water = hydrogen + oxygen in proper proportion.

As detailed in *Traité élémentaire de chimie* (1789), Lavoisier and others transformed the metaphysical arts of alchemy into a reproducible and quantifiable empirical science. The aim was to purify and distill natural substances into their simple component parts so as to describe, classify, catalogue, and analyze the resulting chemical elements. Contrary to the Aristotelean-influenced thinking of the time, elementary substances for Lavoisier were not a set of a priori givens. Rather, "simple bodies" (as he called them) were those that could no longer be decomposable, a state, he noted, that was contingent on the laboratory techniques available. Lavoisier's simple substances were actors, but actors whose performance hinged on the dispositions of a substance, elaborate laboratory instruments, and the skilled manipulation of the scientist in time-consuming trials.

Lavoisier and his team determined that "rapport" (often translated as "affinity") was a "new chemical character." *Tableaus de rapportes*, or tables of affinities, speckle his texts, delineating the descending order of substances obtained by virtue of combustion, dissolution, and distillation when combined with a third substance. The notion of "chemical affinity" sought to capture the relative gradient of force or "elective attraction" between different elements. Thus, not only were new elements identified by one thing they were capable of doing (i.e., oxygène and hidrogène), but elements were also seen as sophisticated agents with capacities in their own right. His "Tableau de la nomenclature chimique" of 1787 listed fifty-five simple substances clustered in groups according to their combining behavior or rapport when combined with oxygen, bases, and acid. This particular table was the rudimentary foundation of the contemporary periodic table.

In the early nineteenth century, John Dalton's theorizing of the atom gave a precision to Lavoisier's simple substances. Dalton hypothesized that matter was made of particles called atoms (from the Greek *atomos*, meaning uncut, unhewn, or indivisible), with each element composed of its own unique atom, always identical in mass and size. This was not the atom of quantum physics; rather, this chemical atom was the simplest unit necessary for combination. Under reactive conditions, Dalton theorized that atoms combined, separated, and rearranged but were never destroyed. By the midcentury, cumulating experimental results indicated that there were particular patterns to how the atoms of distinct elements combined. In 1857, Friedrich August Kekulé—a seminal thinker in the field of organic chemistry—asserted that an element had a fixed capacity to combine with other elements, and he called the measure of this fixed capacity an "affinity unit." For instance, hydrogen had one such unit (or "valence" as it was soon called), while oxygen had two units (or a valence of two). Kekulé concluded that the notation H_2O equaled two monovalent hydrogen atoms combining with one divalent oxygen atom.

About ten years later, in 1869, when Russian chemist Dmitri Mendeleev published his first rendition of how chemical elements fit together, "valence" proved crucial in determining the structure of the periodic table (both the "period" and "group"). Without ever seeing an atom, nineteenth-century chemists determined the valence of a given element on the basis of the number of hydrogen atoms with which that element combined or replaced in a compound. The principle of fixity and exactitude determined that valence was an intrinsic property of each element.

Over time, however, as chemists isolated more elements and laboratory instruments expanded experimental possibilities, chemists became increasingly aware that for most elements their modality and capacity to form bonds also fluctuated. Said another way, the valence of an element, while detailed precisely in the periodic table, was also not a fixed property; it could shift and change depending on its atomic structure, that of the atoms with which it combined, and the particular configuration of the emergent molecular compound. Most elements did not have an absolute valence. Matter, it would seem, was not simply the effect of an element's invariable motive force.

VALENCE

Let me pause here. What is to be learned from this historical-philosophical sketch? And why is it significant for thinking about a legal saga?

As taught in textbooks, an element is a member of a class of 118 pure, essential substances that constitute (either singly or in combination) matter. Their stability accounts for the periodic table—that elegant symphony of precision that orchestrates relations among the fundamental types that make up everything around us. The arrangement of elements in the periodic table coordinates a wealth of knowledge and fixes determinations of mass, weight, oxidation, and valence. Under conditions of experimentation, many of these measures stay constant; valence, however, may very well not.

As has been clear from Lavoisier onward, although elements are the fundamental components of chemical operations, they are not pure facts of nature. Their evincing is an artifact—the consequence of the art of actively engendering facts. They are what Bruno Latour calls a "factish" (1999) and, before him, what French philosopher of science Gaston Bachelard in 1953 called "facticious" (i.e., factitious; Bensaude-Vincent 2014: 70). Actualized through complex processes of chemical purification, the materialization of an element necessitates the intercession of chemical proclivities, skilled scientists, and elaborate instruments. Lavoisier underscored the relativity of elements precisely by defining them as contingent on the analytic techniques at the experimenter's disposal. This in no way undermined the presence of elements and their worldly consequences. It merely emphasized that elements were not passives in a world awaiting discovery.[20] This takes elements not as invariable building blocks of nature but as capacitated simple units "bound to laboratory operations" (Bensaude-Vincent and Simon 2012: 202), whose completeness as a set is indeterminate and whose incompleteness as a unit (expressed through its capacitated and malleable valence) animates openings.

Delving further: the structure of the periodic table emerged from the patterns that Mendeleev perceived after shuffling and coordinating the qualities of simple substances with their simple valence at unit weight. Today, students learn why the structure of the table makes sense in part because of how the electrons of each element differentially reside in atomic orbitals, the very configuration of which proffers to an element its combining power or valence. Explore chemistry beyond the table-derived laws and patterned functions, however, and exceptions abound in the hands of chemists. As a number of chemical philosophers note, an "element" of the periodic table does not exist as a reflection of the real (Bensaude-Vincent 1986, 2008, 2014; Bensaude-Vincent and Simon 2012; Bensaude-Vincent and Stengers 1996; Bernal and Daza 2010). Rather, it is an abstraction, a perfected ideal, that functions as a vital tool in a chemist's operations. As philosophers of chemistry Bernadette Bensaude-Vincent and Jonathan Simon note, "Elements, in contrast to simple and compound substances, have no tangible reality, they are abstract entities that cannot be touched or seen" (2012: 159).

As substance, elements exist only as enactments in relation, not as one in a sequence of essences in juxtaposed isolation, as depicted in the periodic table. And an enacted element's capacities change in relation. That is, capacity hinges on the properties of the emergent compound into which it forms. Consequently, valence (that combining power of chemical agencies) is not absolute, a set numerical index of behavior. Alone, in its regimented order on the periodic table, it is one thing; in different modes of collectivity, it can also be something else. That is because, in association, an element assumes agencies that exceed its behavior in isolation (which is an abstraction). This transformation (in singular, one thing; in assembly, something else) does not simply mean that valence as a quantity changes; this is not the mere numerical amplification of connections. It is that valence, as a capacity and modality, transforms. This means that the way of relating, the tempo and arrangement of combining, and the texture of melding all change. One might think of valence then not as a fixed quantity or expression but as a motive and emotive force that within the merged plurality of collective-becoming—beingness "with relations integrally implied" (de la Cadena 2015: 32)—marks orbitals of coalescence. Tracking de la Cadena (2017) once more, one might say an element is the abstraction that is, *if* it becomes with, and therefore is "not only" what it also is. Said otherwise, rapport, relational-being, is precisely what allows the element to *be*, to exist as the abstraction it not only is.

The task that I have set out for myself is to wade through crude's recombinant valence, the recombining capacities through which multiple truths were made in this lawsuit and countersuit. *Small Matter* takes crude oil as *its* vital element and chemistry as its method for considering crude's valence of truths. Following Lavoisier (and after him a bevy of science and technology scholars), it is clear there is nothing purely natural about crude.[21] It holds no pure fact. Crude oil is begotten of and contingent on a complex of human and nonhuman skill and cunning. It is a vibrant substance imbued with wily capacity, and it only and always exists through the intercession of molecular, chemical, geomorphic, human, and technological processes. It is a sociomaterial composite of atomic intensities, molecular configurations, subterranean geographies, scientific potentials, economic desires, industrial dependencies, ecological horrors, and chemical requirements and promise.

The chapters that follow delve into how what each party said and possibly knew—and presented as evidence and represented as "fact"—about crude was always bound to and by their context of production. Whether the concern was to determine or refute the possibility that crude was toxic, that contamination affected human health, or that a contract foreclosed liability, a constellation of chemical-technical-human operations coalesced the arguments of each legal side. Despite alleging to present pure facts, the parties' arguments were composed, factitious. A plurality of pulls and attachments—variously enrolling transmuting hydrocarbon compounds—differentially deciphered and gave meaning to the chemistry of crude-oil hydrocarbons, or differently devised and inferred epidemiological probabilities, or differently interpreted and site-verified the execution of contractual agreements. Both parties' legal arguments about the molecularly toxic capacity of hydrocarbons, *or* health-risk probability, *or* doctrinal certainties of legal contract can productively be thought of as compositional entities immanent of collectively contingent chemical, disciplinary, industrial, regulatory, legal, and extralegal processes, not of enumerated and sequential essential elements or facts. Conceiving of the legal arguments this way destabilizes any pretense of singular, fixed facticity. And it interrupts the conviction that only one among competing arguments can occupy that role.

In its practice and its doctrine, law demands definitives. Indeed, one of law's constraints (cf. Stengers 1997)—that is, one of the obligations and requirements through which law functions—is an exigency for absolutes. An exigency whose fulfillment demands much work. Parties allege acts, produce evidence to substantiate purported acts, and argue their claim. The court must, after exhausting all possible interpretations and satisfying all procedural criteria, secure a "finding of facts," determine the legal truth, and render a decision. Jurisprudence requires a single, authoritative, final resolution in space and time. What stands out is how within the juris dictio of liberal law, facts are *not* made—a statement that the US district court judge, Lewis Kaplan, pronounced during the RICO countersuit on a number of occasions. Facts simply *are*—the task is to find them. Consequently, each party presents its facts as truths. And in the lawsuits involving Chevron, each side seemed singularly unable to (indeed, could not) trace the production of its own facts and could only address those of the other side by undermining them and calling them corrupt. Facts, however, are rarely so guileless. Extending the work of critical legal scholars, *Small Matter* deploys chemical insights as method devices to grasp facts' compositional enactments and ask what is the effect of claiming otherwise.

When, in prerevolutionary France, Lavoisier decomposed compounds into simple substances, he meticulously monitored and measured his experiments. A curious pattern held: in a chemical reaction, the total mass of the product was the same as the total mass of the material with which he began; that is, an equal quantity of matter existed before and after the reaction. This further confirmed for Lavoisier an "equality" between the body examined and the substances obtained (1789: 141). As he wrote, the relation of a chemical reaction was that of a "faithful mirror" (*miroir fidèle*; 1787: 14). The logic was one of identity; the chemical equation signaled that a compound equaled its constitutive elements—the basis for the law of conservation of matter.

With chemistry's passage through the centuries, theories of atomic structure and molecular architectures finessed the understanding of the chemical compound. An electron's charge, position in energy level, and spin exceeded what a juxtaposed recitation of elements (i.e., H_2O = water; C_6H_6 = benzene) could disclose in accounting for a compound's properties. At the turn of the twentieth century, however, French physicist and mathematician Pierre Duhem worried that structural atomism, much like chemical nomenclature, was an analytical model that led chemists to "imagine that the reactants were actually present in the compounds formed by their reaction" (Bensaude-Vincent and Simon 2012: 196). Indeed, the entire logic of identity encouraged the illusion that hydrogen as such and oxygen as such—as separate entities—actually exist in water. They do not. Duhem wrote, "The chemical formula in no way expresses what really persists in the compound but rather that which is potentially there, that which can be extracted by the appropriate reactions" (2002: 151; Needham 2002: xvii). According to Bensaude-Vincent and Simon, Duhem wanted to hold on to the "enigma of chemical composition" (2012: 127), the conundrum that escaped both equivalent and structural logics.

An analogy made by Aristotle millennia before proved instructive. For Aristotle, the "mixture" was an aggregate in which individual parts come together, retain their identity, and form a new blend—as in barley and wheat in a mixture of grains (Bensaude-Vincent 2014: 67). But the "mixt" was the effect of a chemical reaction in which individual components were no longer decipherable as discrete entities in the constitution of a new body. In Duhem's analysis, the mixture was the combining of entities that retain their qualities (as if, in H_2O, hydrogen and oxygen are present as separate atoms of hydrogen and oxygen). By contrast, the mixt was the combining of entities that, when combined, no longer exist as unique isolates (i.e., hydrogen and oxygen combine to obtain H_2O, but subsequent to their reacting they are no longer present).

The condition of the mixt is "never the simple sum of the properties of its components" (Bensaude-Vincent and Simon 2012: 127). Nor is the mixt analogous to the whole being more than the sum of its parts; this is not the adding together of parts and then enhancing with an extra sprinkle. As Bensaude-Vincent and Simon explain, the quandary of the mixt entails "the conservation of matter accompanied by the emergence of novelty" (127). It is also the dissolution of the prior—as parts cease to exist as clear and distinct units. The conservation of matter implies equivalence. But how can there be "equivalence," really, when the sides neither add up nor exist simultaneously? Either there is the mixt and the properties of its constituents are lost, or there are the decomposed constituent properties and the mixt is lost. But there is never both. And they are not the same (Needham 2002).

Aristotle reconciled this impasse and the paradox of the mixt through a language of pluridimensional, incompatible co-abiding. Although upon chemical reaction, elements no longer exist in *actuality*, they still abide in the mixt in their *potentiality*. Duhem resurrected Aristotle's quandary of the mixt to underscore that, in his mind, "atomism and molecular architecture, the approaches that dominated organic chemistry at the beginning of the twentieth century, were incapable of providing an exhaustive explanation of chemical transformation" (quoted in Bensaude-Vincent and Simon 2012: 126). In the irreducible complexity of chemistry's world, a world "populated by individuals with a range of capacities to put themselves in relation with one another," the work is to understand how the entities at stake "exist not only in the mode of actuality but also in the complementary mode of potentiality" (2012: 209).

THE MIXT

Taking liberties with this chemical concept, and ratcheting up analysis ever so slightly, *Small Matter* takes the controversy and condition of the lived effect of oil contamination as a mixt. The litigation had to reduce the complexity of this mixt. Each legal side deduced facts to establish the truth about crude. Distinct elements in precise arrangements equaled the facts of toxicity, disease, and liability, which when added together sought to equal the truth of the mixt. The error of the chemistry student, however, is to think her equation (H + O + H = water), or molecular model, *is* the compound, *is* a figuration of the real. Similarly, the schooling of legal practice is to project, and then hold, enumerated facts as equal to, as a miroir fidèle of, the controversy at hand.

The Small Matter of Suing Chevron ventures two propositions. First, as noted earlier, facts *were* the processes that produced them. The chapters that follow trace how the facts enrolled by competing legal arguments—of chemistry to determine whether hydrocarbon contamination is toxic, or of epidemiology to determine whether petroleum operations cause cancer, or of contract law to determine whether layered business agreements grant the corporation immunity—were juridico-scientific assertions *and* contingent collective compositions. This does not mean that the facts were not real. Rather, it incites an inquiry into the processes of how these facts were made. Although skilled lawyering rendered facts as isolated absolutes, it was a labyrinth of sociomaterial-juridical techniques and commitments that stabilized them as such. As substance and dynamic forces, facts and their composite elements were already spinning in a constellation of preexisting relations, thwarting any notion that they rendered unmediated truths. This is what the phrase "crude's valence of truths" seeks to elicit: oil's constitutive relationalities that coalesced orbitals of truth.

A second proposition: The facts about toxicity, disease, and liability posed during the litigation were sorely ineffective in capturing the mixt—understood as the controversy and condition of the lived effect of oil contamination. Complex formations are not the sum of their parts—nor more than the sum of their parts—because collective compositions are not the effect of summation. To be complex is not the same as to be computable, despite summation being a preeminent legal method. The mixt exceeds deciphered elements because in-collective elements seep to be other than what they are in isolated narration.

As I suggest in the chapters that follow, the Ecuadorian court procedures, the plaintiffs, and the ruling judge all intuited this; and each variably proffered unconventional modes for effecting an impression of the mixt. By contrast, the US court procedures and ruling curtailed that imagination of facts and their combining. The concept of valence trains attention toward the intensities and emergences of collective coalescence. And it signals the effect of legal authority when unable to accommodate facts' factitious provenance.

In the 1860s, Kekulé and others determined that not all chemical reactions or processes were reversible. Organic compounds (molecules containing a carbon-hydrogen bond) could not easily be decomposed and then recomposed. Indeed, although late nineteenth-century chemists were able to reduce many organic compounds to the varying proportion of their constitutional elements—for example, benzene can be reduced to six carbon and six hydrogen atoms (C_6H_6)—reduction did not inevitably allow for reconstitution. The molecular assemblages derived from animal and plant worlds seemed to afford properties distinct from those derived from mineral worlds.

Organic compounds, it turned out, are composed of a relatively small number of elements (carbon, hydrogen, oxygen, nitrogen, phosphorous, and sulfur). And, most puzzling, the same elements in precisely the same proportion could form numerous distinct compounds. Provoked by this enigma, Kekulé's research on benzene and his theory of its planar and ringed structure prompted organic chemistry, over the subsequent century, to consider the dimensionality of atomic and molecular forms and to rethink valence. How might the spatial dispositions of atomic and molecular collectivities enfold in valence?

Increasingly, a Lavoisierian impulse to understand the properties of a compound by examining the nature of its constituent elements grew problematic. Duhem, for one, rejected this directionality. Seized by Aristotle's quandary, he suggested the reverse: that the mixt engenders an element's properties, not the other way around. Duhem's proposition brilliantly problematized any simple rapport between elements and compounds, thereby "escaping the stifling to-and-fro between simple and composed" (Bensaude-Vincent and Simon 2012: 127). Indeed, the emergence with quantum mechanics of "valence bond theory" and "molecular orbital theory" in chemistry over the subsequent century closely built off of Kekulé's and Duhem's insights. And it did more. The exigencies of irreducible complexity—where the compound cannot be deduced from the elements—called for rethinking agency, causality, and emergence.

For Duhem, valence was neither a fixed nor an intrinsic property of the element. Valence was *of* the molecular configuration; molecular configurations obliged valence. In fact, the mixt—be it a molecule or a multimolecular compound—necessitates that valence be a fluid and relational capacity. Combining tendencies emerge because of and through the larger orchestration. Valence enfolds in enacting transmutations and functions both as an emergent constraint and as possible radical abandon.

The space of valence, in this understanding, does not abide in a realm where atoms and molecules "stand as discrete and isolated entities with permanent properties waiting to be actualized and used" (Bensaude-Vincent 2014: 72). Instead, valence is preeminently of and in the milieu. Its capacities, trajectories, and potentials are derived of association. Said differently, elements and their composite space of valence always "exist as events in a world already furbished with crowds of interacting beings" (72). That the truths of toxicity, disease, and liability—the facts as if discrete essences—never fully accounted for the controversy and condition of contamination is what has both compelled my delving into this legal saga and accompanied my writing. It begs the question, how to enact the complexity of the mixt?

Coalescence III

VALENCE AS A CONCEPT THAT MATTERS

Modestly distilling insights from chemistry, *The Small Matter of Suing Chevron* takes up the challenge that valence affords. If valence is combining power, its potential lies in giving form and texture to orbitals of coalescence: in chemistry, orbitals are cloud realms of connective probability. Indeed, the trace of orbitals of coalescence is what this book seeks to bring into relief. Thus, I deploy valence not as a key to unlock the truth, the truth that escaped or was barred from the court. Rather, I use valence as a conceptual tool "with which one can bring things about by acting in the world" (Bensaude-Vincent and Simon 2012: 206). My hope is to turn valence into a method device that, when exercised, destabilizes the conversation about these lawsuits and expands their capacity to make us pause.

My concern is not dissimilar from that of many scholars of the social: both awareness of the complexity of phenomena and attention to the tools used to "simplify [that] complexity enough to make it visible" (Strathern 2005:

xiii). My effort here is to think this by deploying valence in multiple registers. *Small Matter* enrolls insights from chemical philosophy, and valence in particular, as a method device to demonstrate how legal truths are made (and made differently) of complexity. Simultaneously, it deploys valence as a method device to interrupt a legal truth, devoid of complexity, that both an oil conglomerate and the US court sealed into law. In service of these efforts is the valence of this book's structures. Here, form itself hazards to perform some of that work, just as the form makes self-evident my method (and its limits) for generating interruption.

Two points along these lines. First, if methods do not "discover" the real but partake in generating it, then there is no innocent method. As Law proposes, "method is not . . . a set of procedures for reporting on a given reality. Rather, it is performative. It helps produce realities. It does not do so freely or at a whim" (2004: 143). Conventionally speaking, Law continues, standard social science methods rehearse the "silences of Euro-American metaphysics" (118), by which he means the belief that reality is an out-there, preexisting, independent, definitive, singular fact/truth. The human corollary capable of grasping and predicting this reality is the liberal subject endowed with autonomy, reason, sovereign will, and equal right and obligation—the subject of Euro-American law.

Rather than perform the implicit assumptions of a dominant metaphysics, an alternative method might surface what Bachelard called a "metachemistry" (Bensaude-Vincent 2014: 66)—or, better perhaps, a "metachemics"—an understanding of realities that attends to indeterminacy, openness, and techniques of realization.[22] If, as Law writes, "methods always work not simply by detecting but also by amplifying a reality" (2004: 116), then a metachemics might escape the postulate of givens and engage, despite risks, with a world composed of relational fluxes and generative forces—a world composed through valence. Relations have long been an anthropological concern. Yet, as Strathern reminds, "anthropologists do not pursue connections simply to be ingenious. They route them in specific ways" (1995: 11). To "detect" implies attuning to relational patterns—the tensions of valenced combinations—and to "amplify" implies making those patterned relational tensions consequential. Historically, the ethnographic provocation has been to provincialize conventional Euro-American metaphysics and perhaps, as Godfrey Lienhardt observed, to "further potentialities of our own thought and language" (1953: 270). Toward that end, *Small Matter* provokes questions about the tensions inherent in liberalism and the processes whereby liberal legality partakes (or not) in enacting inequalities in the name of law.

Second, and importantly, I did not set out to follow the method rendered here. But now I am in the brilliant company of innovative social scientists enfolding chemical concerns into their work (Barry 2005, 2015, 2020; Fiske 2020; Hepler-Smith 2019, 2020; Liboiron 2012; Murphy 2006, 2008, 2017; Papadopoulos, Puig de la Bellacasa, and Myers 2022; Puig de la Bellacasa 2022; Shapiro 2015; Wylie 2018). Rather than being premeditated, my method emerged recursively. The deeper I delved into this legal saga's layered controversies, the more questions of chemistry appeared. As such, I studied textbook chemistry as I pored over legal case files; I read chemical philosophy as I pored over legal case files; I researched scientific and legal scholarship on toxicology, epidemiology, and law as I pored over legal case files, all the while learning more about the chemistry of crude. Increasingly, I enrolled insights from chemical philosophy and the chemistry of hydrocarbons to make sense of the legal saga as it unfolded and then folded back on itself. Consequently, over the past decade, my research only at times mirrored the face-to-face encounters that conventionally map anthropology's field method. My writing, however, resonates with the discipline's pliable, but honored, textual soul: ethnography. At its best, ethnographic renderings delve into and extend analytic insights garnered from empirical material, deploying them in such a way as to challenge and transform theory and normative insights. Moved by Isabelle Stengers, my method seeks "not to judge, to critique, . . . but to transform critique into an instrument of modification" (2011: 507).

What is the valence (the transmogrifying, combining capacity) of crude oil? That continuously enacted relational-material effect that affects? That bewildering brew of thousands of carbon and hydrogen atoms? That viscous substance that leaves filmy traces as it permeates soils and slips along streams? That indeterminate affliction and alienation haunting local bodies and ecologies? That stuff of numerous legal arrangements and material infrastructures? That object of intense corporate desire as it surges through the earth's upper crust, swells through pipelines, into corporate headquarters, pixelating in digitized stock pricing? *The Small Matter of Suing Chevron* explores the valenced force of crude: that is, the combining, melding, and repelling forces that converged through petroleum to materialize the objects and collectivities at stake in the lawsuit and countersuits. It homes in on the practices enacted by molecular and scientific, statistical and epidemiological, contractual and para-contractual, sensual and prehensive, and inquisitorial and adversarial agencies, exploring how their associative movements through crude acquired resonance, crystallized facts, dissipated claims, and exuded truths.

Some quick, broad-stroke reflection on this account. During the seven-week bench trial, Judge Kaplan, the US judge who delegitimized the 2011 Ecuadorian ruling, ran a disciplined, tight, and exacting litigation. And of all the complexity before him, he insisted on the absolute nature and transparency of facts. Kaplan's eloquent ruling, nearly five hundred pages long, narrates a seamless logic based on facts that he takes as complete, as absolutes. The unquestioning hubris of his own reasoning leaves no room for doubt, no room for unknowns, no room for any possible opening into that which exceeds him and his knowing. If one were to take his ruling, simply for the sake of a thought experiment, as a chemical compound, it would read as an equation in which isolated and distinct facts or elements, each with a precise quantifiable valence, in summation equaled truth. It would be an unequivocal Lavoisierian compound. That is, Kaplan's ruling would be an assessment in which distinct elements when tethered together—through a periodic table notion of their pure and fixed valence—held the identity of the real. This is the purpose and doctrine of law at work. Indeed, Kaplan's fusing with law—the particular legal doctrine he applied and the particular strictures he imposed on courtroom process—was what directed him to know that his "findings of fact" (i.e., his determination of truth regarding the facts presented in the case) were fixed and absolute. The complexity of evidence before the court, however, was richly valenced, composed of "facts" whose dimensionality and contextual richness were unrecognizable or denied within the constraints Kaplan imposed. It took remarkable legal dexterity to transform complexity into fixed fact and legal truth.

What would it mean to see the litigating process and ruling as a non-Lavoisierian chemical process and compound? How might that shift how we think of the elements as not absolute, less fixed, more provisional, perhaps contrived in their apparent stasis? Would those elements look or even be functioning the same? And, if not, what would their mode of existence be?

The seven-year-plus trial against Chevron in Ecuador (presided over by six judges) and the nearly two-hundred-page ruling (or nearly four hundred pages, when double-spaced) rendered by the final judge, Judge Zambrano, were never so disciplined and clear. If they constituted a compound, the elements were never absolute and isolated. They were muddled, at times murky, in their very accruing and congealing. Clearly, the time span introduced space for irregularities to emerge. As did the courthouse context, a small court on the third floor of a rented concrete building in an Amazonian frontier town that the FARC-EP (Fuerzas Armadas Revolucionarias de Colombia—Ejército del Pueblo) historically used for supplies and rehabilitation.[23] The court had never dealt with a case of this magnitude and was treading water trying to stay on top of correctly inscribing and archiving the proceedings; mistakes happened. Then there were the onsite inspections of alleged contaminated sites, which in all their spectacle and magnificence were replete with difference—for instance, geographic and industrial variation, scientific procedural differences and irregularities, gaping economic disparity, wincing emotive incongruence, and more. In addition, there was the "he said, she said" battle of experts often exacerbated by a corporate intent to meddle in matters and make elements all the murkier. And far from insignificant, corporate litigators and consorts often intimidated through a fawning elitism bordering on racism, an excess of onerous court filings, and instances of outright manipulation through the exchange of favors. Facts were never simply facts.

Curiously, the authoritative humility with which Judge Zambrano's ruling proceeds, while not an elegantly written text, takes care to recognize and provide space for the unknown. In his forthright caution of determinate knowing, he extended an opening for potentialities, for processes that neither science nor the law could know, which the court intuited were more complex and subtle. For Zambrano, there was a clear sense that the controversy and the phenomena at stake (the mixt and its properties) could not be fully deduced from select and isolated parts. The complexity of the mixt exceeded the language of "the sum of its parts." And within the structure of law he allowed for that recognition. Zambrano's judicial rendering lucidly weighs the arguments and precisely delineates wrongdoing, but also it gives weight to ambiguity and phenomena beyond what is known. If it were a chemical compound, it would be a non-Lavoisierian amalgam where the valence of a multiplicity of elements was never absolute, fixed, or even known. Strikingly, despite Zambrano's ruling being upheld on appeal by three Ecuadorian higher courts, this reality (and the inability to understand it) served as a basis on which the US court dismissed and demeaned the Ecuadorian ruling.

Drawing inspiration from chemical philosophy, *The Small Matter of Suing Chevron* delves into this legal saga's distinct controversies and grapples with their intractable complexity. It entrains the capacities of valence to explore how controversy over the toxicity of crude, the health effects of contamination, the question of corporate liability, unorthodox evidence, and judicial procedures exceeded the single-named (because relationally effected) elements that each party proposed in their legal arguments. Enfolded into a mixt, elements transmuted and spun, extending into orbitals of valenced multiplicity with subtending properties, dispersed dependencies, diffuse qualities, unrecognizable substitution, hidden bonds—qualities that often did not register, or alternatively were disavowed, especially by the constraints constitutive of the United States and tribunal litigations. Constraints, simply to remind, being the obligations and requirements (Stengers 1997) through which, in this case, the law functions.

Deploying valence as a method device means attuning to the sympathies of a discipline long focused on "shifting combinations and open systems" (Barry 2017: 8; Barry 2015). As such, it surfaces dimensions of relationality, movement, and transformation that resonate with Strathern's fabulously generative concept of "partial connections" and Deleuzian/Latourian notions of "assemblage." A shared starting point for these, and likeminded, scholars is that entities—human and nonhuman, organic and inorganic—have a relational ontology or, perhaps better, that all phenomena are relational. Strathern (1988, 2005), among others, would sustain that entities do not preexist the relations that enact them, "nor do they exist apart from the relations out of which they are made" (1995: 102). To quote Annemarie Mol, ontology (always relational) "is not given in the order of things . . . ontologies are brought into being, sustained, or allowed to wither in common, day-to-day, sociomaterial practices" (2002: 8).

Strathern generated the concept "partial connections" to incite reflection on the relational form and to interrupt assumed dynamics of part-whole arrangements: atom and molecule, organ and body, individual and community. When "relations" are thought to "exist outside or between these phenomena," there emerges an "image of the interstellar void traversed by the imaginary lines of a 'relationship'" (1995: 52). Strathern's work destabilizes the idea of entities as being unitary—her Melanesian "dividual" (1988) being an exemplar. The seemingly stable nature of entities (be they persons or materials) shifts based on the kinds of bonds produced through and from them. Thus partial connections can interrupt conventional part-whole visions as when entities (either persons or things) are themselves precarious relations between people (2005: 102), or as when severing peoples creatively propagates relations (111), or as when coordinated relation is mutual realization that tenaciously refuses assimilation (39). The latter is Donna Haraway's iconic female-machine-cyborg—"one is too few and two is too many"—who has spawned inspired thinking. Strathern writes of Haraway's cyborg: "It cannot be approached holistically or atomistically, as an entity or as a multiplication of entities. It replicates an interesting complexity" (1995: 54).

If Strathern's proposal for thinking partial connections is, as Law suggests, "'this' (whatever 'this' may be) is included in 'that,' but 'this' cannot be reduced to 'that'" (2004: 64), then the imagination of valence I deploy coincides with and also diverges from and extends this figuration. A parallel chemical corollary might be that hydrogen-bonded H_2O molecules constitute water, but H_2O cannot be reduced to water. But other chemical corollaries would deviate, such that "this" is never included in "that" because once "that" is composed, "this" no longer exists (H and O are not in H_2O); "this" partakes in constituting "that" but "this" cannot predict "that" (C in C_6H_6); and, more puzzling, "that," but also "that," and "that," partake in determining "this," but none of those "thats" contain "this" (C_6H_6, CH_4, CO_2 do not include C). This is the felicitous complexifying of partial connections.

Valence as a method device might more explicitly allow for the movement and transformation that Deleuze's concept of *agencement* (Deleuze and Guattari 1987) sought to elicit. If "assemblage" as a concept has come to mean a far too fixed and determined arrangement than was intended, valence might infuse uncertain passage, agitated process, indeterminate unfolding. Like assemblage, valence as combining force is attentive to materiality and it affords a grammar through which to honor materiality's continuously relational changing advent. Reverberations here invoke Deleuze's "virtual": the "material force field" of which complex relations generate an actualization that has "no similarity to an original form" such that "proper novelty" is realized (Jensen 2018: 36). This is the condition of the "mixt."

Valence then, perhaps, speaks less of ontology than "of movement in movement" (Rees 2018: 82) in which pieces (as discrete and isolated entities) have no place. Rather, there are intensities of compositional transmutation that give something else while always keeping the elements as abstractions *in potentia*. Processes of reduction capture components although those components are never essences, never origins. And they are never the compositional movement—the mixt—because the latter is always not what prior was. Thinking with valence is an experiment that hopes "to evoke new modes of relatedness" (Jensen 2012: 52). This is an instance, perhaps, of Andrew Barry's "chemical geography": a worlding composed of "events and situations . . . contingent, contested and frequently inexplicable" (2017: 2, 4).

LATTICED CONFIGURATION

In an attempt to respect the complexity of the lawsuits, *The Small Matter of Suing Chevron* unfolds in different registers. The book is divided into three parts, each containing two chapters accompanied by short interstices. Each "part" ("Dissociating Bonds," "Spectral Radicals," and "Delocalized Stabilities") begins with a short text that, in extending insights from the "Chemical Philosophies" sections above, brings forth a particular texture of chemistry's grammar of valence—that transmogrifying force of combining agencies. This invites the subsequent chapters to perform a figurative chemical fugue. As one analytic movement explores the constellation of material practices that rendered knowledge truths crucial to key dimensions of litigating the case and countercases, another movement manifests a particular mode of complex relationality that valence as a method device might yield. And, with the intention of further holding space for complexity, the interstices obliquely extend or unsettle concerns explored in the adjacent chapters, possibly serving as "interstitial forms that are generative of emergent effects" (Barry 2015: 120).

Part I, "Dissociating Bonds," explores two controversies suffusing the Ecuadorian lawsuit and the bonding orbitals that configured opposing legal arguments. Two movements are at play here in crude's valence of truth. First, that the science on which legal arguments were based—the elements derived and the facts obtained—were bound by and contingent on research practice. They were factitious. A combination of technology and technique, protocol and expertise, and material proximities and propensity effected unique fabrications. Second, far from discrete facts in juxtaposed isolation, the elements of scientific expertise were already enrolled in distant institutional, regulatory, and ethical predicaments.

Chapter 1, "Chemical Agency: Of Hydrocarbons and Toxicity," explores this theme by delving into the controversy over whether crude oil is toxic. Tracing the North/South networks among corporate, regulatory, and academic science, this chapter shows how and why distinct techniques for dissecting the molecular structure of hydrocarbons resulted in conflicting chemical determinations of toxicity. Here the temporal/spatial complexity of hydrocarbon molecules combined with distinct scientific, industrial, and regulatory processes to coconstitute toxicity as a sociomaterial accomplishment. Far from providing certainty, the chemistry of oil proved deeply contentious. Chapter 2, "Exposure's Orbitals: Of Epidemiology and Calculation," probes this theme differently by examining the epidemiological studies indicating or disavowing an association between oil operations and cancer. It evinces a long-standing strategy by US industries (tobacco, most infamously) to avoid liability by producing doubt over the effects of their activities. The specific form this strategy took on Chevron's behalf was unique to the temporality and materiality of oil extraction. But just as importantly—and unrecognized in the mixt—tensions within the field of epidemiology unwittingly gave traction to the doubt produced, and further entrenched an indeterminacy in the link between crude and ill health.

Part II, "Spectral Radicals," borrows from a form of chemical relationality of triggered chain reactions. It explores two very different moments of legal process that transmuted and spun in ways never anticipated. In one instance, questionable corporate arrangements threatened to foil the very instrument legally created to sustain the corporate form: the legal contract. In the second instance, a bewildering cascade of experiential events haunting the judicial inspections transformed into exceptional and singular forms of prehensive knowing, such that phenomena subsumed experience and transformed beingness.

Chapter 3, "Alchemical Deals: Of Contracts and Their Seepage," examines the concept and practice of the contract form as it unfolded in the litigation. The more the corporation invoked layers of legal contract in order to bring closure to the dispute and preclude its liability, the further the dispute extended. Data and testimony acquired during the judicial inspections suggested that layers of corporate and state contracts were imprecise, spinning the dispute off into parallel espionage inquiries, contract-fraud indictments, and international arbitrations. Chapter 4, "Radical Inspections: Of Sensorium as Toxic Proposition," explores how sensorial processes coalesced to consequential effect in the lawsuit. Over the five-year judicial inspection process, oil's hydrophilic propensity, the design and ubiquity of industrial waste pits, and hundreds of affect-laden testimonies converged with experiential evidence (e.g., the felt slickness of shimmering, oil-laced matter; the visceral recoiling at the smell of crude; the empathic receptiveness to compromised human, animal, and plant life) to generate for the court consistent forms of sensorial knowing not readily available within US courts.

Part III, "Delocalized Stabilities," draws inspiration from the singular form of molecular bonding present in aromatic compounds. It explores the dramatically different process, structure, and tenor of the legal reasoning that stabilized the 2011 Ecuadorian and 2014 US court rulings. Chapter 5, "Plurivalent Rendering: Of Prehension Becoming Precaution," dissects the 2011 $9 billion Ecuadorian judgment. It takes the ruling's argument (easily dismissed in the US court as convoluted and unconventional) seriously and analyzes the statutory foundations and legal logic for how and why the court rendered the largest international liability in environmental litigation history. Faced with scientifically indeterminate yet materially and sensorily uniform conditions across oil-extraction sites, the court joined a cluster of recent civil law rulings in Latin America and beyond and invoked the precautionary principle as a guiding legal ethic with statutory obligation. The Ecuadorian ruling established a legal precedent that induced extreme unease among extractive industries. And for this very reason, it has been fiercely fought. Chapter 6, "Bonding Veredictum: Of Corporate Capacity and Technique," examines Chevron's successful civil RICO countersuit against the Ecuadorian plaintiffs and their lawyers. In March 2014, a US district judge unversed in Ecuador's legal procedures determined that the 2011 ruling was procured through fraud and thus illegitimate. Filed under the Racketeer Influenced and Corrupt Organizations Act (RICO)—a federal law enacted to prosecute organized crime—Chevron's countersuit represents a novel corporate legal strategy for responding to adverse foreign judicial opinions. It also raises fundamental questions about juris dictum, legal ethics, and translation writ large.

The conclusion, "Metamorphic Reprise: Valence in the Mixt," brings this book to a close by reflecting on the international arbitration that Chevron filed against the Republic of Ecuador before the PCA in The Hague. It captures how an entire legal-fraud worlding sustaining Chevron's corruption narrative rested on a reductive understanding of chemistry and how a constrictively valenced enactment of contract transfigured a national environmental contamination dispute into an international investment dispute in the name of upholding the sacrosanct disposition of the contract form. Despite compelling evidence to the contrary, the arbitral panel of the PCA—an intergovernmental dispute resolution body—found that the republic had breached its bilateral investment treaty and impeached the republic's sovereign judiciary for having denied Chevron justice.

Writing's Orbitals

The Amazonian town hosting the Ecuadorian lawsuit is Nueva Loja, but everyone calls it Lago Agrio. Sour Lake was the birthplace of the Texas Company (i.e., Texaco), the site where, in 1903, two mavericks struck crude in the backlands of Texas and turned their partnership into a major oil producer. Lago Agrio was Texaco's Latin American Sour Lake. The first time I traveled to Lago Agrio was in 1988, at the tail end of Texaco's operations. At the time, decades after its founding, Lago Agrio felt like a raw oil town. Although much bigger and more bustling than when Texaco established it as the company's base of operations, Lago Agrio was still rough-and-tumble. Ensconced in an oil-rich region, the town was marked by potholed, crude-strewn, muddy streets; semi-open sewers; and mildew-ridden, half-built, cinder-block houses whose rebar, jutting to the sky, hinted at the slow corrosion of soaring dreams. That the town is known as Lago Agrio underscores the entrenched sway that US petrocapital has held there. So strong and confident was that sway that, during the near-decade of pretrial hearings (1993–2002) in US federal court chambers of the Southern District of New York, Texaco extolled the virtues of Ecuador's judicial system and fiercely petitioned that the lawsuit be sent to Lago Agrio for trial.

Over the course of thirty-odd years, Texaco's operations indelibly transformed the northern Ecuadorian rainforest, scoring it with thousands of miles of seismic grids, hundreds of oil wells and waste pits, numerous separation and pumping stations, an oil refinery, and the bare-bones infrastructure essential for petroleum operations. By the early 1970s, a network of roads linked oil wells and facilitated the homesteading of the region by over one hundred thousand humble mestizo farmers, or *colonos* (colonists) (see Center for Economic and Social Rights 1994; Trujillo 1992; Uquillas 1985, 1989, 1993; Vickers 1984; Zevallos 1989). As their lands dwindled from encroachment, many northern indígenas retreated eastward deeper into the forest, while others joined the economic ranks of the non-Indigenous, semi-urbanized, and rural peasants. It is these Indigenous and non-Indigenous people on whose behalf the lawsuit was filed.

I began following this legal saga at its inception. When the case was first filed in the New York federal court in November 1993, I happened to be in Ecuador. At the time, I was conducting my dissertation research in the Ecuadorian Amazon on a separate conflict over oil extraction roughly three hundred kilometers south of Texaco's core operations, working with what was then one of the most consequential Indigenous movements in Ecuador and Latin America in general (Sawyer 2004a). Because of the effectiveness of the political and environmental organizing of this Indigenous opposition, a number of campesino and Indigenous leaders who formed the plaintiffs' group in the lawsuit against Texaco approached Indigenous leaders with whom I then worked. They sought guidance in organizing local residents around the collective effects of oil operations. It was through my collaboration with local Indigenous and environmental groups in the Ecuadorian Amazon that I became connected in the 1990s with individuals who increasingly became key actors in the lawsuit as it progressed over the subsequent twenty-five years.

At the start of the new millennium, I published a few articles on the early stage of the lawsuit. One cluster examined the case during its decade-long life between 1993 and 2002 of pretrial hearings in New York federal courts (Sawyer 2001, 2002, 2009), and another analyzed the opening of the trial in the Amazonian provincial court in 2003 (Sawyer 2006, 2007). But it wasn't until 2010 that I began to read the exploding case file concertedly and began to conceptualize the project that ultimately became this book. That emergent analytical form shifted my research methodology. Rather than being grounded in day-to-day fieldwork, my research entailed years of studying legal documents and consort scientific literatures in an ever-transforming legal saga. This meant that, although I sustained connections with the plaintiffs' legal team and a number of the plaintiffs over the decades, I was not caught up in the ever-piling intricacies and intrigues surrounding the cases. This distance afforded me space to distill and reassemble this legal saga otherwise.

My compulsion to attend to oil relations exceeds, of course, academic concerns and enfolds consideration of how the exigencies of oil give form to a being. Beyond the complicity that living in the hyper-consumptive North brings, the legacy of oil in South America and North Africa largely shaped the trajectory of my paternal family over the greater part of the twentieth century. My grandfather, uncle, and father each lived and worked for decades in Latin America (although not Ecuador) for Standard Oil of New Jersey qua Esso qua Exxon qua ExxonMobil as an engineer, geophysicist, and geologist, respectively. And I, in turn, grew up around oil.[24] Significant portions of what I know about oil operations come from my uncle (whose research I cite in this book) and from extended conversations and travels over the years with my now long-retired, social justice–oriented father. Paradox, emergence, and surprise suffuse these orbitals of coalescence; they partake in forming the research I have done, just as that research has modestly partaken in altering their trajectory.

Writing any book is no easy task. And this one has presented its own unique challenges that deserve comment. First, the mass of information in this legal saga is staggering and unwieldy—with multiple actions producing immense case files and associated scientific, technical, legal, and social debates together being of equal magnitude. With reason, Chevron hired more than two thousand lawyers from sixty law firms in order to launch its RICO countersuit. I am a team of one. Second, writing about a legal case in a way that fundamentally breaks with the decisions of the US district court and court of appeals is daunting. Ours is a judicial system I respect, despite its flaws, and tracing the persuasive threads of skilled legal maneuvering within it has been sobering. Lastly, writing critically about the practices of the second largest US oil company—with its multiple tentacles—is not for the faint of heart. Litigation is a tool that Chevron has used relentlessly to debilitate, both financially and emotionally, individuals and organizations associated with positions it does not like. Chevron's crushing offensive against the LAP's

US advisor counsel, Steven Donziger, is disturbingly the most egregious. But he is not the only target of the corporation's reprisal tactics.[25] At multiple junctures along the way, these concerns have given me pause. None are to be taken lightly. And, consequently, much care and deliberation have accompanied me in writing this book. I now relinquish that attention to you, dear reader, to judge.

PART I
DISSOCIATING
BONDS

IN CHEMISTRY, a "bond" signals an occasion of relationality among elements and between molecules. In models of chemical structure, a bond often appears as a discrete point of transaction between discrete electrons surrounding discrete nuclei (e.g., methane, CH_4).

FIGURES 4 & 5

This is the molecule as the sequential juxtaposition of isolated elements—good for offering explanations and triggering further experimentation but poor as a figuration of a real. A better proximate, perhaps, is thinking of a bond as an orbital of coalescence. In chemistry, "orbitals" are cloud formations (limitless in expanse) in which the probability of locating an electron exists. This is not an "orbit" that delineates the trajectory of a planet-like electron encircling its sun-like nucleus. Rather, orbitals, as mathematical functions, are amorphous distributions of condensing and dissipating probability that signal an electron's agitated disposition and capacity to move and be moved, to join and be joined, to affect and be affected, such that valence emerges, generating new compositional effect.

Over the twentieth century, an emerging electronic theory of atoms developed elaborate models of the workings of orbitals, understood as the varied force fields in which different electrons move. The models encoded and made visible the orbital realms through which electrons differentially suffuse atoms and how those realms adjust and morph through their enactment and participation in inter- and intra-atomic forces. What has resulted are models of molecular bonding with multidimensional architectures and a multiplicity of modes of interaction. There are electrons that bond through sharing, donating, repositioning, displacing, substituting, hybridizing, delocalizing, and so on, within and across energy levels, or force fields. The arrangement of electrons within architectures of energy signals an element's and a molecule's valence, its relational capacity and ingenuity of combining force. As an expression of energies and intensities, valence is a molecular complex's puissance (strength, influence, its power to act). Extending poetic liberties, valence affords modes of "attractions and repulsions, sympathies, and antipathies, alterations, amalgamations, penetrations, and expressions" (Deleuze and Guattari 1987: 90) that transform bodies into collective ecologies. As transforming expressions of valence, orbitals sustain what Deleuze following Spinoza called a body's susceptibility toward absorbing the traces of another (affection, the affections of a body) and a body's continuously varying capacity for acting (affectus, its affects) (Deleuze 1978). Incompleteness compels the process. Metamorphosis is what transpires.

Taking two controversies suffusing this lawsuit—"is crude toxic?" and "does contamination undermine human health?"—this section explores the tenuous and contingent orbitals that formed to configure opposing and repelling legal arguments. The orbitals conjoining each party's argument formed dissociating bonds; they simultaneously coalesced a multivalenced legal-scientific order and sundered association with the other. Two movements are in play. First, the science on which legal arguments were based—the elements derived and the facts obtained—was bound by and contingent on laboratory practices and research regimes. It was "facticious" (cf. Bachelard 1953).

A combination of technology and technique, protocol and expertise, and material proximities and propensities effected unique compositions. Second, far from being composed of discrete facts in juxtaposed isolation, scientific arguments were already enrolled in distant institutional, regulatory, and ethical predicaments. Rather than being mirrors of truth, arguments reflected dense orbitals of coalescence—effected contingent composites in which relations were always already integrally implied. In line with the chemist, the Deleuzian question is not one of judgment. Attentive to how orbital encounters evince a body's affects and affections, it asks, what can a composite body do? Of what is it capable?

HEARING

On October 21, 2003, the court proceedings in the lawsuit against Chevron began. This text is from my field notes and experience at the Lago Agrio courthouse that day.

"The morning dawned with rain," Pablo Fajardo bellowed. He stood on the flatbed of a truck seconding as a stage before a crowd of approximately five hundred Amazonian peasants and indígenas. "But as we say here," he winked, "water cleanses. And the rain is a sign—our decadelong struggle demands that ChevronTexaco cleanse our rivers and lands." Standing under umbrellas, huddled beneath overhangs, or welcomingly drenched, expectant onlookers laughed and cheered. "Today, October 21st, is a momentous day," Pablo continued, his words blaring through loudspeakers outside the court: "For today begins the lawsuit that holds our greatest hope for cleaning up the criminal destruction that Texaco left in our Amazonian communities. After ten years of organized legal battle and solidarity work with our national and international allies, we the defenders of life and rights have succeeded in making ChevronTexaco submit to our laws." Those gathered filled the street. "ChevronTexaco, ya viste," he rallied, and the crowd joined in: "La justicia sí existe!" Unperturbed by the rain, men and women, young and old, had traveled to Lago Agrio to mark what they called "the trial of the century." At that point, the century was only three years old.

Past the guarded metal gates, on the third floor of the superior court, one hundred people packed a muggy courtroom. The opening hearing of the law-

FIGURE 6 Frente de Defensa de la Amazonía (FDA) / Unión de los Afectados por Texaco (UDAPT) rally outside the Superior Court of Justice of Nuevo Loja (aka Lago Agrio). Photo by author.

suit against Chevron had just begun. The crowd on the street below represented a fraction of those pressing the suit.[1]

Chevron's lawyers had arrived at the courthouse early, escorted from their private jet in armored cars with a security detail. The Ecuadorians' lawyers headed the march of the hundreds of plaintiffs determined to hold vigil outside the court. In the single-chamber court, the judge sat behind the dais flanked on either side by opposing legal teams, each comprising Ecuadorian- and US-based lawyers. Among the spectators, a collection of plaintiffs listened, periodically relaying news to the assembly outside; dozens of human-rights and environmental activists watched attentively; and national and international reporters set up their video cameras and microphones as security police and bodyguards watched over the crowd. All focused their attention on Chevron's chief lawyer, Dr. Adolfo Callejas Ribadeneira, as he responded to the plaintiffs' claims.

On the street below, Pablo intoned a collective catharsis. "Here at the very scene of the crime, we can anticipate that ChevronTexaco will incessantly

FIGURE 7 Judge Alberto Guerra, the first judge to preside over the lawsuit against Chevron, talking with reporters in the single-chamber court. Photo by author.

present the court with papers, documents, rhetoric, and motions to delay; they have nothing else to show. We, however, have overwhelming proof to present to justice: our poisoned rivers, our ailing forest, our disappeared animals, our dead. Raise your hands, compañeros! Because today we have come to say 'NO' to Texaco, 'NO' to Chevron!" Some in the crowd shouted in agreement. Pablo yelled: "Las pruebas les dimos, con estos te jodimos!... Muerte y destrucción, dejó la Texaco en toda la región." The protest chants echoed on. And from the crowd emerged another: "Fluye el petróleo; sangra la selva."

At the time, Pablo was a human-rights advocate heading up a legal-complaints desk in Shushufindi (another Amazonian oil town). Pablo knew the effects of crude oil all too well. As a teen, he had worked in oil fields. Over the previous ten years, he had advocated unflinchingly to seek reparations for poor farmers confounded by the damage, death, and distrust that oil had inflicted on their lives. Pablo was a deeply respected presence. He was also, at the time, studying law. And, although no one would have predicted it, within two years, he became the stalwart and enduring chief lawyer for the plaintiffs.

"This lawsuit is for everyone," Pablo continued, "because the harm has spread far beyond the oil fields. Chevron has offended our national dignity. We call upon our local, national, and international civil society and authorities to be vigilant and ensure our rights are fulfilled. And to make sure that never again will an oil company use criminal technology that threatens life and nature." The thinly veiled concern? That this small provincial court would be overwhelmed and succumb to the power and pressure that the oil industry can exert. Having national and international eyes watch over the legal process in Lago Agrio gave hope that this would not easily be the case.

A swell of statistics flowed. Pablo read from a sheet:

Between 1964 and 1990, Texaco operated 15 oil fields, 22 production stations, and 339 wells. . . . The company extracted approximately 1,434,000,000 million barrels of crude . . . across a region of approximately 500,000 hectares using technology rooted in a principle of minimal investment and maximum profit. The harm has been devastating. The number of gallons of crude Texaco spilled in rivers and streams: 16.8 million gallons. The number of gallons of toxic water Texaco poured in rivers and streams: 20 billion gallons. The number of waste pits Texaco left: over 600. The number of indigenous groups affected: six. The number of persons affected by Texaco: approximately 30,000. Most common illnesses suffered by those affected by Texaco: skin and stomach infections, dizziness, headaches, spontaneous abortions, and cancers. Most common damages suffered by those affected by Texaco: the dying of cattle and domestic animals, the loss of crops from poisoned water and oil spills, the dying of fish in rivers.

Carrying homemade placards and swaddled infants, burdened by their stories and ailing bodies, protesters bore witness to the despair that Texaco's oil operations had brought them. Some joined Pablo and his colleagues and spoke into the microphone. Silvia was one.

"First, my sister fell sick," Silvia started. She trembled from nerves and the cold of the rain: "It's been a year and eight months now. The doctors say she has stomach cancer. Everyone in my family is feeling some pain, but my sister suffers the most." In her late twenties, Silvia was from an Indigenous Kichwa community not too far from Coca (another oil town).

This disease, it ravages. We sold everything to heal her and even doing that we haven't been able to win her back. What she has you can't cure, that's what the doctors say. She can't eat. She can only take small amounts of liquids. And the only thing she gives back is blood, bloody vomit from

FIGURE 8 Plaintiffs holding a banner reading "TEXACO = DEATH. IMPLORING JUSTICE FOR AN EXPLOITED AND WOUNDED PEOPLE." Photo by author.

inside. The doctors did an exam and all her insides are eaten away. From one moment to the next, she fell ill. She was healthy, strong. Now she lives only in bed. She's dreadfully thin, pure bones, that's what she is. There is no cure, no hope. There is no recuperation. No one can heal from this.

From his ersatz stage, Pablo spoke forcefully: "The hour of justice has arrived." Behind him a huge black banner spelled *JUSTICIA* in bold lettering. Both Pablo's words and the banner signaled that many—previously wary of Ecuador's judicial system—increasingly believed that the courts might treat them fairly.

Upstairs in the court, Chevron's chief lawyer proceeded to read the corporation's ninety-six-page written response, which denounced the plaintiffs' "false accusations" and "negat[ed] all claims."[2] Within minutes, Callejas asserted that the then president of the superior court, Judge Alberto Guerra, "lack[ed] jurisdiction and competence over ChevronTexaco Corporation" and that the plaintiffs sued the wrong entity. Texaco Inc., which "never lost its legal personhood," was the corporation directed by the US court to submit to Ecuadorian law, not ChevronTexaco.[3]

Callejas continued: it is "not true that ChevronTexaco Corporation replaced Texaco Inc. in all its 'obligations and rights'" as the plaintiffs claimed.

And it is "also not true," Callejas asserted, "that ChevronTexaco Corporation accepted in any manner to submit to the jurisdiction and competence of the Ecuadorian courts and tribunals. . . . ChevronTexaco Corporation was never the operator nor a party to the Concession Contract, which existed since 1973, nor did it replace nor is it the successor of Texaco Inc. nor of Texaco Petroleum Company (Texpet). . . . Therefore, I repeat that ChevronTexaco Corporation is not subject to your jurisdiction or competency, Mr. President, and is not a legitimate opponent in this case."[4]

Furthermore, Callejas asserted, in 1995 Texpet (Texaco's subsidiary operating in Ecuador) entered into a "Contract for Implementing of Environmental Remedial Work and Release from Obligations, Liability and Claims" for which Texpet spent $40 million on environmental remediation work (an arrangement that is the focus of chapter 3). When in 1998 the work was completed, the government of Ecuador "released Texpet, Texaco Inc., their successors and predecessors from any additional liability for environmental impact arising from [its former] operations." Because, Chevron lawyers claimed, this arrangement was a legal settlement, the lawsuit represented "an attack against the immutability of judicial decisions" and "a violation of all legal principles of any civilized society governed by laws" in attempting "once again to debate an issue that was concluded to the satisfaction of the Government of Ecuador."

As for the substance of the allegations, Callejas asserted that Texpet "did not use or implement obsolete exploration and exploitation procedures and techniques. . . . Texpet always worked with the leading technology at the time when it performed its operations in Ecuador, using techniques and procedures generally accepted in the petroleum industry at the time." He continued: "The procedures and methods [used] do not have and did not have," as the lawsuit claims, "lethal effects on the environment." All "techniques and procedures such as those used by [the] Consortium complied with hydrocarbon industry standards."

Callejas argued that "accusations regarding the supposed negative effect of the Consortium's operations on the health of the local population have not been proven by any scientific or factual evidence [the focus of chapters 1 and 2]. Indeed, an analysis of general health indices do not reveal any negative effect directly caused by the oil operations. . . . The true causes of the problems" faced by residents of the region—"which are falsely deemed to be attributed to the oil operation"—include colonization, pesticide use, and sanitation, all factors "deliberately and maliciously ignored by the plaintiffs."

In response to these claims, the plaintiffs' then chief lawyer, Dr. Alberto Wray, a prominent Ecuadorian legal scholar and former member of Ecuador's supreme court, spoke slowly and clearly: "Regardless of the name used, regardless of the legal disguise deployed, Texaco caused environmental damages to the Ecuadorian Amazon. And the poison is still there today. It is simplistic to claim it is not. The plaintiffs are not against the exploitation of petroleum. Rather they are against the act of pursuing it aggressively, solely for the purpose of economic gain, and toward that end using production technologies that released toxic elements into the environment—and which in turn caused harm to the people, fauna, and flora there—all the while knowing that there were less toxic ways of working. . . . To speak of contaminating elements is not to speak of a myth from the past. We are talking about a present danger that is still harming and causing injury to local people, animals, and the environment, and to you, Honorable Judge."

Chemical Agency

Of Hydrocarbons and Toxicity

BETWEEN JANUARY 2004 AND MARCH 2009, fifty-four judicial inspections of alleged contaminated sites unfolded in the Ecuadorian Amazon as part of the litigation against Chevron.[1] The inspections involved the judge, the legal teams from both sides, various scientific experts, the press, interested observers, and local inhabitants trekking through the secondary rainforest surrounding former Texaco oil wells, processing stations, and exposed or purportedly remediated waste pits. At each site the plaintiffs' and the defendant's teams of technical experts extracted soil and water samples; examined their texture, color, and smell; noted the coordinates and depth of their extraction; and placed these samples in airtight, labeled containers ultimately to be sent off to faraway laboratories that would analyze their chemical content. Two written reports—one from the plaintiffs and one from the defendant—with multiple parts and appendices resulted from each site inspection. Ranging in the hundreds of pages, each report maps an inspection site, its geologic and hydrocarbon history, the biophysics of crude oil degradation, and the position of waste pits and oil spills relative to waterways, human habitation, and other Texaco infrastructure (wells, pipelines, and pumping and processing stations). Most importantly, each report methodically detailed the geomorphic and chemical composition of the samples taken.

The judicial inspections constituted a significant part of the "evidentiary phase" of the trial and, as such, they were crucial events for garnering or dispelling proof of contamination. The scientific reports emerging from them formed an integral part of the evidence that the superior court judge, Nicolás Zambrano, ruled on in February 2011. Among the issues at the heart of the legal proceedings was the capacity to materialize or dematerialize the presence of toxic elements, derivative of Texaco's operations, in the region's soil and water systems forty-odd years after petroleum extraction began. Although the presence of crude and its by-products in the environment was not in question, the very *toxicity* of this substance was.

Virtually all the technical reports submitted to the superior court on the plaintiffs' behalf contained language to this effect: "Soils, dispersed at various points [at the site], are severely contaminated with the presence of petroleum residues and toxic heavy metals." And the presence of these toxins "represents a real present and future risk to the [local] population."[2] By contrast, virtually all the reports that the defendant submitted reached opposite conclusions. Chevron's scientific analyses asserted that alleged contaminated sites pose "no oil-related risk to public health or the environment" and that collected samples of water and soil "contained no hydrocarbons—BTEX, PAH, and metal concentrations—that pose risks to human health."[3]

How was that possible? During these litigation proceedings, Chevron admitted that Texaco's operations dumped over 16 billion gallons of formation waters directly into the environment, burned roughly 230 million cubic feet of natural gas, and dumped heavy oil from exploratory and producing wells into open waste pits. Surely Chevron experts must have distorted, manipulated, or concocted purifying evidence for the corporation to receive such clean reports.[4] Much was at stake. There was, of course, a potential corporate liability. But more ominous loomed the potential to set precedent: that a parent corporation be found accountable for the negligent actions of its subsidiaries in distant lands.

In most cases, however, the chemical constituents and concentration levels detailed in reports submitted to the Ecuadorian court on behalf of Chevron were not radically dissimilar from the chemical results obtained from the soil- and water-sample analyses that the plaintiffs' experts submitted to the court in their reports. This was especially the case in the early years of the inspections.[5] Differences, of course, did exist, which I will detail shortly. But for field samples taken from or near former waste pits, the laboratory results of chemical analyses from each side broadly corroborate and coincide with one another. That is, the raw data that each legal party generated de-

FIGURE 9 One of the petroleum waste pits adjacent to the oil well Sacha-77.
Photo © Frente de Defensa de la Amazonía / Unión de Afectados por Texaco.
Used with kind permission.

tailing the molecular compounds found in soil and water samples were not significantly dissimilar. Indeed, one of the plaintiffs' key pieces of evidence to prove Chevron contaminated the environment was the fact that molecular hydrocarbon levels—as measured by both the plaintiffs *and* the defendant— exceed Ecuadorian standards (by tens to hundreds of times) in 97 percent of the sites examined during the judicial inspection. Clearly—so the lawyers for the plaintiffs maintained—this was the present toxic materialization of past negligent practices.

So how did diametrically opposed interpretations of contamination emerge from the same material reality?

This chapter explores how molecular, technological, and regulatory orbitals coalesced to make possible opposing scientific arguments over the same contaminated conditions. Recall that, as substance, compounds exist only as relational enactments, not as a sequence of essences in juxtaposed isolation. Being-ness as relation—that is, with relations integrally implied—means that, far from being fixed, ontology emerges from within encounter. In chemistry, valence captures this emanating and subsuming combining capacity, and in quantum theory (although engaged with greater sophistication), valence is the immanence of orbital realms. In this way, a compound's capacities hinge not on constituent essences but on the valenced force of emergent configurations.

What follows below leans on this theory in a double movement. First, taking the parties' arguments *as* compounds, this chapter unfolds the chemical, scientific, and governing orbitals that coalesced to materialize or dematerialize crude as toxic. Rather than being defined by isolated essences—a truth of the chemical, a truth of the scientific, a truth of monitoring agencies—the parties' arguments emerged from orbital realms dependent on, as they transmuted, each other's combining force. Second, taking the compositional force of valence seriously, this chapter disconcerts by tracing how partial the understanding of the chemistry of crude is. As will become clear throughout this book, this partial knowing, and partiality, makes crude's chemistry vulnerable to recombinant appropriation and signification. Here, unknowns and unknowability become a presence, not absence, and constitutive of the matrix through which novel compositions emerge.

And, so, the context. Yes, there was crude oil present in the environment. Yet there was no consensus as to whether the hydrocarbons present were toxic. Disagreements over the intricate chemistry of crude—the temporal, spatial, and quantum dynamics of thousands of distinct hydrocarbon compounds—unfolded in a morass of indeterminacy. Intriguingly, yet disconcertingly, opposing scientific claims were not invalid. That is, neither side was proclaiming scientific untruths. Rather, conflicting toxicological knowledges produced divergent ways of deciphering the chemistry of hydrocarbons. Opposing reports described how the fates of distinct hydrocarbon compounds isolated through distinct assays enacted contrary trajectories. To be clear, I (along with many others) have no doubt that the plaintiffs' lawyers and experts more relevantly captured the precarious consequences of contamination in the rainforest. But what Chevron argued is not incorrect. Rather, corporate science was incongruously constrained and deceptive in portraying that constraint as truth.

Contrary to common assumptions, toxicity is far from natural.[6] Rather, the mattering of toxins—whether from seepage, spills, or combustion—is suspended in orbital cloudscapes enfolding industrial practices and chemical bonds, corporate profit and failing bodies, and scientific procedures and regulatory standards. This chapter focuses on interpermeating molecular, laboratory, and regulatory processes that made toxins matter, or not. Understanding determinations of toxicity in the Lago Agrio lawsuit demands, I suggest, considering how the production of scientific knowledge, the spatial/temporal complexity of hydrocarbon compounds, and the structure of regulatory reasoning allowed for multiple determinations of crude oil that indexed distinct toxic and nontoxic realities. Material, temporal, and spatial relationalities catalyzed the atomic-scientific-statutory formation of toxins in disparate ways, animating concern and exacting action, or securing righteous conceit and generating disavowal.

Following Antoine-Laurent Lavoisier's footsteps, the work on "historical ontology" (Hacking 2004) further inspires my analysis. Historical ontology claims that "what counts as 'truth' is the result of historically specific practices of truth-telling—laboratory techniques, instruments, methods of observing, etc.—and the objects that are apprehended through that truth-telling are also historical" (Murphy 2006: 7–8). As noted earlier, what comes to matter as reality, key scholars claim, is the product of historically precisioned instruments, techniques, protocols, nonhuman capacities, and human dispositions (Latour 1993, 2005; Law 2004; Mol 2002; Shapin and Schaeffer 1985). Bruno Latour and Steve Woolgar showed how various arrangements of machines and experimental procedures act as "inscription devices" that "transform a material substance" into a decipherable and manipulable entity (1987: 51). Of concern was the relational capacity among laboratory instruments capable of detecting a substance's determined properties and the predilections of a substance to enable that detection to occur. Latour and Woolgar's attention demonstrated how lab assemblages entrained instruments and substances such that particular relations changed "from non-trace-like to trace-like form" (Law 2004: 29), thus transposing them into recognition.

In the context of the litigation, these insights are consequential to understanding how a crude substance was deemed scientifically to be toxic or not. The way each party produced and then dissected and/or collapsed the chemical constituents of petroleum distinctively determined whether crude oil had lethal and less-than-lethal impairing capacities. Leaning on chemical philosophy, this chapter makes three subtending arguments. First, that the "chemical substances" (i.e., distinct hydrocarbon readings) detailed in experts' reports were "the product of technique rather than bodies found in reality"—a discernment that led Gaston Bachelard to state that "the real in chemistry is a realization" (cited in Bensaude-Vincent 2014: 66). Second, "to be toxic" is not simply a property or essence of particular chemical compounds; "to be toxic" is a valenced capacity triggered by and surfacing through conditioned chemical-metabolic-laboratory-statutory arrangements. And, third, toxicity functions like a mixt: a transforming, enigmatic phenomenon whose complexity exceeds the ability to understand it by pulling apart its molecular and atomic components.

Texaco's Oil Operations

As Pablo Fajardo, the young lawyer for the plaintiffs, explained during the judicial inspection of the Sacha Sur Production Station, the lawsuit demanded environmental remediation. The plaintiffs claimed that, between 1964 and

1992, Texaco (which, remember, had merged with Chevron in 2001) system-atically used substandard technology in designing and engineering its oil operations in the northern Ecuadorian Amazon. Shoddy equipment and practices spewed industrial wastes throughout its oil concession, devastat-ing the local ecology and endangering the health of local inhabitants. At a quick tally, Texaco drilled (at a minimum) 325 oil wells, excavated over 900 waste pits, operated 22 production stations, and built a maze of primary and secondary pipelines transporting crude oil along the trans-Andean pipeline (Sistema de Oleoducto Trans-Ecuatoriano; SOTE) to a Pacific port where it was readied for export. The mechanics of oil that bred these infrastructural associations were messy, changing as they spanned over space and time. So, let me delve a bit into Texaco's oil operations.[7]

When in 1964 Texaco first acquired rights to explore and exploit oil in Ec-uador, its concession was inaccessible except by air—or by trekking for weeks through the forest. As such, Texaco carried out its early geophysical explora-tion largely by air, flying and shooting thousands of miles of air-magnetome-ter profile and air photography before starting its seismic exploration. At the time, seismic work entailed cutting a grid of thousands of miles of swaths through the rainforest, drilling two- to five-meter-deep holes every one hun-dred meters along the swaths, and detonating ten to twenty kilos of dyna-mite in every hole. Cables connecting the charges of dynamite in each hole transmitted the soundwaves once detonated to geophone equipment. The geophone read the soundwaves and conjured a vertical map that pictured the subterranean stratigraphy. Airfields, heliports, and worker camps dot-ted the rainforest landscape as short take-off planes and helicopters hauled the equipment and bodies, and tents housed the laborers, necessary for this work.

Once studied by geophysicists, the aerial imagery, seismic maps, and sur-face geology presented locations for exploratory drilling. Drilling for oil is no easy task. And even less so when boring wells two miles deep in the neo-tropics of the 1960s, 1970s, and 1980s. Heavier equipment was aired in to erect derricks for drilling and build longer-term worker camps. Geologists and en-gineers monitored the drilling; as the bit penetrated deeper through rock, compression, density, and heat shifted, demanding that the earth's subter-ranean pressure be equalized with calibrated quantities of synthetic drilling muds that also lubricated and cooled the drill bit. With deeper boring into the earth, a mixture of cuttings (rock and sands), chemical muds, and waters emerged from the well. Advancing upon an oil formation, crude and highly salinized subterranean fluids, called "formation waters," intermingled with the drilling sludge. These industrial by-products needed to be dealt with, and toward that end Texaco excavated two to three (or more) pits alongside each exploratory well in which it dumped the sludge, formation waters, and unusable heavy crude that surfaced during the drilling process—along with the chemical muds and industrial solvents essential for maintaining and repairing drilling equipment.

Were an exploratory well to spud crude oil, geologists would run "production tests" to estimate the future flow of the well. That is, they would let the well run through a pipe and measure the quantity, consistency, and constancy of the flow as it shot into the excavated pit. These production tests would run at half-hour or hour-long intervals, spewing thousands of gallons of crude into the pits. If Texaco determined the well to be commercially viable, it drilled additional wells in the area to estimate the size of the discovered petroleum reserve. Beside each of these drilled wells, Texaco similarly excavated purpose-built earth pits.

Once further infrastructure was in place, the crude from the successful wells passed through pipelines or "flow lines" to what is called a "production station," where it converged with crude from other oil wells. Crude oil, of course, does not surface from subterranean depths as a pure substance. Miles deep in the earth, it resides in interstices of rock, together with waters, gas, and sands. Consequently, when crude rises through the earth, it emerges as a matrix. The production stations were engineered to isolate crude oil from its associated substances so that the crude alone traveled along Ecuador's principal pipeline across the Andes to the Pacific coast.

Texaco designed its production stations to separate crude principally by means of specific gravity—that is, the propensity of different substances to act differently according to their molecular weight. Because of their molecular density, the gases dislodged from association by floating up, the molecularly heavier formation waters and solids did so by sinking down, and the crude oil floated and flowed in the middle. Catalyzed by heat (in the case of the "separator") or an emulsifier (in the case of the "wash tank"), once "separated" each substance had a distinct fate. The natural gas was funneled through pipes ultimately to be flared. The floating crude was shunted off along central manifolds ultimately to reach the Pacific. And the heavier formation waters and sands were drained from pipes into excavated waste pits at the stations.

According to the plaintiffs, Texaco took negligible environmental precautions during its twenty-eight years of oil activities (1964–92). More precisely, the lawsuit claimed that the company engaged in practices that, while illegal

FIGURE 10 Ignited petroleum waste pit. Photo © Amazon Watch.
Used with kind permission.

at home, boosted corporate revenues by the billions, reducing the per-barrel costs of production by three to four dollars (Sawyer 2001). Cost-cutting practices pervaded all aspects of Texaco's operations, the plaintiffs alleged: seismic exploration, exploratory drilling, extractive drilling, processing facilities, pipeline maintenance, and pumping stations. Each of these aspects of Texaco's operations disrupted lifeways—both human and nonhuman—and altered their possibilities. The company's seismic lines ripped thousands of miles of forest and detonated thousands of pounds of dynamite, irrespective of the presence of waterways, forest ecologies, dwelling sites, agriculture, or livestock. And its exploratory and extraction wells, and separation and pumping stations, raised forests, moved earth, dug trenches, and excavated craters similarly causing damage, irrespective of surrounding life collectivities.

But what most worried plaintiffs were the open, unlined waste pits that, they claimed, left an inhabited landscape festering in chemical industrial wastes. These waste pits—many still open to the elements in the 2000s, others covered with dirt or remediated—and their effects were a primary focus of the judicial inspections.[8] In theory, the excavated pits were meant to facilitate extraction unproblematically: the heavier, solid debris (drilling cuttings, muds, and solvents) were to sink to the bottom; the residual hydrocarbons would float to the top; and the heavier, watery fluids would rest in the middle. Time, heat, and gravity would do their work; volatile components would evaporate or biodegrade, and other elements would settle into their designated space according to molecular weight, all ensuring that the watery fluids, when siphoned into the environment, were just fine. Amazonian clay soils, which formed the base and walls of the pits, were, purportedly, impermeable, and crude molecules nestled in muds or coagulated in surface caps were, purportedly, immobile. Underground leaching was nil. However, the plaintiffs claimed this was not the case and that the mechanics and physics of Texaco's facilities were hardly benign. Dug out of the earth, unlined and open, the pits served as holding receptacles for slow-motion toxic seepage and overflow.

When they design Texaco's Amazonian oil operations, engineers and geologists face a fundamental problem: how to maintain the structural integrity of waste pits. A substantial amount of fluid flowed into these pits, whether from exploratory wells or at production stations. To give a sense, the ratio of formation water to crude oil produced from an operating Amazonian well ranged from 3:1 to 10:1, depending on the age of the well and the geological formation from which the crude emerged. That is, for each barrel of crude oil produced, between three to ten barrels of formation waters were also released from the earth.[9] When combined with the region's annual rainfall, averaging three to five meters a year, this posed challenges. Changes in fluid levels and temperature could cause pit walls to collapse or crack. Consequently, virtually all pits were designed with a gooysenecked pipe engineered into the side, one meter down from the rim, to prevent the pits from reaching capacity and the walls from imploding.

As Texaco logs indicated and Chevron lawyers acknowledged, the goosenecked overflow pipes decanted formation waters down embankments into adjacent gullies inevitably coursing along local creeks. And these waters, despite the chemical theory of molecular separation, were never pure. They were emulsions of hydrocarbons and subterranean fluids with influxes of industrial chemicals from drilling muds and solvents. These admixtures, the plaintiffs claimed, contaminated surface waters, where rivulets flowed into streams that flowed into rivers, all essential water sources for human and nonhuman ecologies.

Even during the early years of Texaco's operations in Ecuador, it was standard oil practice in the United States not to store hydrocarbon-laced brine in open waste pits. Indeed, open waste pits had been illegal in Texas since 1939.[10] At that time, it was becoming the norm to reinject formation waters and subterranean sands back into a rock structure well below the water table, to process chemical solvents until they were environmentally safe, and to refine subterranean gases. On the US Gulf Coast, states prohibited the release of formation waters and industrial wastes into water systems by the early 1930s and mandated that it be reinjected at least one mile below the surface of the earth.[11] In Ecuador, Texaco never reinjected waste fluids, despite doing so in the United States, despite retaining a patent for reinjection technology, and despite corporate executives knowing about the dangers that formation waters posed.[12]

As a consequence, the plaintiffs' lawyers claimed, the wastes from crude leached into subterranean soil and water systems, oozed along hydraulic flows through streams and swamps, seeped into water sources used for household consumption, and interrupted the metabolic processes of plants, animals, and people. The soils adjacent to the pits in particular suffered from contamination that, over time, the plaintiffs argued, percolated slowly through the soils, affecting both the surface and underground water networks.

Operational oil spills were similarly a concern. Between 1972 and 1990, the Texaco-operated trans-Andean pipeline spilled an estimated 16.8 million gallons of crude into Amazonian headwaters—over one and a half times the amount spilled by the *Exxon Valdez* oil tanker (DINAPA 1993; Kimerling 1991a, 1993, 1996). Estimates for secondary pipelines compete with that figure. Oil spills are an inherent risk in the petroleum business, and Texaco claimed that "natural" factors—especially the 1987 earthquake—were responsible for the major portion of crude that was spilled. Yet economic factors determined how the company chose to mitigate oil slicks should they occur. Texaco invested minimal resources in maintaining deteriorating pipelines, and "cleanup" of spills often took the form of covering them with dirt.

Routinely, Texaco workers siphoned the coagulated crude floating atop waste pits into tanker trucks that then sprayed the heavy crude on local roads as faux asphalt. Just as frequently, Texaco set ablaze the coagulated hydrocarbons that formed a thick layer on the surface of the pits. Or, similarly, it set fire to accidental spills, big or small, in an attempt to contain them. The burning off of crude led to the phenomenon many called "black rain"—what some Indigenous people called the "bleeding of the skies"—as hydrocarbon soot mixed with atmospheric mists and spat down. Traces of this rain appeared on the ubiquitous black-speckled clothes and pitted tin roofs. Similarly, volatile gases flamed freely into the atmosphere twenty-four hours a day. Researchers estimate that Texaco's operations generated up to 4.3 million gallons of hazardous waste daily over a period of twenty years (Center for Economic and Social Rights 1994; Kimerling 1991a).[13]

Chevron is correct in asserting that Ecuador was not the only place where such practices were employed.[14] My father recalls growing up near similar operations in Argentina and Venezuela during the 1930s and 1940s. But ubiquity hardly indicates soundness of a practice for life processes writ large. And such was the plaintiffs' claim in Lago Agrio.

Valenced Dissociation

As noted, the judicial inspections produced numerous technical studies and appendices regarding each alleged contaminated site. The appendices included reams of data on the chemical composition of soil and water samples— 2,817 unique samples (2,371 from the defendant and 446 from the plaintiffs) and many times more analytes—whose values included levels of total petroleum hydrocarbons (TPH) and various delimited constituents. Numerical values were to provide the empirical evidence necessary to determine whether or not chemicals derived from Texaco's former petroleum operations were still present in the environment in sufficient quantity to cause harm. Yet far from being indicative, these data were a source of disagreement. Determining what exactly specific chemical values signified roiled in controversy, as scientific experts differed on how to isolate, measure, and assess the presence of hydrocarbons and their effect.

Methodology posed a first level of divergence. Crucially, distinct analytics— theories and instruments themselves developed in parallel with quantum theory and thermodynamics—assured each party distinct (yet related) results. The effect was valenced dissociation: where combining powers performed allied dissimilarity. That is, distinct laboratory protocol, assays, and equipment uniquely converged with the specific recalcitrance of crude oil to configure two distinctly bonded methods that muddled comparison just as they dissolved the significance that distinct methods made. This methodological divergence formed the first step in consolidating the scientific basis from which Chevron claimed that Texaco operations neither contaminated the environment nor posed a risk to human health.

As Antoine Lavoisier affirmed in the late 1700s, the chemical compounds evinced in a laboratory are not the capture of a pure nature. Rather, chemical constituents derived from a lab are the function of an available and deployed analytical method and its instruments as it encounters (with the help

FIGURE 11 An auger soil core from the judicial inspections at Sacha Sur. Photo © Amazon Watch. Used with kind permission.

of human skill) a substance and its unique proclivities and properties. The capacity to materialize and dematerialize measured concentrations of a compound hinges on molecular *and* technical work. Any reading obtained is a consequence of both a substance's tractable qualities *and* the analyst's care in manipulating the technologies and protocol proper to the specific method used to produce it.

So, consider TPH, one of the primary constituents analyzed for in the water and soil samples extracted from Texaco's former oil concession. As will become clearer in the subsequent section, crude oil is composed of a complex brew of thousands of different hydrocarbons—molecules or compounds composed of carbon and hydrogen atoms. The complexity is mindboggling, but suffice it to say at this juncture that TPH is the umbrella term used to point toward that chemical complexity. It is the measured concentration of hydrocarbons in a specific matrix at a specific time and place.

In the 1970s, the US Environmental Protection Agency (EPA) developed a couple of methods for extracting TPH from a matrix, and these methods are recognized and used around the world. But because each method uses vary-

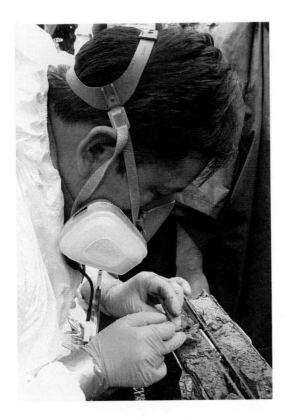

FIGURE 12 Scientist examining extraction samples from a soil core. Photo © Amazon Watch. Used with kind permission.

ing techniques to extract and detect hydrocarbons, they measure slightly different sets and subsets of the petroleum-derived hydrocarbons present in a sample (Total Petroleum Hydrocarbon Criteria Working Group [TPHCWG] 1998). That is, different analytic tests register only those hydrocarbons for which that specific method was designed to discern, extract, and record. A TPH "value," then, depends on the method and on the analytic technique used. Even the same sample when analyzed by different TPH methods may produce different TPH values (TPHCWG 1998: 13, 14). The take-home message is that because different methods detect different ranges of hydrocarbons, a TPH reading measures the hydrocarbons within that range. Which hydrocarbons and their concentrations are measured differ depending on the test used.

Understandably, the parties involved in the judicial inspections worried about how distinct laboratory methods might influence analytic results. And, indeed, the parties could not agree on a single method. The plaintiffs' experts used EPA Method 418.1.[15] This method provides a "one-number" value of TPH in an environmental medium. And although it does not provide in-

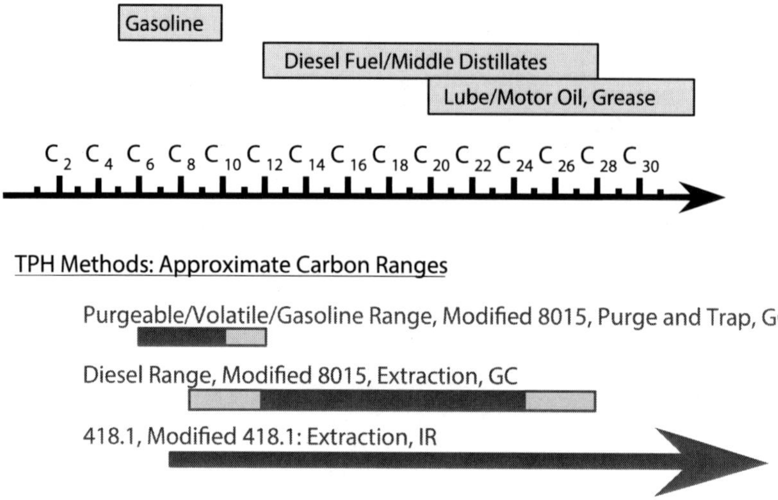

FIGURE 13 Carbon number ranges addressed by different analytical methods. Source: McMillen, Rhodes, et al. 2001: 60.

formation about the individual constituent hydrocarbons behind that one number, it captures the broadest range of the hydrocarbon compounds in a matrix. The amount of TPH measured "depends on the ability of the solvent used to extract the hydrocarbon from the environmental media and [the ability of the hydrocarbons in the solvent extract] to absorb . . . infrared (IR) light" (Agency for Toxic Substances and Disease Registry [ATSDR] 1999: 11).[16] This method extracts and registers the greatest range of hydrocarbons present in a sample. "Range" here refers to the number of equivalent carbon numbers. As figure 13 indicates, EPA Method 418.1 will register concentrations of molecules that contain eight to forty-five carbon atoms (or more depending on possible tweaking). This is the analytical method most widely used (and required by many regulatory agencies around the world) when assessing years-old exploration and extraction sites.

Chevron, by contrast, used analytical methods that captured only specific ranges of hydrocarbons in the complex brew that makes up crude. One method, EPA Method 8015B adjusted for gasoline range organics (GRO), is designed to detect molecules with six to twelve carbon atoms (C_6 to C_{12}). The other, EPA Method 8015B adjusted for diesel range organics (DRO), is designed to detect hydrocarbons with ten to twenty-eight carbon atoms (C_{10} to C_{28}). These methods are designated as GRO and DRO, because the boiling point ranges of the hydrocarbons in each group roughly correspond to those of gasoline (C_6 to C_{10-12}) and diesel fuel (C_{8-13} to C_{24-26}). These methods are great for assessing hydrocarbon contamination during the drilling and testing of wells, or testing for well leakage during extraction if the integrity of the well is compromised, or testing for underground storage leaks at gas stations. But at old and abandoned sites, these methods are less effective. They cannot capture and register the higher end of the range of hydrocarbon compounds found in crude oil.[17]

The difference between these analytical methods was not a guarded secret. In fact, Sara McMillen, of Chevron Research and Technology, coedited an influential industry book in which figure 13 was published. More will be said about this later, but McMillen was Chevron's senior scientific advisor in the lawsuit against the corporation. In explaining this figure, McMillen and colleagues wrote, "Lastly, TPH Method 418.1 covers the complete range from gasoline through lube oil, motor oil, and grease (i.e., C_8 to C_{40})" (McMillen, Rhodes, et al. 2001: 60). The point is that Chevron experts knowingly used analytic methods that would underestimate the total amount of petroleum hydrocarbon that was present in the samples they extracted. Even when the chemical concentrations for DRO-TPH and GRO-TPH were added together, they would be lower than any analysis registering blanket TPH. At a number of sites, Chevron collected and analyzed what were termed "split samples," meaning its experts extracted and tested samples from the exact location and depth as the plaintiffs' samples. When comparing the analytes produced from split samples, in virtually all cases, Chevron's GRO-TPH plus DRO-TPH values were lower than the TPH values obtained by the Lago Agrio plaintiffs (LAP).

But using a laboratory method that can detect only C_{28} hydrocarbons prompts a further concern. In practice, there is no generic crude. There are only site-specific crudes. In fact, crude is a noun best thought in the plural and multiple rather than singular and unified. Around the world, crude oil's molecular composition varies dramatically. This is because a particular crude's chemistry is an effect of geology: geography and history, space and time. According to Euro-American geology, crude oil is derived from single-celled life-forms that, through hundreds of millions of years of shifting seas and land formations, heat, and pressure, coagulated into semidecomposed matter that embedded into mineral source rock. The unique geophysics of this sequestering deep within what is now the earth's crust determines the complex composition of the hydrocarbons that make up any crude (ATSDR 1999: 146). Each unique geology affects the crude oil's molecular density, which, in turn, affects its commercial value as much as its cleanup costs. Crude oils from different regions (even different rock formations within a region) are evaluated according to molecular density—or the American Petro-leum Institute (API) gravity degree.[18] The general geoformation rule is that crude oils with a higher API gravity (or lower density) are more valuable and contain fewer heavy hydrocarbons. While, by contrast, crude oil with a lower API gravity (or greater density) contains a greater concentration of large hy-drocarbons, or hydrocarbon compounds composed of more than thirty car-bon atoms (C_{30}). And let it be noted that crude oils can have molecules with as many as one hundred carbon atoms (C_{100}) (TPHCWG 1998: 23).

Chevron's senior scientist made this correlation visible in one of her co-authored chapters in her coedited book. Analyzing the relationship between API gravity and the volume of hydrocarbon molecules over C_{44}, McMil-len and collaborators demonstrated a striking association. The percentage of "vacuum residuum" (i.e., the volume composed of $C_{44}+$) rises as the API gravity declines. The crude oil coming out of Ecuador ("Napo crude" and "Oriente crude") leans toward the lower end of the API gravity spectrum. With the global range for crude oils stretching from 12 to 70 degrees, Ecua-dor's hovers around 19 to 24 degrees API. According to McMillen's graph, the vacuum residuum would account for roughly 30 to 60 percent of Ecuador's crude. When considering old weathered crude at former Texaco exploration and extraction sites, that percentage, in all likelihood, would be higher.

Because Chevron took so many more samples than the plaintiffs, the range of TPH levels (as measured in parts per million [ppm]) that the corporation encountered far and wide subsumed the range of TPH levels encountered by the plaintiffs. And this despite these methodological differences. Further sa-lient, however, is that not only were Chevron's laboratory assays incapable of discerning the presence of hydrocarbon compounds exceeding C_{28}, but they were helpless in the face of that vacuum residuum of which weathered crude is imbued. This methodological choice is disquieting, given that Sara McMil-len's own research underscored its problematic nature.

Fractioning Risk

This first-tier analytical choice for extracting TPH from a sample did not, however, provide the scientific basis for claiming that oil operations caused no harm to humans or the environment. Rather, it formed the substrate for further chemical and laboratory work.

As noted, crude oil is a complex brew of thousands of different hydrocar-bon compounds. Its complexity thwarts any simple understanding of what crude oil is. Consequently, scientists have devised various analytic techniques to make sense of and give meaning to the concoctions that make up rock oil. TPH is the umbrella term used to gesture toward that chemical complexity. As the US Agency for Toxic Substances and Disease Registry (ATSDR, a division of the US Department of Health and Human Services) notes in its toxic profile of crude, TPH is "the measurable amount of petroleum-based hydrocarbon in an environmental medium" (ATSDR 1999: 9). The amount of TPH found in a sample is "useful as a general indicator of petroleum contamination at [a] site" (2).

TPH as an indicator of toxicity, however, is controversial, especially in the United States. This largely stems from TPH being a multifarious concoction "of hydrocarbons of varying chemical composition" (ATSDR 1999: 146). First, as noted above, TPH is a method-dependent reading. Second, although levels of TPH are general indicators of petroleum contamination in soil, water, or air, the amount of TPH calculated tells little about the particular petroleum hydrocarbons in a given sample, let alone how they may affect human, animal, and biotic life. Meaning: a number does not equal harm. A favorite industry example is household petroleum jelly, which has a TPH content of nearly 750,000 ppm yet is harmless to humans. Deciphering the actual constituents of a TPH measure demands further chemical isolation and analysis.

Thus, following scientific wisdom in the United States, a TPH measure does not determine risk. Being a gross assessment, a TPH value does not, and cannot, proffer any meaningful information on the multiple chemical compounds within that measure. Nor does it offer insight into how chemical components have in the past or will in the future interact with each other or the medium in which they exist—both key concerns in understanding toxicity (ATSDR 1999; TPHCWG 1998). As noted by the Total Petroleum Hydrocarbon Criteria Working Group (TPHCWG)—a US industry-sponsored network of scientists concerned with soil contamination and cleanup—TPH analysis is useful as a preliminary reading to assess whether a potential problem exists (or how remediation efforts are progressing), but it is not directly indicative of toxicity or environmental impact.

Within the United States, TPH levels are not regulated by the federal government, and they never have been. Rather, beginning in the 1970s, individual US states engaged in monitoring TPH by establishing regulatory cleanup levels—that is, levels beyond which contamination is not permissible. These cleanup standards (measured in ppm) have varied dramatically from state to state, ranging from 10 ppm to 10,000 ppm TPH depending on proximity to human habitation, with regulation in the preponderance of states hovering around 100–200 ppm TPH (Michelsen and Boyce 1993: 3; Staats, Mattie, and Fisher 1997: 660). This discrepancy from state to state on tolerable TPH levels was thought by many, especially the industry, to be a problem. Consequently, in 1993, the oil and gas industry—together with the consulting community, the US military, state regulatory agencies, and the University of Massachusetts—established the TPHCWG. This was the year the lawsuit against Texaco was initially filed in New York. The consortium's mandate was to address the disparity in regulatory directives. It reasoned that by standardizing the science behind cleanup levels, regulatory disparities among states would diminish.

In one of five volumes outlining a method for understanding the complexity of petroleum hydrocarbons, the TPHCWG noted: "TPH concentration data cannot be used to quantitatively estimate human health risk. The same concentration of TPH may represent very different compositions and very different risks to human health and the environment. For example, two sites may have TPH measurement of 500 ppm but constituents at one site may include carcinogenic compounds while these compounds may be absent at the other site" (1998: 5).[19] With this paradox being of primary concern, the TPHCWG was formed "to develop scientifically defensible information for establishing soil cleanup levels protective of human health at hydrocarbon-contaminated sites" (TPHCWG 1997a: ix; Twerdok 1999).

In pursuing this goal, the TPHCWG enacted a crucial transformation in how to assess contamination. This transformation shifted the focus from measuring the gross value of hydrocarbons in a matrix to assessing the risk to human health that constitutive groupings of hydrocarbon compounds might cause. In explaining the science behind assessing the toxicity of crude oil, Sara McMillen—again, Chevron's senior scientist for the lawsuit—stated, "It used to be, in encountering petroleum in the environment, that we would ask, 'how much of the contamination do we need to clean up.' Now we ask, 'how much do we need to clean up to make the area safe for humans?'" (quoted in Berlinger 2009). Unpacking this statement provides insight into the shift among oil-producing US states to regulate for specific compounds, instead of gross TPH levels as they previously had. And it offers insights into the capacity to dematerialize toxins in the Ecuadorian Amazon. Such an unpacking requires a brief excursion through the chemistry of crude.

Hydrocarbons are a class of organic chemical compounds composed of the elements carbon (C) and hydrogen (H), which account for roughly 95 to 99 percent of what makes up crude. The carbon atoms join together to form an architecture for the compound and the hydrogen atoms variously attach themselves in a plurality of configurations. Scientists think the chemistry (the properties and capacities) of an individual hydrocarbon compound depends in large part on the *structure* and *type* of chemical bonds that form between and within constituent carbon and hydrogen atoms (ATSDR 1999; TPHCWG 1997a; TPHCWG 1998: 54).

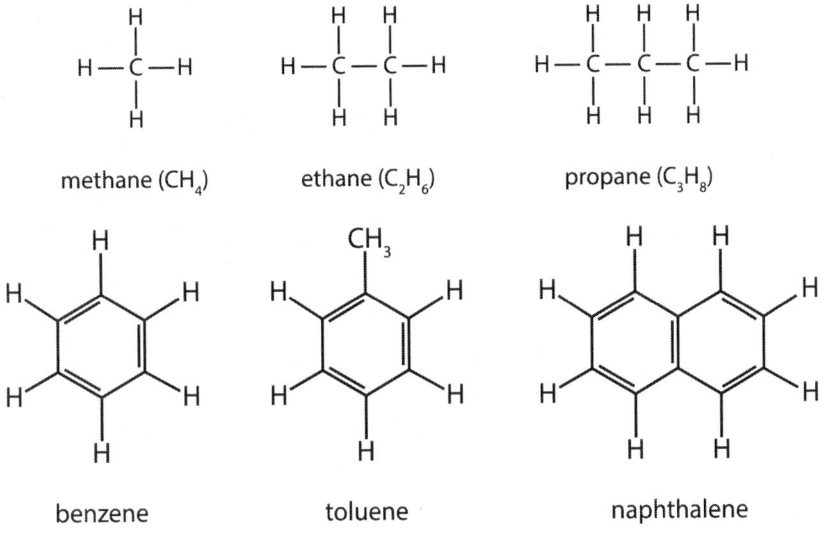

methane (CH_4) ethane (C_2H_6) propane (C_3H_8)

benzene toluene naphthalene

FIGURES 14 & 15 Simple aliphatic hydrocarbons

In teaching the chemistry of crude, hydrocarbons are schematically broken into two groups on the basis of structure: *aliphatic* and *aromatic* (figures 14 and 15). Aliphatic hydrocarbons are chains or branching chains of single bonds, or carbon-carbon double bonds, or carbon-carbon triple bonds. Aromatic hydrocarbons are ringed compounds, with benzene being the purest form. This ringed structure of schematically alternating carbon-carbon double and carbon-carbon single bonds is said to possess a "special stability" due to the movement (or what in chemistry is called "dislocation") of valence electrons in the bonding of the six carbon atoms forming the ring. This special dislocation-dependent stability (to which I return toward the end of this book) creates the uniqueness of an aromatic compound such that the ringed structure as composite (not simply the carbon-carbon links) is what solidifies the compound, making it stronger than would be mathematically anticipated. Chemically this means that aromatic rings are more stable, less reactive, and contain greater thermodynamic force. This structure and its qualities give aromatics the capacity to transform into different compounds under the right conditions while maintaining their potency.

In trying to assess the toxic constituents of a TPH value, scientists classify constituent compounds into groupings—or "fractions"—that cluster hydrocarbons according to their structure (i.e., aliphatic versus aromatic), their equivalent carbon number, their boiling point, and their "fate" (meaning how they will react and move in an environment depending on their solubility, vapor pressure, and propensity to bind with geomorphic and organic particles). Together these properties serve to establish the proclivity of a hydrocarbon to be volatile, to leach, or to persist in a matrix (ATSDR 1999: 13; McMillen, Magaw, and Carovillano 2001; TPHCWG 1998). The assumption is that hydrocarbon chemicals grouped by transport fraction have similar toxicological properties, although this is not always the case (ATSDR 1999: 13–14).[20]

The ATSDR has determined minimal risk levels (MRLs) and the TPHCWG developed fraction-specific toxicity values (reference dose [RfD] and reference concentration [RfC]) for a collection of hydrocarbon fractions, indicating at what point negative consequences may emerge from inhalation or oral or dermal contact. But, as noted earlier, crude oil contains thousands of hydrocarbon compounds—some consisting of many dozen carbon atoms—and the majority of these compounds have never been analyzed. By 2001, scientists globally had identified the physical and chemical properties of only 250 hydrocarbon compounds (ATSDR 1999: 9, appendix D; McMillen, Rhodes, et al. 2001: 58; TPHCWG 1999: 3). And of those 250, only 25 hydrocarbons had been sufficiently studied and characterized to determine their potential toxicity (ATSDR 1999: 94; TPHCWG 1997a).

Of these twenty-five hydrocarbons, two classes of aromatic hydrocarbons—benzene, toluene, ethlybenzene, and xylene (BTEX) and seventeen polycyclic aromatic hydrocarbons (PAHs)—were of particular concern. BTEX are crude oil's lightest aromatic compounds—all based on one benzene ring—while PAHs (a class consisting of several hundreds of compounds [Sanders and Wise 2011]) are hydrocarbons composed of two or more fused benzene rings. BTEX and the seventeen light PAHs are known to be carcinogenic, mutagenic, and/or teratogenic; they intervene in the cellular development of life-forms. Beginning in the 1970s with the Clean Water Act, the US EPA incrementally included these hydrocarbons among its "priority pollutants"—a set of chemicals that the agency regulates and for which it has developed testing methods, given their potential harm.

TABLE I Representative Physical Parameters for TPH Analytical Fractions Based on Correlation to Relative Boiling Point Index

Fraction	Solubility (mg/L)	Vapor Pressure (atm)	Henry's Law Constant (cm^3/cm^3)	log K_{oc}
AROMATICS				
EC_5–EC_7[a]	220	0.11	1.5	3
$EC_{>7}$–EC_8[b]	130	0.035	0.86	3.1
$EC_{>8}$–EC_{10}	65	0.0063	0.39	3.2
$EC_{>10}$–EC_{12}	25	0.00063	0.13	3.4
$EC_{>12}$–EC_{16}	5.8	0.000048	0.028	3.7
$EC_{>16}$–EC_{21}	0.65	0.0000011	0.0025	4.2
$EC_{>21}$–EC_{35}	0.0066	0.00000000044	0.000017	5.1
ALIPHATICS				
EC_5–EC_6	36	0.35	47	2.9
$EC_{>6}$–EC_8	5.4	0.063	50	3.6
$EC_{>8}$–EC_{10}	0.43	0.0063	55	4.5
$EC_{>10}$–EC_{12}	0.034	0.00063	60	5.4
$EC_{>12}$–EC_{16}	0.00076	0.000076	69	6.7
$EC_{>16}$–EC_{35}	0.0000025	0.0000011	85	8.8

a The only compound contained in this fraction is benzene.

b The only compound contained in this fraction is toluene.

EC = EQUIVALENT CARBON NUMBER.

Source: TPHCWG 1997a.

Within the technical reports emerging from the judicial inspections were tables that enumerated concentrations of TPH, BTEX, equivalent-carbon fractions for aliphatic and aromatic hydrocarbons, and sixteen PAHs within a given soil core or water sample. Chevron extracted 2,371 samples that produced 50,939 test results, and the plaintiffs extracted 466 samples that produced 6,239 test results. Together the parties produced hundreds upon hundreds of tables.

By and large, the TPH values recorded in these tables by both the defendant and plaintiffs coincide—ranging from low to astronomical levels.[21] However, the technical reports presented by both parties registered only a scant presence of BTEX and light PAHs (including the EPA priority pollutants)—the hydrocarbon compounds within crude oil understood universally to be detrimental to life-forms. It is on this basis that Chevron's counsel and experts were able to state that former Texaco operations posed no present oil-related risk to public health or the local ecosystem.

When crude oil is released from the earth its composition changes, at first abruptly, then more slowly over time, but always irreversibly, as a result of various physical and biological processes known collectively as degradation. BTEX (the four aromatic compounds composed of one benzene ring) is volatile (meaning it precipitates from a liquid to a gas state easily) and tends to dissipate from oil relatively quickly. As oil weathers, BTEX evaporates within days to weeks to months, depending on conditions once exposed to the air; when underground, BTEX dissolves in groundwater and can even slightly evaporate. The same is true of light PAHs (i.e., those with two to four benzene rings and a relatively low molecular weight). Consequently, Chevron is correct when it says "all our test results demonstrate the virtual absence of BTEX, and the disappearance of light and mobile fractions of PAH."[22] By fractioning risk, toxins disappear. In those isolated cases in which Chevron's sampling did indicate the presence of one or two known toxic compounds, this occurred at well sites still under production. Quite legitimately, the corporation argued that, given these compounds' volatility, their appearance could not be the result of Texaco activities between 1964 and 1990.

The capacity to evince crude oil's gift to harm radically differed depending on the matrix of legibility— TPH levels versus BTEX and light PAHs— into which it was placed. The plaintiffs focused on TPH; Chevron focused on BTEX, PAHs, and hydrocarbon fractions. Each matrix of legibility rested on, as it fused, a distinct constellation of molecular, technical, and social processes. Differently assembled, they invested hydrocarbon compounds with unique meaning, evincing and foreclosing qualities, capacities, and the possibilities of their effect. In the late twentieth century, the spatial and temporal volatility of light hydrocarbon compounds conjoined with a scientific impulse to standardize and a corporate compulsion to obviate undue regulation. Disturbingly, this atomic-scientific-industrial-legal assemblage allowed toxins (narrowly defined— BTEX, light PAHs) to be dematerialized and culpability to disappear, precisely at the moment when the industry advocated an "accurate" (rather than "general") science for making a contaminated "area safe for humans" (McMillen, Magaw, et al. 2001: 125). An emergence of unique integral relations, precision narrowly defined had scant resonance with the type of veridiction that lived experience implied.

Knowledge Production and Risk Management

In reworking a methodology for assessing hydrocarbon contamination, the TPHCWG was motivated by one "truth": "there is no single TPH toxicity criterion for developing human health risk–based cleanup goals" (TPHCWG 1999: 2). Meaning: there is no scientific basis for why a permissible TPH level in one place might be half that in another. Indeed, the introduction section of each TPHCWG volume that I referred to affirmed that the range of TPH standards used across different US states to assess the need for hydrocarbon cleanup was "not based on a scientific assessment of human health risk" (TPHCWG 1998: ix). Although these "sometimes arbitrary TPH standards" may "reduce human health risk," this is "by an unknown amount" (TPHCWG 1999: 2), and standards may be overly conservative and costly. Upon compiling and reviewing hydrocarbon chemical and toxicological data in five extensive volumes, the TPHCWG "developed an approach for calculating RBSL [risk-based screening levels] that provides a quantifiable degree of health protection" (1999: 2).

TPHCWG risk-based cleanup goals emerged from a tiered, risk-based decision-making framework: (1) determining the specific fraction composition of the particular hydrocarbon contaminant at a site; (2) executing mathematical calculations that establish (for ingestion, inhalation, and dermal contact) the RBSL for thirteen hydrocarbon fractions (in soil/groundwater/surface water) based on the TPHCWG's assigned toxicity criteria; and (3) assessing the hydrogeological conditions, history, possible exposure pathways and receptors of contamination, and present and future land use of the site.

Since the late 1990s, scientists working for oil companies, the API, and a number of environmental consulting firms have avidly promoted, disseminated, and extended the TPHCWG's work in peer-reviewed scientific journals and books. As noted in the preface to Sara McMillen's coedited volume, the work of the TPHCWG was formative in transforming how hydrocarbon-contaminated sites are understood in the United States: "The most desirable environmental goal" is "risk reduction . . . not achieving generic hydrocarbon concentration limits" (Loehr 2001: 4). "Risk assessment" and "risk management" based on scientific knowledge are what will "achieve an environmentally protective endpoint, i.e., a concentration of a chemical in such soils below which there is no expected adverse effect to human health and the environment" (2). This reasoning was powerful. Whereas from the 1970s through the 1990s all US oil-producing states regulated contaminated hydrocarbon sites by using gross TPH measures, by the mid-2000s virtually no oil-producing state regulated or determined cleanup in this way. Rather, following and adapting TPHCWG guidelines, state regulatory agencies set new cleanup standards based on dividing hydrocarbons into the thirteen constituent fractions. In theory, breaking up hydrocarbons into these fractions provides a more accurate understanding of risk. Yet it was at the expense of worlds of unknowns that risk management claimed to mitigate harm.

The publications of a cohort of industry-related scientists that built on the TPHCWG's work sounded a recurrent take-home message: after a contamination event, those hydrocarbon compounds known to detrimentally affect human health and the environment—light aromatic compounds (BTEX and PAHs $\leq C_{24}$)—for the most part will dissipate or biodegrade in the environment. Heavy PAHs (with more than five rings) remain in the environment but are immobile, inert, and, as such, cannot threaten human and nonhuman ecologies (Alexander 1995; Bobra, Shiu, and Mackay 1983; Claff 1999; Heath, Koblis, and Sager 1993; McMillen, Magaw, and Carovillano 2001; O'Reilly and Thorsen 2010; Staats, Mattie, and Fisher 1997; Twerdok 1999; Vorhees and Butler 1999). These studies reach this conclusion by using scientific methodologies and reasoning: they complete a review of "the literature" and perform the requisite tests and equations (K coefficient for sorption, Henry's Law coefficient for volatility, vapor pressure, and water solubility) to demonstrate that the petroleum hydrocarbons that remain after seepage, discharge, or a spill—that crude which contains heavy PAHs—do not pose a risk to human and environmental well-being.

The research published by Sara McMillen and colleagues came out of the multiyear joint-industry Petroleum Environmental Research Forum (PERF) project, which, working in tandem, extended the TPHCWG's findings to the specific concerns emerging from oil exploration and extraction sites. Two main concerns drove the research. First, it sought to discredit the use of gross TPH standards to regulate contamination (especially TPH levels that it believed to be unjustifiably onerous). The goal here was to affirm the science behind assessing contamination by means of equivalent carbon (EC) fractions.[23] Second, in the event a state agency were to set a gross TPH cleanup standard—as is the case for many countries across the globe—the research sought to establish a TPH level that was sound scientifically. Toward this end, McMillen and colleagues (McMillen, Magaw, et al. 2001) analyzed seventy different crude oils representative of the chemical composition of the types of crude around the world and devised a risk-based TPH screening level for the crude by adapting the TPHCWG's EC fraction method. Their analysis indicated that a composite TPH level of 41,300 ppm (derived from a range of 35,000 ppm to 67,300 ppm) at production and exploration sites was "protective of human health." They determined this level valid "because most of the equivalent carbon fractions found in crude oils are either not soluble or volatile enough to cause a concern" (McMillen, Magaw, et al. 2001: 126). This conclusion made the industry-devised 10,000-ppm TPH standard for all oil-exploration sites and oil-production sites more than adequate.

Much of this industry-sponsored work rests on particular understandings of degradation and biodegradation. Hydrocarbons biodegrade when soil microbes metabolize and convert them to carbon dioxide, water, and biomass, and they degrade when a compound is decomposed by light, heat, or an element, such as oxygen. A number of factors affect these processes, especially in soil, including temperature, moisture, aeration, pH, mineral content, and specific hydrocarbon characteristics. But the general pattern is for degradation to "attenuate the more mobile, light-end aromatic and water-soluble petroleum hydrocarbons, leaving behind the more recalcitrant hydrocarbons with little potential for contaminant migration" (Hamilton, Sewell, and Deeley 2001: 41). The refrain is echoed over and over: "Aging and weathering of anthropogenic hydrocarbons in soil can result in greater sequestering and less release and leachability of such chemicals" (Loehr 2001: 3); "The higher molecular weight compounds are generally less mobile and stay near the source location, while the lighter weight compounds migrate deeper into the subsurface because of greater aqueous solubility" (Hamilton, Sewell, and Deeley 2001: 40); "Leaching to groundwater and volatilization to outdoor air are of lower concern for the complex mixtures as a whole" (McMillen, Magaw, and Carovillano 2001: 16); "From a mobility perspective, the high molecular weight hydrocarbons $>C_{44}$ will not move significantly from the area of release via groundwater" (Edwards, Tveit, and Emerson 2001: 117).

Bringing this concern back to the litigation in Ecuador, Kirk O'Reilly (a former Chevron employee) and Waverly Thorsen (2010) analyzed data that

Chevron's technical experts collected during the Ecuadorian judicial inspections to examine if weathering affects the solubility of complex hydrocarbons, especially those large "recalcitrant" PAHs. They concluded that, given the "rapid weathering of the more soluble aromatics and the low effective solubility of larger PAHs," soils impacted by Ecuadorian crude would "not . . . result in dissolved [PAH] concentrations that exceed health-based drinking water goals" (402).

Alternative Chemical Orbitals

Within other spheres of science, a growing literature suggested different conclusions than those of industry-related studies. For a number of decades, scientists understood that the aging of crude reduces *acute* toxicity—as single-ringed (BTEX) and two- to three-ringed (light PAHs) aromatic compounds evaporate and decompose (Griffin and Calder 1977; Mackay and McAuliffe 1989). Similarly, scientists have understood that aromatics with a greater molecular weight are more toxic, and increasingly so by an order of magnitude per carbon ring, than their lighter, more volatile counterparts (Black et al. 1983). Because of their low solubility and tendency to sequester in the micropores of soil particles, however, it was thought that heavier aromatic compounds were not biologically available and thus posed little concern.

But research following the 1989 *Exxon Valdez* disaster and other oil spills challenged this assumption. Spurred by declines in fisheries populations, a number of scientists document the negative effect that long-term exposure to low concentrations of weathered crude has had on fish embryos and larvae (Incardona et al. 2005, 2012; Marty et al. 1997; Peterson et al. 2003; Rice et al. 2001). Contrary to prior assumptions, these scientists discovered that many of the multiringed PAHs in weathered oil are bioavailable and that chronic exposure to weathered crude can result in long-term negative effects. In Alaska's Prince William Sound, heavy aromatics passed through the porous membranes of fish embryos and lodged in lipophilic yolk reservoirs during cellular differentiation and development. Observed long-term toxic consequences on larvae and fish—although often not expressed until long after exposure ended—included cranial and spinal malformations, cardiac dysfunction, decreased size, slowed development, inhibited swimming, increased mortality, reduced marine survival, and reproductive impairment (Bue et al. 1996; Incardona et al. 2005). Most likely these effects were the result of PAH clastogenesis: three- to five-ringed PAHs metabolized as clastogens—or agents that added, deleted, or rearranged sections of chromosomes—inducing chromosomal disruption (Incardona et al. 2005; Rice et al. 2001). Research from more recent oil spills in the Gulf of Mexico (Brette et al. 2014; Incardona et al. 2014), San Francisco Bay (Incardona et al. 2012), and Norway (Sørensen et al. 2017) has substantiated and extended these findings. Significantly, these scientists' capacity to materialize the toxic effects of multiringed PAHs depended on experimental designs whose locus of analysis and laboratory techniques differed significantly from that of industry-associated science.

Overall, these studies on molecular and genetic toxicity suggest that weathering does not *necessarily* mean becoming more benign. While the acute toxicity of crude may quickly dissipate through degradation, the chronic, long-term, sublethal effects of crude have been shown to increase with time; and toxicity can intensify, rather than diminish—especially for tricyclic PAHs (Amat et al. 2006; Heintz, Short, and Rice 1999; Incardona et al. 2005, 2012). This confirms Griffin and Calder's (1977) early research suggesting that weathering magnifies toxicity. The mechanisms by which intensification occurs have yet to be understood. But a number of studies suggest that grouping PAHs into fractions by carbon number (or carbon equivalence) and assuming that the individual compounds within a fraction share similar properties (in terms of transport and fate) may not be an effective way to assess risk to life-forms. A number of scientists (ATSDR 1995; Environmental Protection Agency 2010; Incardona et al. 2005, 2012; Jacob 2008) underscore the complexity of polycyclic aromatic hydrocarbons: that PAHs with similar molecular weight (i.e., the same number of carbon atoms) but different ring arrangements have different capacities for solubility and uptake; that the pathways that enable PAHs to bind to receptors that control genes encoding enzymes (converting PAHs to water-soluble derivates) may metabolize and eliminate xenobiotic compounds *or* they may intensify toxic capacity and effect; and that the metabolites of distinct PAHs vary in their toxicity depending on the organism, the tissue, and the stage of development of the entity that has metabolized the PAH. The assertion by industry-related science that complex hydrocarbons are immobile and harmless increasingly appears premature.

Knowing and Unknowing

Differing opinions among industry and nonindustry scientists as to the behavior and consequence of multiringed petroleum compounds raise questions about the production of scientific knowledge. As several scholars (Markowitz and Rosner 2002; Michaels 2008; Proctor 1995) have explored, generating scientific uncertainty is a time-honored strategy by an assortment of industries in an attempt to preclude unwanted regulation and/or postpone liability. The tobacco industry—with its "doubt-is-our-product" infamy—has been perhaps the most egregious (Proctor 2011). But the petroleum industry is hardly innocent in this regard. Despite an association recognized between chemicals in fossil fuels and cancer in 1775 (Pickering 1999), and medical research having documented a causal link between benzene and fatality beginning in the 1920s (ATSDR 2007: 39), and between benzene and leukemia beginning in the 1930s, the oil and gas industry effectively forestalled federal regulation of benzene for fifty years by "manufacturing uncertainty" (Markowitz and Rosner 2002; Michaels 2008: 70–78).

With respect to hydrocarbons more broadly, the industry has more recently been engaged in a quest to produce certainty—a certainty that has delegitimized prior TPH regulatory standards in the United States, facilitated more lenient cleanup directives, and sought to foreclose the need for further research. Toward this end, corporate and consulting science has pursued a double tactic. On the one hand, it has repeatedly demonstrated that gross TPH measures are meaningless; that the best way to assess risk from hydrocarbon contamination is by measuring the concentration of thirteen distinct fractions and assessing toxicity from them; and that multiringed PAHs are inert and pose no risk to human health. On the other, it has demonstrated that soils contaminated with a gross TPH concentration of 41,000 ppm are not deleterious to humans or the environment. The former is an effort to control the science and assert truths in the face of ambiguity. The latter is an effort to reduce the need to assess and analyze exploration and extraction sites. Transforming uncertainty into alleged certainty, both, in turn, promote cost reduction—in terms of analysis, restoration, and reparations. In a self-referencing citational loop, the corporate-consulting science of hydrocarbons has sought to forge the scientific legitimacy and technical protection of oil operations and their collateral damage.

This production of truth claims serves to hide the controversy around petroleum hydrocarbons and the partiality of the industry's own assertions. At one level, industry-related scientists must depict the state of scientific knowledge in constrained and limited terms in order to magnify corporate certainty; that is, they misrepresent by selectively ignoring, even censoring, other science—those alternative chemical orbitals that don't suit industry interests. Not one of the industry-sponsored or -associated studies I examined cited research outside its bubbled industry-science world. More insidiously, the laboratory techniques and protocol that industry scientists have standardized are unable to apprehend, let alone capture, the harmful effects of PAHs with more than four or five benzene rings. That is, their experiments cannot register a consequential value. The life-debilitating capacity of heavy PAHs is imperceptible, and that imperceptibility is produced.

Developing methodologies to understand exposure risks from specific hydrocarbon fractions is not wrong. It simply has limitations. The shift in US regulatory policy in the mid-2000s to assess petroleum in the environment on the basis of discrete hydrocarbon fractions (instead of gross TPH measures) was arguably an industry strategy to contain and stabilize what can be understood as contamination. The historical context is notable. The TPHCWG was officially formed in 1993, only a few years after the *Exxon Valdez* spilled 11,088,000 gallons of crude oil in Alaska's Prince William Sound in 1989. The magnitude of the spill—the largest at that time in US history, spreading over 750 kilometers—and the ensuing scientific investigations, cleanup operations, and legal actions over the following decade gave witness to the impressive financial liability that assessments of contamination could wreak on the oil industry. Among those states with TPH cleanup regulations on their books at the time, most tended toward the more conservative side (action at 100 ppm to 1,000 ppm TPH) for sensitive areas—with the majority of those hovering around 100–200 ppm, and a handful of other states extending beyond 1,000 ppm (Hamilton, Sewell, and Deeley 2001: 38; Staats, Mattie, and Fisher 1997: 660; Tomlinson and Ruby 2016: 916). Were Alaska to have been among those states mandating regulation at TPH levels of 100 ppm or even 1,000 ppm, the consequences of post-spill environmental politics and remediation might not have been so devastating for Alaskan aquatic and terrestrial ecologies.

Valenced Capacities

In a written rebuttal to a Chevron technical report, Alberto Wray, the plaintiffs' chief lawyer during the first years of the litigation and an eminent Ecuadorian legal scholar, noted, "[Chevron's] expert is trying to confuse and distort the very concept of contamination. He seeks to relativize it, when the very concept of contamination is absolute. The contamination either exists, or it does not exist."[24] One might argue that Ecuador's 2001 Executive Decree 1215—"Environmental Regulations for Hydrocarbon Operations in Ecuador"— concurred. In a move contrary to regulatory science in the United States, Decree 1215 legislated that contamination be assessed and regulated using a gross TPH measure, not hydrocarbon fractions. The 2001 law established national limits for the permissible quantity of TPH in soils according to distinct land uses: that is, 1,000 ppm for sensitive ecosystems; 2,500 ppm for agricultural land; 4,000 ppm for industrial-use land; and 10,000 ppm in areas of industrial wastes.[25] The legal limit for the Amazon—a sensitive ecosystem—was 1,000 ppm TPH. With 900,000 ppm TPH (90 percent TPH content) being the highest concentration obtained from both Chevron and plaintiff samples during the judicial inspections and all sites exhibiting samples well in excess of 1,000 ppm TPH, the plaintiffs concluded there was no argument. Clearly the contamination was real.

By contrast, Chevron sustained the conviction that crude toxicity was not directly related to TPH readings and that a corporate science of risk management—based on specific hydrocarbon fractions and whose logic was powerful enough to shift regulatory process in the United States—best determined the extent to which crude contamination posed a risk to human and environmental health. Indeed, the industry expended significant energy establishing certainty about hazard. Industry-promoted science claimed that only science-based risk assessment (not a gross TPH measure) could determine danger, legitimize regulation, direct remedial action, and establish liability.

But given the complexity of crude oil, to claim that the most scientifically accurate way to determine risk is to compare the concentrations of thirteen EC fractions (with assigned toxic-concentration levels) is deeply misleading. Much is left out of this corporate-sponsored initiative to determine scientific truths. In 1999, the ATSDR stated: "Despite the large number of hydrocarbons found in petroleum products and the widespread nature of petroleum use and contamination, only a relatively small number of the compounds are well characterized for toxicity. The health effects of some fractions can be well characterized, based on their components . . . (e.g., light aromatic fraction— BTEX . . .). However, heavier TPH fractions have far fewer well-characterized compounds. Systemic and carcinogenic effects are known to be associated with petroleum hydrocarbons, but ATSDR does not develop health guidance values for carcinogenic end points" (1999: 16).[26] This scenario has little changed.

Left out of the story told by the industry-sponsored TPHCWG is that many of the unstudied or understudied hydrocarbon compounds within its hypothetical 500-ppm TPH sample could also be deleterious to well-being and health. We just don't know. And many of the compounds in the excessively high levels of TPH in samples gathered during the judicial inspections from old, weathered crude could likewise be harmful to life-forms. We just don't absolutely know. The belief that only a relatively small collection of isolated molecular structures can be deemed toxic or can pose a hazard to life-forms may not be the appropriate reference for understanding contamination in the Amazon.

As historian Christopher Sellers (1997) writes, the field of toxicology was founded on a critical move in the early twentieth century that shifted investigation from the chaotic world of disease to the sanitary world of the laboratory. Shifting the site of investigation away from ailing bodies and toward dose-response trials, industrial hygienists labored to decipher the causes of industrial and urban disease through controlled experiments. Within the lab, their credibility hinged on the technical ability to make visible what was invisible to clinicians and the population at large: the specific chain of effects that a chemical consistently caused in animal physiology. By passing the reaction between toxic environments and human bodies *through* the laboratory, industrial hygiene promised to discover the chemical causes of industrial diseases (Murphy 2006; Sellers 1997). A key dimension of the judicial inspections shared this conceit. To ascertain the relationship between a contaminated landscape and ailing bodies, laboratories that analyzed unique soil and water samples for their chemical content and toxicological consequence would provide the connecting links.

Historically, toxicology has been singular in its capacity to make visible the relationship between certain chemicals and their bodily effects: think lead and lead poisoning, asbestos and asbestosis, mercury and neurological disorders, and so on. The capacity to render such laboratory achievements, however, is circumscribed. As others have argued, the science of chemical exposure is inherently inexact (Fortun and Fortun 2005; Jasanoff 1995, 2002, 2005; Murphy 2006; Petryna 2002; Sellers 1997; Strauss 2013; Wylie 2018). Conventionally, toxicological experiments eliciting dose-response curves graphed a "threshold-limit value" for a unique chemical—the point at which a substance becomes toxic and negatively impinges on a body. The aim was to materialize "physiological reactions to chemicals that were both regular— that is predictable—and specific—that is, a signature physiological reaction for that chemical" (Murphy 2006: 90).[27] In parallel fashion, industry and regulatory toxicologists studying crude oil and its products have determined an MRL or RfD for each of the thirteen EC hydrocarbon fractions (ATSDR 1999; TPHCWG 1997b). A wonderful feat. Yet the vast majority of hydrocarbons have never been characterized.

Outside the world of the laboratory, however, discrete compounds or hydrocarbon fractions do not exist in isolation. As substance, they abide only as enactments in relational collectivity. This begs a slew of questions—of which toxicologists themselves are fully aware—that are not easy to register in a laboratory setting that conceives of a toxicological event narrowly. What toxic capacities ensue from combinations of chemicals? Or appear with temporal and spatial molecular change? Or manifest in due course at concentrations previously deemed insignificant? And the galaxies of chemicals that have yet to be studied? They are unregistrable nonevents (Strauss 2013; Wylie 2018).

Tackling these questions with regard to crude oil is daunting. As noted, hydrocarbons are mind-bogglingly complex; I've provided only a glimpse. To give a hint of further shifting complexity, consider this: a hydrocarbon compound containing twenty-five carbon atoms can transform into 36,797,588 different isomers (McMillen, Rhodes, et al. 2001: 61), a number that increases with the size of the hydrocarbon.[28] This only underscores the folly of Chevron's analytical method. Not only was it unable to detect that "vacuum residuum" (of which weathered crude is largely composed) but it also was unable to account for that amalgam's variable mobility and bioavailability. This vacuum, then, was a present absence conditioning how harm might be recognized.

Widespread, long-term, low-level exposure to hydrocarbon elements (some studied and many not) that change over time and space is precisely what fisheries scientists are confronting as they study the consequences of oil spills decades later (Incardona et al. 2005, 2012; Jacob 2008; Peterson et al. 2003). Widespread, long-term, low-level exposure to hydrocarbon elements (some studied and many not) is precisely the condition that haunts lived realities in Ecuador's northern Amazon. And this is the predicament that Ecuador's twenty-first-century environmental regulations recognized.

Thus, in setting a legal limit for TPH at 1,000 ppm in sensitive ecosystems (i.e., the Amazon), an emergent Ecuadorian regulatory science legislated by Decree 1215 was *not* asserting that "the very concept of contamination is absolute—that contamination either exists, or does not exist." Rather, Ecuador's 2001 environmental law reflected a recognition that scientific knowledge of crude toxicity is open-ended. Indeed, it would seem that an emergent environmental expertise understood that relying on limited toxicological knowledge (that is the certainty that BTEX and light PAHs are toxic) might not be the best way to secure care for life-forms. Hydrocarbons pose deep uncertainty to soils and waters and, by extension, to all human and other-than-human ecologies dependent on them. An emergent Ecuadorian environmental reasoning enfolded knowledge, curiosity, and imprecision whereby indeterminacy, uncertainty, and probability proffer an expanded platform for those who share a stake in its epistemological rules (cf. Petryna 2002: 28, 117–18). Of concern were not discrete chemical structures and fractions, but complex material substances over which apprehension hung. These were not isolated chemical compounds but ever-transforming chemical complexes subsumed "in a world already furbished with crowds of interacting beings" (Bensaude-Vincent 2014: 72). The profound ignorance of and deep curiosity in this predicament compelled Ecuador's regulatory reason toward a principled precaution.

As the ATSDR notes, "Petroleum hydrocarbons are commonly found environmental contaminants, though they are not usually classified as *hazardous* wastes. . . . The volume of crude oil or petroleum products that is used today dwarfs all other chemicals of environmental health concern. Due to the number of facilities, individuals, and processes and the various ways the products are stored and handled, environmental contamination is potentially widespread" (1999: 10). Hydrocarbons are not classified as "hazardous" in the United States because crude oil was, curiously, exempted from being considered "hazardous waste" under the 1976 Resource Conservation and Recovery Act (RCRA) (the federal law that governs the disposal of solid and hazardous wastes) and the 1980 US Comprehensive Environmental Response, Compensation, and Liability Act (CERCLA)—or Superfund legislation (the federal law that makes liable those responsible for the release of a hazardous substance into the environment).[29]

In the literature, crude oil is said to be "complex mixture." But perhaps we might more productively call it a mixt. Its hydrogen and carbon compounds are not inherently isolated and bounded. They are made to be that. As a mixt, they are fusing, melding, transforming temporally, spatially, and energetically—obliging a complexity that exceeds the ability to understand the effects of crude oil by pulling hydrocarbons apart. A few years before the Ecuadorian litigation began, the ATSDR warned, "It is extremely difficult to make general statements about typical TPH or TPH-component levels in environmental media. Environmental fate and transport processes of TPH mixtures are complex. Interactions of the chemicals within bulk oil typically result in different environmental fate and transport than would be predicted for individual components" (1999: 81). That is, Chevron's EC fractions, once extracted, might exhibit fates, properties, and transport capacities that differ from their mode of existence in the mixt. Interactions may become synergistic and/or antagonistic (95), their effects may bioaccumulate and biomagnify (95), and they may biodegrade into biocidal transitory compounds (75). Toxicity cannot really be understood to emanate from a bounded, sealed, pure molecular composite. Yes, benzene clearly has fatal effects. But this molecular structure also supports life. Thus, isolating it and seventeen other aromatic molecules (those "priority pollutants") sorely misses the complexity of hydrocarbon worlds, let alone their permeating seepage into other worlds. Compelling both death and life, benzene is multiple. Toxicity does not dwell in molecular structure. Rather its valence is preeminently of and in the milieu.

As this chapter has shown, toxicity is not an inherent property. Rather, it subsumes relational capacities, trajectories, and potentials derived of association. And these associations exceed purely chemical worlds. Enfolding molecular, technological, and regulatory orbitals coalesce to configure the truths of toxicity. That is, the ability to claim toxicity and chemical hazard is uniquely dependent on scientific apparatuses of knowledge production (i.e., ecology or industry associated), differently materialized depending on methods (i.e., gross TPH versus equivalent carbon fractions), distinctly apprehended according to endpoints (i.e., chemical versus molecular versus genetic), *and* variously adjudicated in accordance with regulatory regimes (i.e., national and jurisdictional standards). Yet these realms are not distinct, for they interpenetrate and shape one another. That is the immanence of valence— of affect and affections—to transmute collectively enfolding chemical, technical, and legal work.

The mattering of toxins—whether from seepage, spills, or combustion— rests suspended in orbital cloudscapes enfolding atoms and industry, molecular fates and share value, chemical bonds and legal contracts, failing bodies and corporate profit. From the Love Canal to Bhopal to Chernobyl, from the *Amoco Cádiz* to the *Exxon Valdez* to the *Deep Horizon*, chemical toxicity is a contentious concern. Historians of science suggest that what counts as truth is the result of historically specific practices of truth-telling—laboratory techniques, methods of observing, modes of calculating, regimes of classification. And those truth-telling practices, what could be called technological orbitals, coalesce with—that is, co-emerge and mutually transform in relation with— molecular and statutory orbitals. Science holds no pure truth.

In the lawsuit against Chevron, these orbitals of coalescence situated numerical data within specific economic, political, and social constellations of possibilities: investing or divesting them with qualities and capacities, and endowing them with or stripping them of meaning. In so doing, molecular-technical-juridico orbital cloudscapes variously intervened in this unstable realm, animating concern and exacting action, or securing righteous conceit and generating disavowal. Unquestionably, however, the scientific understandings produced by LAP experts more fully captured the pervasive threat posed by a landscape tainted with heavy hydrocarbons. The science emerging from Chevron's expert served to dismiss that threat and the precarity it wrought. As you will see in chapter 6, misunderstanding the profundity of this distinction led the US district court in 2014 to mistakenly determine that conspiracy riddled the LAP's science in an attempt to exaggerate the contamination. Conspiracy had no part in LAP's science. And far from exaggerating, their method best rendered the conditions of lands compromised by weathered crude.

INSPECTION

The following is taken from court transcripts of legal arguments before Judge German Yanez Ruiz, president of the Superior Court of Justice of Nueva Loja, on March 8, 2006, during the judicial inspection of the Sacha Sur Production Station, San Carlos Parish, Canton Joya de los Sachas, Orellana Province. I have translated and gently amended the language here.

FIGURE 16 Pablo Fajardo, chief lawyer for the Ecuadorian plaintiffs, during the judicial inspection at Sacha Sur. Photo © Amazon Watch. Used with kind permission.

Pablo Fajardo, lawyer for the plaintiffs: "Your Honor, the plaintiffs requested this judicial inspection so that the court of justice can attest to and verify the present state of the damages left by Texaco, damages that have never been eliminated. We are standing, Your Honor, on what was one of the pits or pools that Texaco built when it began operating the Sacha Sur station. I would like us to briefly examine these aerial photographs from 1976 and 1985. Here we can see the road on which we came and the township of San Carlos. . . . At this moment, we are located at this black-colored blotch that was one of the pits that Texaco built at that time. Looking at this other photograph, we can verify that this pit indeed existed. Right here is the 'wash tank' from which poured formation waters that then flowed into the pit. From there, the waste passed into a channel and onward to a local stream, which we will visit in the course of this judicial inspection. The defendant has always claimed that the [formation] waters were treated through oxygenation and aeration by passing through a series of pits that you can see in the [aerial] photographs. . . . Your Honor, please stipulate that the parties' technical experts conduct an investigation into whether these successive pools, these pools that the defendant claims served to treat formation waters, eliminated TPHs [total petroleum hydrocarbons], heavy metals, and other minerals that exist in production wastes. . . . [1]

"According to the information we have, this station was built solely by Texaco in 1972. Since that time (until its management was relinquished to Petroecuador [in 1990]), it released more than 1,100,000,000 gallons of toxic water. In addition, it burned approximately 1,870,710,000 cubic feet of gas. Because this combustion was incomplete, Your Honor, and because of the gas's sulfur content, the burned hydrocarbons generated carbon dioxide and sulfur dioxide, or what is known as 'acid rain.' . . . All these waste elements have been mixing together, exacerbating each other and putting at risk the lives of those who reside in this area. The defendant claims that these installations reflect the operating practices of the time. Totally false. Chevron wants to deceive public opinion, you, Your Honor, and all of us here at this judicial inspection, by convincing us that this was how operations were carried out around the world. That is a total lie, Your Honor. The Texaco corporation held [at the time] a patent for reinjecting [formation] waters. It had a [legal] obligation not to contaminate natural water. Despite that, they did, recklessly disregarding what would happen to the people of this area. . . . [2]

"Your Honor, as we indicated at the previous location containing the pits we just visited . . . over 1,100,000,000 gallons of toxic waters were dumped at this very station. Consequently, I ask that you now observe and place in the court record that here, Your Honor, before us is a canal that most likely was built by the defendant, formerly Texaco and today Chevron. This canal was used to funnel all the formation waters [from the pits] into the stream below that we will visit presently. Please observe from whence this channel flows. It originates at the pits. This means that the discharge of toxic waters was done in a premeditated fashion. In fact, this installation was designed and engineered in such a way that the formation waters would converge with natural water sources, which directly served the local population. . . .[3]

"In addition to noting the origin of this canal, Your Honor, I ask that we walk briefly and follow the canal outside [the station] where these waters were discharged into one of the community's rivers. We are briefly tracing the trajectory of the canal as it flows towards the south. And here, under the roots of this breadfruit tree, the canal continues and empties its waters into this small stream or spring of natural water. This is where the canal built by Texaco converged and dumped toxic waters that ultimately affect other water sources that cross the community. . . .[4]

"As we follow the stream down this ravine, I ask Your Honor that you instruct the technical experts to take the samples necessary to determine the extension of this substance that we see and inform the court of the elements that exist along its length and in the swampland beyond. . . .[5]

"These toxic elements are still active. I ask that one of the assistants please use an auger to bore a small hole in the soil. This way, Your Honor, you will observe what is unearthed and indeed concealed below the surface layer of the earth. . . . One encounters, scattered throughout this area, this crude substance embedded deep in the soil. And although, as you can see, there is absolutely no evidence of hydrocarbons on the surface, they are hidden there. Hydrocarbons stay buried for decades and continue to have an effect, migrating and leaching toward this stream that is scarcely ten meters away and obviously affecting the human population. As a consequence, dozens of men and women are suffering from cancer and dying at this very moment. This hydrocarbon substance is toxic, despite the defendant not wanting to recognize it as such. And despite not wanting to say that it kills human beings, it does exactly that. . . .[6]

"One cannot hide the truth that we are able to perceive, Your Honor. Hundreds of people live in the township of San Carlos. They are human beings who have a right to a healthy and pollution-free environment. This lawsuit,

FIGURE 17 Adolfo Callejas (*far right*), chief lawyer for Chevron, during the judicial inspection at Sacha Sur. Photo © Amazon Watch. Used with kind permission.

Your Honor, is not motivated by economic ambitions as the defendant insistently contends. Rather, the motivation is life, the life of the people. It is their dignity. It is their health. It is their relatives who are slowly dying. Those who are affected know we cannot return their health or life. But they do want the environment to be completely remediated so that, at the very least, there is a future where one can live in dignity and with transparency, in a healthy environment exactly as the Constitution guarantees. Your Honor, the people do not demand money. They demand dignity, that there be a healthy environment suitable for human life."[7]

Dr. Adolfo Callejas, chief lawyer for the defendant: "The plaintiffs claim that the production waters that the Consortium Petroecuador-Texpet discharged have brought about a number of adverse impacts to the local environment and the health of its inhabitants. I argue that this assertion is false, lacking proven scientific foundations for the following reasons: (1) As I previously mentioned, the formation water was always treated by removing the crude oil and other materials before being discharged into the environment. (2) Many hydrocarbon components rapidly biodegrade when in contact with the environment, as is the case with benzene or toluene. (3) Due to processes of evaporation, given the high prevailing temperatures in this zone the small quantity of hydrocarbons with a low molecular weight like benzene and toluene that could have existed in formation waters would have been rapidly diluted and biodegraded in the rivers and water currents in concentrations so low or nonexistent . . . that ultimately they do not effectively pose a risk to either human health or the environment. (4) In areas like those we encountered today with a high index of rainfall and whose rivers have a high flow of rapids and movement, and as such a high dilution potential, the impact of formation water is low and reversible. (5) The environmental impacts that could eventually occur by the discharge of production water would be caused basically in areas downstream very close to the discharge and usually would be clean tens of meters from that geographic point. (6) Adverse impacts are also reversible once the discharging of production water ceases. . . . According to the composition of the formation water collected in this station and the aforementioned considerations, my client affirms that there is no risk to human health or the environment directly associated with the technology that the Consortium Petroecuador-Texaco used until June 1990 to treat and discharge production waters in this and the other production stations that were built and operated in the concession area."[8]

Exposure's Orbitals

Of Epidemiology and Calculation

2

IN LATE OCTOBER 2003, amid the trial's two-week opening conciliatory hearings in Lago Agrio, Chevron lawyers grilled Dr. Miguel San Sebastián, a medical doctor and faculty member in the Department of Epidemiology and Global Health at Umeå University, Sweden. Over the course of twelve years, San Sebastián had practiced clinical medicine and conducted epidemiological research among Indigenous and non-Indigenous peoples in the northern Ecuadorian Amazon, research that constituted the core of his PhD investigations in environmental epidemiology from the London School of Hygiene and Tropical Medicine. The plaintiffs' lawyers used journal articles and reports from this research to substantiate their claim before the court that oil operations negatively affected human health. Chevron's lawyer's questions were uncompromising, insisting that the good doctor's research hardly proved anything. In response, San Sebastián retorted:

> In environmental epidemiology, it is extremely rare to speak of conclusive proof. . . . But petroleum contains, among others, two components—benzene and polycyclic aromatic hydrocarbons—which have been conclusively proven to be carcinogenic. Our studies suggest a relation between exposure to these chemical components of petroleum and cancer. . . . Although environmental epidemiology cannot speak of "cause and effect" . . .

this does not mean we cannot affirm that the incidence or risk of suffering from cancer is much higher in cantons where petroleum exploitation exists in Ecuador than in cantons where no petroleum exploitation exists. . . . These findings are corroborated by other researchers examining the relationship between acute exposure to petroleum and health in which similar symptoms obtain.[1]

Four years later, after the completion of over three years of judicial inspections, Chevron produced a document titled "Affirming the Truth: About Texaco's Past Operations and Questions of Health" (Chevron Corporation 2007). Printed on glossy paper with vibrant images and graphs, the pamphlet affirmed the corporation's commitment to human and ecological well-being. Moreover, it sought to address the "unfounded allegations . . . that our former operations in Ecuador have led to health problems among people of the region" (2). Chevron assured readers that "facts based on scientific evidence" established the "incontrovertible truth that the claims of health problems caused by Texpet's [Texaco Petroleum's] oil operations are unsubstantiated" (1–2). Moreover, San Sebastián's scientific rigor was wanting; quoting two world-renowned epidemiologists whom Chevron had contracted as experts, the text states, "As a body of data, [San Sebastián's] reports collectively contain little material information about the relationship between oil development in the Ecuadorian Amazon region and health effects among residents of that region . . . making it seem that [San Sebastián's] role is closer to that of advocate than that of a skeptical scientist" (Rothman and Arellano, quoted in Chevron Corporation 2007: 4).

This chapter delves into a second controversy animating the lawsuit: did oil operations detrimentally affect human health? It examines the controversy surrounding the reality, or truth, of an association between oil extraction and human illness and, specifically, how experts sought to signal or refute an environmental health crisis. As part of the evidence presented to the Ecuadorian court, each party to the lawsuit submitted epidemiological analyses that alternatively avowed, or disavowed, an association between oil extraction and bodily disease among local residents in a region.

In examining these epidemiological analyses, I seek to further tease out a theoretical concern at the core of this book: not to determine or arbitrate truths, but to demonstrate how truths were made. As in chapter 1, at issue here is not that one side or the other simply manipulated data to argue its case. Rather, of concern is how the legal parties created worlds with the capacity to generate seemingly inevitable truths. Extending Lavoisier's

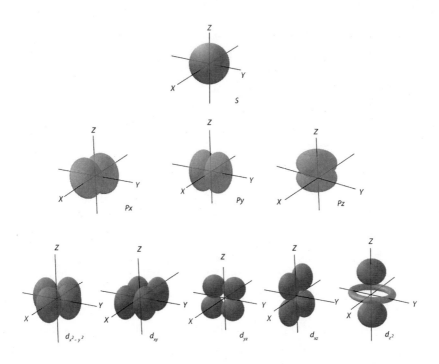

FIGURE 18 Shapes of the *s*, *p*, and *d* orbitals. There are more, and each shape replicates itself subsequent to the force field in which it first evinces.

intuition, this chapter explores how modes of observation, regimes of classification, techniques of calculation, and logics for extrapolation generated singular truths. I call these world-forming capacities "exposure's orbitals."

In chemistry, remember, an orbital is the vibrant, multidimensional, overlapping cloud space in which electrons dwell. Within the idealized atom, differentiated energy levels accommodate increasingly complex orbital shapes. These shapes, in turn, transform and hybridize, redistributing an electron's province of probability as a unique atom or molecule forms. This makes orbitals conveyors of valence, whose very capacities they share in relationally generating. Orbitals' shape, energy, and magnetism usher forth the panorama of combining potentials—valence's affective capacities to form bonds and create worlds.

In the lawsuit against Chevron, distinct epistemic commitments coalesced unique orbital worlds through which to ascertain exposure. These epistemic commitments similarly reflected distinct tendencies or ways of making knowledge within the discipline of epidemiology. Contrasting modes of reasoning, design methodologies, and analytical techniques among a handful of scientific investigations and assessments accounted for (or discounted) a connection between the presence of disease in the area and Texaco's activities over the last quarter of the twentieth century. Guided by a broad public-health imperative, the epidemiological studies presented by the plaintiffs demonstrated scientifically an association between elevated incidence of human disease and the presence of long-term oil operations. By contrast, circumscribed by positivist imperatives, the epidemiological analyses commissioned by Chevron demonstrated the impossibility of claiming scientifically any link between oil operations and human ailments in Ecuador.

Mixed into the fray were elements of a time-honored practice—long rehearsed by prominent industries—of funding counterstudies and analyses that sow uncertainty (see Markowitz and Rosner 2002; McGarity and Wagner 2008; Michaels 2008; Proctor 1995, 2011). But this was not merely a story of corporate-influenced science inflicting the wound of doubt and forestalling liability. Other attachments were also in the mixt. The capacity to forge bonds between contamination and disease hinged in part on disciplinary tensions within the study of public health. These tensions reflected conflicting convictions about what science is and what makes for good science.

Importantly, unique attachments, commitments, and compulsions within epidemiology distinctly determined what constituted exposure's orbital—both giving form to competing epidemiological analyses in the lawsuit and imputing them with meaning. For epidemiology—an observational science of the patterns, effects, and etiology of health and disease across a given population (whose porous disciplinary boundaries are infused by social medicine, biostatistics, and toxicology)—certainty is achieved only rarely. Yet Science (with a capital S) and Epidemiology (with a capital E) congealed the principal orbitals that allowed Chevron's counteranalyses to proclaim certainty. As this chapter demonstrates, Chevron's fixation with scientific and methodological rigor triggered a disconnect between analytic exactitude and contextually grounded process. That is, an emphasis on quantitative formula divorced precision from veracity. To wit, Chevron's counterstudy generated analytically precise, trust-conjuring numbers that had scant referents in the historical process of oil operations. Ignoring the spatial/temporal singularities of petroleum extraction and its capacity to affect allowed analytically rigorous epidemiological science to index crude oil as a statistically improbable etiology of cancer in the Amazon. Yet this calculation had a distorted connection with the historical reality of Texaco's oil operations.

Further deploying the purported exigencies of scientific rigor, Chevron's experts critiqued San Sebastián's work in relation to "causal criteria"—criteria allegedly foundational to any sound epidemiological study. The degree to which San Sebastián's research conformed, or not, to these methodological criteria deemed his work more, or less, scientifically valid. Claiming, however, that a handful of "causal criteria" were essential for constituting true science imbued limited quantitative and epistemological concerns with primacy, while simultaneously bracketing ethics from the realm of science. That is, granting "causal criteria" deliberative epistemic capacity as *the* arbiter tended to divorce epidemiology from its broader mission to promote and ensure public health.

Chevron's move to reduce the complexity of the crude-health compound to the constraints of methodological rigor (and thus limit exposure's orbitals) sanctioned dissociating science from historically and contextually lived process in the name of objectivity. The question "what is to be done" about the relationship between petroleum extraction and human illness remained wrenchingly elided. And, in the process, all the phenomena that constituted the mixt, that exceeded the categories of "quantification" and "causation," volatized in Chevron's method. All too often, the authority invested in positivist rigor foiled efforts to gain traction on the bonds between contamination and disease. So let's see how that unfolded in the Ecuadorian litigation.

Valencing Crude and Disease

Since the lawsuit's inception, the plaintiffs alleged that Texaco's oil operations had led to severe health repercussions. Their lawyers grounded this claim in a cluster of epidemiological studies and analyses authored by Miguel San Sebastián and various colleagues, published in peer-reviewed journals between 2001 and 2005 (Hurtig and San Sebastián 2002b, 2004; San Sebastián et al. 2001; San Sebastián, Armstrong, and Stephens 2001, 2002; San Sebastián and Hurtig 2004a, 2004b, 2005).[2] The studies were designed to assess the presence of a relationship between petroleum extraction and health in the northern Ecuadorian Amazon. Given their scientific credentials, these epidemiological studies, the plaintiffs' lawyers argued, carried significant weight and represented proof that oil operations led to disease.[3] Here I focus on the cancer studies that San Sebastián and his colleagues conducted, as these studies garnered the most attention during the Ecuadorian litigation and were the target of Chevron's counteranalyses.

SAN SEBASTIÁN: ASSOCIATING CANCER
INCIDENCE AND OIL EXTRACTION

In a study of adult cancer, Anna-Karin Hurtig and Miguel San Sebastián (2002b) collected data from SOLCA (Ecuador's national cancer registry, located in Quito) of all the cancer cases diagnosed and reported between 1985 and 1998 for individuals living in fourteen Amazonian cantons (administrative units for local government). They then classified the collected data according to where the individual diagnosed with cancer reported that they lived. If that person lived in a community in one of four cantons where oil extraction had been ongoing for at least twenty years, the authors classified that case as "exposed." Those cancer cases in which the individual resided in one of the eleven counties without oil development activities were classified as "nonexposed." The authors analyzed 473 cancer cases in the exposed counties and 512 in the nonexposed counties, concluding that the incidence of cancers was higher in the "exposed group." Higher rates were statistically significant for stomach cancer, rectal cancer, melanoma, soft-tissue cancer, and kidney cancer in men and for cervical cancer and lymph-node cancer in women. Recognizing that their "ecologic study cannot lead to a causal inference," Hurtig and San Sebastián suggest that the "exposed" group was likely affected by waters contaminated by volatile organic compounds—that is, BTEX (benzene, toluene, ethylbenzene, xylene) and light polycyclic aromatic hydrocarbons (PAHs) that, as we've learned, can dissipate over time—and that Texaco's petroleum operations had routinely discarded into the environment.

In a study on childhood leukemia, Hurtig and San Sebastián (2004) collected data from SOLCA on cancer cases for children up to fourteen years old from fourteen Amazonian cantons between 1985 and 2000. They then classified this data with respect to the presence or absence of oil activity in the county of residence. They wanted to see whether the incidences of leukemia were different between children living in proximity to oil fields and those living in areas free of oil exploitation. Like with the adult cancer study, the childhood leukemia study is "ecological" (in the epidemiological sense) and defines exposure "on a county level." That is, "exposed children were defined as those living in a county where oil development had been ongoing for at least twenty years at the time of study. Non-exposed [children] were identified as those living [in] counties without oil development activities" (247). Their analysis found that children in exposed cantons below four years of age were three and a half times more likely to have leukemia than children in nonexposed counties, and children below fourteen years of age were two and a half times more likely to have leukemia than children in nonexposed counties. As with the adult cancer study, the authors suggest that contaminated water represents a likely exposure pathway by which children could be in contact with hydrocarbons of toxicological import: BTEX and light PAHs. In particular, they note, benzene is a "well-known cause of leukemia and maybe other hematologic neoplasms and disorders," and, to date, "no adequate data on the incidences of cancers after human exposures to the other volatile organic chemicals exist" (248). Recognizing, again, that their "ecologic study cannot lead to a causal inference," Hurtig and San Sebastián affirm, however, that "the results suggest a relationship between leukemia incidence in children and living in the proximity of oil fields" (249).

KELSH: DISSOCIATING CANCER MORTALITY AND OIL EXTRACTION

While for the plaintiffs' legal team the Hurtig and San Sebastián cancer studies confirmed that oil operations detrimentally impacted human health, for Chevron's lawyers these studies underscored the inconclusiveness of such a claim. Indeed, Chevron's legal team expended considerable energy examining the credibility and scientific underpinnings of these epidemiological studies. They hired a collection of consultants to analyze the peer-reviewed studies (Christopher 2010a, 2010b; Green 2005; Hewitt 2005; Kelsh 2006, 2009, 2010, 2011; McHugh 2008, 2011; Rothman and Arellano 2005) and also one consultant to conduct a detailed counterstudy (Kelsh, Morimoto, and Lau 2009).

Published in a peer-reviewed journal, Michael Kelsh and colleagues' counterstudy reads like a thoughtful intervention into the "considerable controversy" surrounding the "potential adverse effects on health" that oil operations have generated in the Ecuadorian Amazon—where, like in other developing regions, socioeconomics, sanitation, education, and lack of public-health infrastructure confound the ability to clearly ascertain the effect of industrial activity on people's lives (Kelsh, Morimoto, and Lau 2009: 391). Furthermore, an array of problems in Hurtig and San Sebastián's work—connected to data quality, inability to document exposure (let alone cancer cases), method of interpretation, and study reproducibility—"limit the reliability and accuracy of [their] results" (382). The question of how hydrocarbon extraction and public health are connected remained indeterminate.

To provide answers, Kelsh and colleagues conducted a study comparable to Hurtig and San Sebastián's two cancer studies. But instead of looking at morbidity (cancer incidence), the study authors looked at mortality (cancer death). Although cancer mortality rates would, of course, be lower than cancer incidence, Kelsh reasoned that mortality data "still provide[d] reasonably accurate estimates of relative risk" (Kelsh, Morimoto, and Lau 2009: 382). And the results? Kelsh's mortality study contradicted San Sebastián's work. Kelsh's research concluded that his analysis "does not provide evidence for an excess cancer risk in regions of the Amazon with long-term oil production" and study findings were "not . . . supportive of earlier studies in this region that suggested [an] increased cancer risk" (381).

When comparing cancer mortality rates in Quito with Amazonian counties in which long-term oil activity had taken place, mortality rates in those Amazonian cantons were lower for all causes, overall cancer, and all site-specific cancers. When exclusively considering the Amazon region, analysis indicated that "mortality in cantons with long-term oil production activities were similar, or lower, compared to those without such activity for overall mortality, overall cancer, circulatory disease, infectious disease, and respiratory diseases, and for many site-specific cancers, with relative risk estimates generally near 1.00 (RR [relative risk] range = 0.40–1.52)" (Kelsh, Morimoto, and Lau 2009: 391). The only statistically significant elevated cancer mortality in oil-producing cantons was death from liver cancer among males. But this high level was met with skepticism. Liver cancer, the researchers note, "is relatively high in developing countries with rapid population growth"— most "probably attributable" to liver-cancer-causing "infectious agents (e.g., hepatitis virus B or C)"—and "is the most common cause of cancer death in several Latin American countries" (391). Furthermore, liver cancer "has not been previously associated with exposures to crude oil, oil refining, or benzene in previous epidemiologic studies (Wong and Raabe [2000a, 2000b])" (391). Consequently, "it is unlikely that the observed elevation is related to potential exposure to oil extracting activities" (391).

So, what accounted for such contrasting conclusions?

Orbitals of Calculation

The research studies by both San Sebastián and Kelsh are examples of "ecological" studies. In epidemiology, ecological studies are studies that attempt to ascertain whether there is a relationship between a risk factor (i.e., oil operations) and an outcome (i.e., health), where both the presence of a risk factor and health outcome are determinates for a population over a distinct geographic area and then analyzed statistically. Ecological studies differ from a number of standard epidemiological methodologies—such as cohort studies or case-control studies—that look at specific individuals and prospectively or retrospectively assess their risk for disease due to exposure levels. Ecological studies neither follow individuals nor calculate exposure levels.

Consequently, defining a given geographic unit as containing (or not) a given risk factor importantly determines the overall capacity of a health outcome to register as statistically significant or not. That is, how a geographic unit is classified dramatically affects the capacity to associate a risk factor with a disease. Significantly, San Sebastián and Kelsh used different criteria to define whether a canton was "exposed" or "unexposed." This difference enabled cancer incidence or mortality to register as either consequential or not.[4] These differences bear witness to how the world-making of technique, method of calculation, and trust in numbers (Porter 1995) impute or purge meaning and effect. As exposure's orbitals, classification, extrapolation, and calculation assumed unique combining capacities in the service of coalescing a truth.

EXPOSURE CLASSIFICATION

San Sebastián and Kelsh sought to understand whether geographically defined exposure status corresponded with cancer incidence or mortality. That is, in both cases, "exposure" was not a quantified measure of a discrete individual's contact (be it dermal, digestive, or inhalation) with a suspected noxious substance. Rather, "exposed" and "nonexposed" was a gross categorization for a canton. Both research teams defined the "exposed" population as those individuals residing in an Amazonian canton in which "long-term oil operations" had taken place. By contrast, "unexposed" populations were those individuals residing in a canton in which oil operations had never occurred.

According to Hurtig and San Sebastián (2002b, 2004), at the time of their study in 1999, there were four cantons in the Ecuadorian Amazon whose inhabitants had been exposed to long-term hydrocarbon activity: Lago Agrio, Shushufindi, Orellana, and Sachas (see map 1). In each of these counties, the researchers state that oil exploitation and production had been ongoing for at least twenty years prior to their study date. As map 3 indicates, Texaco's concession (in its final configuration) was situated within the confines of these four cantons. According to the National Institute of Statistics and Census, 118,264 people (55 percent male) resided in these four cantons in 1992. Similarly, there were eleven Amazonian counties whose inhabitants had not been exposed to oil activity through the late 1990s, "excluding seismic stud-

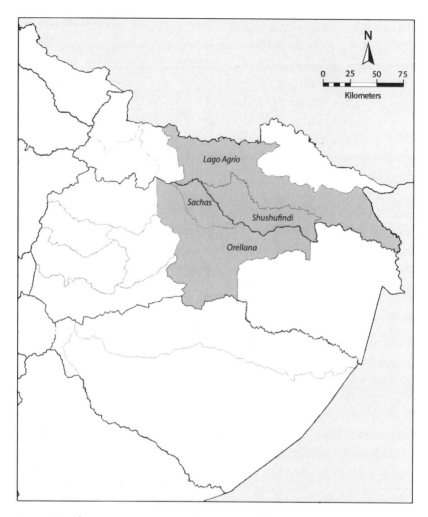

MAP 1 Map showing counties included in the study. The counties whose populations were deemed "exposed" are in gray. Source: Hurtig and San Sebastián 2002b: 1022.

ies during the late 1990s with no exploitation activities" (Hurtig and San Sebastián 2002b: 1023). Those cantons are Sucumbíos, Cascales, Putumayo, Gonzalo Pizarro, El Chaco, Baeza, Archidona, Tena, Aguarico, Mera, and Puyo. According to the National Institute of Statistics and Census, 155,710 people (52.4 percent male) resided in those eleven cantons in 1992. In Hurtig and San Sebastián's map (see map 1), the "exposed" population was everyone living in the shaded areas on the map; the "unexposed" population was represented by everyone living in the unshaded areas (1023).

By contrast, Kelsh and colleagues maintained that in addition to the four cantons that Hurtig and San Sebastián (2002b, 2004) classified as "exposed," three more cantons (Cuyabeno, Cascales, and Putumayo) also fit this category. Kelsh and coauthors offered a map (see map 3) as evidence that oil activity occurred in more than the four cantons selected by San Sebastián's teams. With each solid circle representing an oil well, the map indicates (in line with Hurtig and San Sebastián) that the cantons of Lago Agrio, Shushufindi, Sachas, and Orellana have a concentrated number of wells within them. But a significant concentration of oil wells also occurs in Cuyabeno, Cascales, Putumayo, and Aguarico.

Kelsh and coauthors added mathematical rigor to give further meaning to the visual presentation of oil wells on their map. Compiling data for twenty northern and central Amazonian cantons, they calculated the relative density of oil wells for each canton over time.[5] The algorithm is as follows: "Well-years/100 km^2 = (# of wells in canton × duration from spud date through 2005) / (area of canton in square kilometers × 100)" (Kelsh, Morimoto, and Lau 2009: 385). The data, apparently, revealed "natural breakpoints for oil well density across cantons based on their distribution of well-years/100 km^2" (387). Three categories—high density, moderate density, and low density or no wells—emerged to designate the "exposure" status of different cantons (see table 2).

EXTRAPOLATING EXPOSURE

Kelsh and colleagues' map (see map 3) and table (see table 2), however, deserve closer inspection. Kelsh used data from IHS Energy (a data analytics company) to produce these representations. The data for twenty north and central Amazonian cantons detailed: (1) the number of oil wells drilled and (2) the spud date (the day on which drilling was initiated) of each well. Because, the authors explain, "well operation dates were not available," they "estimated the working duration of each oil well as the number of years between spud date (date drilling began) through 2005 (the last year of death data follow-up)" (Kelsh, Morimoto, and Lau 2009: 384–85).

Consequently, in devising their "exposure classification" index (represented in table 2), Kelsh considered all wells equivalent: as being the same materially and as having the same material effects. That is, all wells were treated as the same regardless of whether geophysical instruments registered a well to be "dry" (i.e., contained no hydrocarbons), commercially unviable (i.e., the grade or API gravity indicated the crude was too heavy), noncommercially viable (i.e., the estimated quantity of crude was insufficient), com-

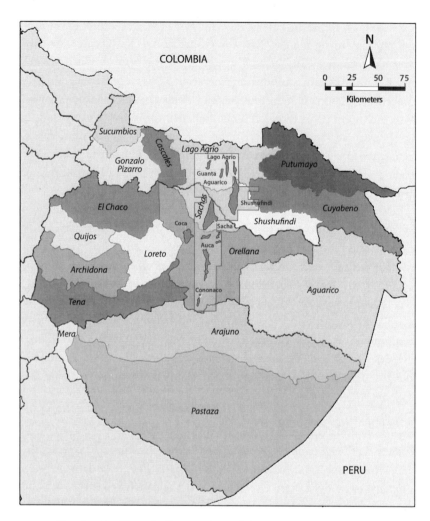

MAP 2 Map showing Texaco's oil concession and Amazonian counties.
Source: https://theamazonpost.com/chevron-ecuador/wp-content/uploads
/1996-canton-map.pdf.

mercially viable but never developed or only developed at a much later date
(i.e., because of the capital investment needed), or commercially viable and
producing oil soon after its spud date. The effect is that Kelsh's index assumes
that all wells produce comparable exposure, irrespective of a well's life trajectory and the fate of any hydrocarbons it contained.

A number of problems ensue. Principally, if the goal is to understand the
relationship between oil operations and death due to cancer, then counting

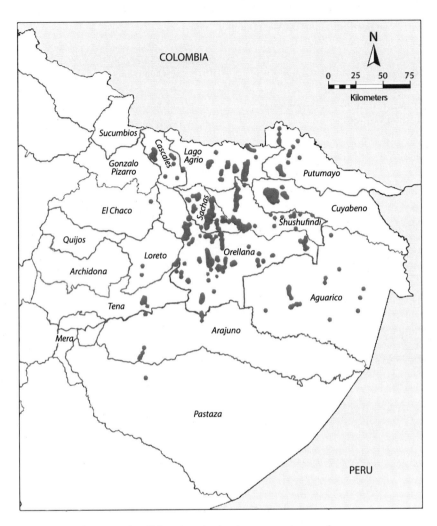

MAP 3 Map showing oil-well locations in the Amazon region as of 1990.
Source: Kelsh, Morimoto, and Lau 2009: 386.

the mere existence of drilled wells will confound, not clarify, that relation-ship. The presence of drilled oil wells does not signify the presence of long-term oil operations. So, let's look at the three cantons that Hurtig and San Sebastián (2002b, 2004) purportedly neglected to take into account.

CUYABENO — With respect to Cuyabeno, Kelsh's claim is incorrect. As com-paring maps 1 and 3 indicates, Cuyabeno was formerly the eastern half of

TABLE 2 **Exposure Classification of Cantons,
Based on Distribution of Oil-Well Density**

Category	Canton	Oil Well-Years per 100 square kilometers
"Low or none"= 0 to <1 well-year per 100 km^2	Archidona	0.9
	El Chaco	0.0
	Mera	0.0
	Gonzalo Pizarro	0.0
	Sucumbíos	0.0
"Moderate" = 1–100 well-years per 100 km^2	Tena	10.5
	Aguarico	33.0
	Pastaza	9.4
	Putumayo	38.0
"High" = 100+ well-years per 100 km^2	Orellana	145.3
	Sachas	360.5
	Cascales	116.9
	Lago Agrio	102.2
	Shushufindi	190.1

Source: Kelsh, Morimoto, and Lau 2009: 387.

Lago Agrio, and as such, Hurtig and San Sebastián did include it in their research.[6] The fact that Kelsh does the same and melds Cuyabeno into its former Lago Agrio canton has, however, a quelling effect. The cluster of wells in Cuyabeno was drilled by a company named Cayman from 1972 to 1979, and approximately one-fourth of these wells were dry.[7] Consequently, adding the wells and landmass of Cuyabeno when calculating the "oil-well density" classification for Lago Agrio significantly diminishes an understanding of the density of oil operations in Lago Agrio.

CASCALES — Kelsh's determination of "oil-well density" in Cascales is more problematic. When looking at Kelsh's map (map 3), those wells toward the west of the canton form part of what is known as the Bermejo oil field. In 1967, Texaco had drilled and capped three exploratory wells in Bermejo (*American Association of Petroleum Geologists Bulletin* [*AAPGB*] 1972: 1651–52), but the discovered oil would remain undisturbed two miles below the surface of the earth for nearly two decades. As you will learn in chapter 3, the exact configuration of Texaco's oil concession changed multiple times over its first decade. By 1973, Texaco's concession no longer included the Bermejo field, and Texaco lost control of the wells it had drilled. In 1984, Corporación Estatal Petrolera Ecuatoriana (CEPE; the then state oil company) began producing oil from the Bermejo field (Martz 1987: 364; Rivadeneira 2004: 211–13). Consequently, any long-term effects of oil contamination in western Cascales could only begin to be triggered seventeen years after the Bermejo wells were first drilled.

The oil wells toward eastern Cascales on Kelsh's map had a different history. In early 1970, the Ecuadorian state granted OKC Petroleum International (an Oklahoma-based drilling firm) a small concession that wrapped around the Bermejo field on three sides. OKC drilled four wells between 1971 and 1973 (Martz 1987: 59, 109; Rivadeneira 2004: 217; Sawyer 1975: 1151). Each of these wells proved dry and no hydrocarbons were ever extracted.[8] Thus, in contrast to Kelsch's assumption, these wells in eastern Cascales never produced any long-term (or short-term) effects of oil contamination. However, by using nonoperating wells to calculate oil-well density (well-years/100 km²) in a small landmass, Kelsch substantially skewed an understanding of the effect of oil contamination in Cascales (see table 2).

PUTUMAYO — The wells in Putumayo similarly have an erratic trajectory. In 1971 and 1972, Texaco drilled and capped five wells (Cuyabeno-1 and Vinita-1 encountered oil; Margaret-1, Confane, and Farfan were each dry) (*AAPGB* 1973; Rivadeneira 2004: 212). As such, Texaco never brought the wells into production. With the redrawing of its concession in 1973, Texaco relinquished control of these Putumayo wells. In the early 1980s, CEPE drilled fifteen wells to determine the extent of what became known as the Cuyabeno, Sansahuari, and Tetete oil fields.[9] Four of these wells were dry (or indicated scant subterranean hydrocarbons) and the other eleven did not come into production until 1986, after further drilling (*AAPGB* 1987). CEPE drilled a subsequent Putumayo oil field—the Victor Hugo Ruales (formerly Cantagallo)—in 1988, which Petroecuador (the then state oil company) brought into production in 1991.

Here, again, Kelsh's oil-well density calculation is off. Including forever-dormant wells or wells from which oil was exploited a decade-plus after their "spud date" in his calculus dilutes any capacity to clearly determine a relationship between oil extraction and disease.

CALCULUS OF TECHNIQUE

By 1972—the year the trans-Andean pipeline was completed—Texaco extracted crude oil from nearly one hundred wells in the cantons of Lago Agrio, Sachas, and Shushufindi.[10] And while Kelsh correctly notes that oil operations occurred outside these cantons, his map (map 3) and exposure-index grid (table 2) misrecognize—and in so doing impute new meaning into—their object. Indeed, the technoindustrial history and material geography of oil operations in the Amazon belie Kelsh's artifice of mathematical precision. Devising an algorithm based on the number of wells drilled created a world distinct from lived practice. It presupposed that the dots on a map—all drilled wells—were equal and represented comparable levels of exposure to hydrocarbon activity: an oil well is an oil well is an oil well. However, as demonstrated above, long-capped or dry wells drilled in Cascales, Putumayo, and Cuyabeno do not signal the same intensity of oil activity. Residents in these cantons had not been exposed to "long-term oil production activities" (Kelsh, Morimoto, and Lau 2009: 383). This other-worlding is subtle. But its effects were sufficient to produce, indeed to ensure, a negligible association between oil extraction and cancer deaths.

Let me explain.

As table 2 indicates, Kelsh determined that oil activity occurred at a greater level of intensity in Cascales than it did in Lago Agrio. But anyone familiar with the region knows that this calculation flies in the face of a lived geo-historical reality. Lago Agrio—the place in which Texaco first discovered petroleum and in which no dry well was drilled—was a high-volume production region by the early 1970s. In Cascales roughly one-third of all wells drilled were dry and the others did not produce until nearly two decades after that initial drilling. The difference in physical operations led to radically distinct material geographies than those that Kelsh's numbers-worlding imputes. Kelsh's quantitative conjuring is consequential: transforming Cascales into a high-density oil region dilutes the capacity to statistically register a relationship between oil operations and ill health in Lago Agrio.

A complementary technique of calculation further distorts efforts to measure the health impact of oil operations. Kelsh is clear on the need to include Cuyabeno (like Cascales and Putumayo) as a canton in which long-term oil operations took place. However, to determine his exposure index (well-year/100 km²), Kelsh enfolds Cuyabeno into Lago Agrio without explaining why (see table 2). The consequence is a diminished calculated exposure effect. Because the wells in Lago Agrio were drilled in the late 1960s and early 1970s, they add up to a substantial number of "oil well-years." But if the index is well-years divided by one hundred square kilometers, then the larger the area over which those well-years are divided, the lower the calculated effect. In other words, collapsing the area of Cuyabeno into that of Lago Agrio reduces the number of well-years per square kilometer, contributing to the notion that the oil-well density in Lago Agrio was less significant than that in Cascales. This is inaccurate.

Clearly, Kelsh's study leaned on an analytic rigor for generating the numbers essential for statistical analysis: absolute numbers on oil wells drilled, the "well-years/100 km²" algorithm, mortality statistics, and yearly population calculated off two national censuses (instead of one). This reasoned precision was consequential but not because it resonated with the technoindustrial trajectories of oil operations. Rather, it was consequential because it worlded otherwise. Indeed, how Kelsh determined "exposed" versus "unexposed" areas figured critically into whether mortality rates due to cancer were statistically significant. By including the populations of Cascales and Putumayo among the "exposed" group, their study made impossible the ability to measure a statistically significant association between "exposure" and cancer mortality. Lumping together populations as if they shared exposures, when they had not, served to dilute any measurable effect.[11]

Here a calculus of technique—an exposure index that conflated an object (a drilled well) and a process (long-term oil extraction)—displaces a concern for veracity with a semblance of analytic precision. One line in the Kelsh article hints at a glitch: the IHS Energy oil-well data on which his analysis depended include wells designated by the operator as "exploratory" (Kelsh, Morimoto, and Lau 2009: 384–85)—meaning, wells drilled but not developed. This triggered, however, no concern. As Kelsh and colleagues conclude, "Our analyses of national mortality data of the Amazon Region in Ecuador does [sic] not provide evidence for an excess cancer risk in regions of the Amazon with long-term oil production" (381). The article's acknowledgments state "This research was funded by Chevron" (393). And the conflation of object (drilled well) with process (oil extraction) fabricated a methodology that buttressed Chevron's claims.

"Doubt Is Our Product"

In 1969, an executive of Brown and Williamson, a tobacco firm, wrote a now-infamous memo: "Doubt is our product since it is the best means of competing with the 'body of fact' that exists in the minds of the general public. It is also the means of establishing a controversy."[12] Although not voiced in such brazen terms, the Chevron-sponsored research and analysis had a similar effect. It created doubt, stirred up uncertainty, and sowed confusion—strategies familiar among industries confronted with legal proceedings or regulatory actions. But perhaps more valuable for the corporation, this research kept the debate open and bought time, invaluable time.

Indeed, Kelsh's work (and that of other Chevron experts) resonates with practices that Michaels (2008) calls "tricks of the trade," Proctor calls "agnotology in action" (2011), McGarity and Wagner call "bending science" (2008), and Markowitz and Rosner call "deceit and denial" (2002). As Michaels writes, "Epidemiology is a sitting duck for uncertainty campaigns" (2008: 61). Uncertainty is inherent to its practice; except under rare circumstances, an epidemiologist cannot state definitively that exposure to X chemical caused Y disease in a specific patient. This is especially the case when it comes to cancer, given the complexity of its etiology and its long latency period (Jain 2013; Michaels 2008; Proctor 1995). What an epidemiologist can do, instead, is provide a skilled probability statement—that is, an assessment that establishes reliable probabilities of a given population manifesting said ailments under certain conditions. Given this, it is remarkably easy to produce contrary results to most epidemiological studies. By challenging and changing some of the parameters, Kelsh, for instance, managed to demonstrate that the statistically significant elevation of cancer risks found by San Sebastián was no longer statistically significant.[13] Drawing once more from Michaels: "Re-analyses are a specialty of some of the product defense firms, whereby one epidemiologist reanalyzes another's raw data in ways that almost always exonerate the chemical, toxin, or product in question. The studies are carefully designed to do just that. Statistically significant differences disappear; estimates of risk are reduced" (2008: 50).[14] As a number of scholars argue, much industry-conducted and industry-funded research serves to mask rather than reveal exposure-disease relationships—thereby protecting the corporate entity and not those affected by it (Proctor 1995, 2011; Bohme, Zorabedian, and Egilman 2005). This is precisely the effect of Kelsh's research. And it is the effect of another study that found no association between petroleum and cancer in the Ecuadorian Amazon, published by the same science-for-hire firm (Moolgavkar et al. 2014) after the Ecuadorian litigation had concluded, but used by Chevron in its international arbitration in The Hague.

A brief survey of the petroleum industry's relationship to benzene is instructive along these lines. Benzene—that confoundingly complex and wily, single-ringed hydrocarbon molecule within the brew that makes up crude oil—forms the basis of all aromatic petroleum hydrocarbons. The uniqueness of both its structure and its dislocated electron bonds gives benzene a paradoxical volatility and potency. By the 1930s and 1940s, benzene was cited in the medical literature to be linked with leukemia—cancer of the blood-forming tissues, especially bone marrow. In 1948, an American Petroleum Institute (API) review (prepared by the Harvard School of Public Health) noted that there were "reasonably well documented instances of the development of leukemia as a result of chronic benzene exposure" (3). And in discussing "safe limits" the same review noted, "Inasmuch as the body develops no tolerance to benzene, and as there is a wide variation in individual susceptibility, it is generally considered that the only absolutely safe concentration for benzene is zero" (4). Twenty-five years later, in 1973, a director of the Medical Research Division for Esso Research and Engineering, Dr. Robert Eckardt, wrote, "[The] accumulated literature of cases of leukemia following benzene exposure leads to the inevitable conclusion that benzene is a leukemogenic agent" (906).

Yet when the specter of regulation loomed, the oil industry employed a different tactic. In the early 1970s, the US National Institute for Occupational Safety and Health (NIOSH) collaborated with the US Occupational Safety and Health Administration (OSHA) to establish regulatory standards. Concerned about the health of workers exposed to benzene, NIOSH conducted a study of the effects of benzene on workers in a Goodyear factory producing synthetic rubber. Using Goodyear's production logs, which recorded the amount of benzene used to produce synthetic rubber, the study quantified workers' leukemia risk and found that with four years of exposure the risk doubled, in five to nine years the risk magnified by fourteen-fold, and in ten years and more the risk magnified thirty-three-fold. Convinced that their study demonstrated how dangerous it was for workers to be exposed to benzene, the study's authors strongly advocated for the regulation of benzene— "an agent known for almost a century to be a powerful bone-marrow poison" (Infante, Risnsky, and Wagoner 1977: 78, in Michaels 2008: 71). Soon after the study results were published, OSHA lowered the legal level of benzene to which a worker could be exposed over the course of an eight-hour day from 10 parts per million (ppm) to 1 ppm.

Immediately, the API challenged the new standard in court. Three years later, in 1980, the US Supreme Court ruled against the legality of the 1-ppm standard, arguing that OSHA had exceeded its authority as defined by the Occupational Safety and Health Act of 1970.[15] The majority opinion stated: "When OSHA acts to reduce existing national consensus standards . . . it must find that (i) currently permissible exposure levels create a significant risk of material health impairment; and (ii) a reduction of those levels would significantly reduce the hazard."[16] In the court's view, the agency did neither: "OSHA's rationale for lowering the permissible exposure limit from 10 ppm to 1 ppm was based, not on any finding that leukemia has ever been caused by exposure to 10 ppm of benzene and that it will *not* be caused by exposure to 1 ppm."[17] The court stated that, in order to regulate a workplace chemical, OSHA needed to demonstrate a significant risk associated with each chemical and that the proposed standard would reduce that risk. Since OSHA had done neither, according to the court, the new standard was not lawful. It took NIOSH scientists a decade to garner the research evidence to support their 1977 legal limit for benzene exposure. In 1987, OSHA reissued the new exposure standard of 1 ppm of benzene.

The industry has not stopped investing millions in analyzing and reanalyzing epidemiology on the effects of crude oil.[18] Industry-sponsored meta-analyses, in particular, have demonstrated that exposure to benzene can cause no harm.[19] The industry has commissioned studies determining that ambient benzene concentrations do not pose a risk of leukemia to the general population, as well as studies that refute the possibility that benzene exposure can cause non-Hodgkin's lymphoma. Among the epidemiologists heading this research is Otto Wong, whose meta-analyses of refinery and petroleum workers have shown no incidence of excess leukemia (Wong and Raabe 1995, 1997, 2000a, 2000b). Given that neither San Sebastián's study nor Kelsh's study could definitively prove or disprove a connection between oil operations and cancer, this industry-generated secondary literature serves to extend a scientific credibility.[20]

Hurtig and San Sebastián (2002a, 2002b, 2002c, 2004) situate the discussion of their findings, which suggests there is an *association* between increased cancer-incidence rates and "exposure," within a literature exploring the relationship between cancer and hydrocarbons. The studies they cite employ a range of methodologies to grapple with how hydrocarbons might be carcinogenic. Among others, Hurtig and San Sebastián discuss laboratory studies of mice who developed tumors after crude oil was applied to their skin; epidemiology studies that document evidence of the carcinogenic effect of

PAHs on petroleum and petrochemical workers; and ecological studies that register an association between various cancers and residence in proximity to petroleum instillations in the United States, Britain, and Taiwan. And they cite studies that refute the latter findings. Broadly, Hurtig and San Sebastián underscore the "considerable attention devoted to the biological mechanism by which some of the components of crude oil (benzene, PAH) could increase cancer risk" (2002b: 1025).

In stark contrast, Kelsh and coauthors contextualize their study by citing meta-analyses of previous research on the carcinogenicity of petroleum chemicals by Wong and Raabe (Wong and Raabe 1995, 2000a). The epidemiological studies that Wong and Raabe review are cohort mortality studies that include over 350,000 petroleum workers. As Kelsh and coauthors report, "Wong and Raabe observed no increases [among petroleum workers] in mortality from most cancers, including digestive, lung, bladder, kidney, and brain cancer" (Kelsh, Morimoto, and Lau 2009: 382). Buttressing this conclusion, the Kelsh report also highlighted that "other occupational studies have produced similarly null findings" (382). The authors do cite one study whose outcome conflicts with these analyses; however, that study, of a cohort of Australian petroleum workers with a higher-than-anticipated rate of leukemia (Glass and colleagues, cited in Kelsh, Morimoto, and Lau 2009: 382), is neatly sorted out as anomalous. Seeming to close the question of the relationship between cancer and hydrocarbons, Kelsh and others note that the International Agency for Research on Cancer, a division of the World Health Organization, long ago concluded that "crude oil was not classifiable as to its carcinogenicity in humans" (Kelsh, Morimoto, and Lau 2009: 382).

Thus, while the Kelsh research finds solace in the IARC's conclusion that crude oil is not carcinogenic, San Sebastián and his colleagues' research invokes the constituent complexity of what makes up crude, attentive to research on hydrocarbon compounds known to undermine life-forms. They write, "Crude oil is a complex mixture of many chemical compounds, mostly hydrocarbons. The petroleum hydrocarbons of most toxicological interest are volatile organic compounds (benzene, xylene and toluene) and PAH. . . . Benzene is a well-known cause of leukaemia, and perhaps other haematological neoplasms and disorders" (San Sebastián et al. 2001: 517; see also Hurtig and San Sebastián 2002b: 1023–24; Hurtig and San Sebastián 2004: 248). All fresh Amazonian crude oil contains benzene.

In 2012, the IARC published a monograph, *Chemical Agents and Related Occupations, Volume 100F: A Review of Human Carcinogens.* The publication was part of a program initiated in 1969 to critically evaluate data on the carcinogenicity of agents to which humans are known to be exposed. Among the chemicals and compounds that the IARC volume examined was benzene. According to the IARC's cancer taxonomy, benzene is a Group-1 carcinogen—meaning that it is carcinogenic to humans (2012: 257, 249–95). In reviewing the epidemiological literature on benzene, the publication called into question the research of specific scientists: the IARC "decided not to take into consideration a series of meta-analyses of studies of petroleum workers (Wong and Raabe 1995, 1997, 2000a, 2000b)" (2012: 258).[21] In a move seldom taken, the IARC declared the meta-analyses generated by Wong and Raabe to be methodologically untrustworthy. Disturbingly, Wong's work had in part been sponsored by the industry, and it is precisely Wong and Raabe's research that Kelsh and his coauthors invoked to support the conclusion that a connection between crude and disease is unsubstantiated.

Chevron's Experts

In addition to funding epidemiological counterresearch, Chevron commissioned a number of experts to examine the work of San Sebastián and colleagues.[22] Hailing from the academy and consultant firms (or straddling both), Chevron's experts independently and consistently declared that San Sebastián's studies lacked scientific rigor and did not establish any causal link between crude oil and disease—between alleged chemical exposure and observed health conditions. Indeed, the basis for delegitimizing San Sebastián's research was that it was unable to satisfy basic epidemiological criteria for establishing causation. The irony is that San Sebastián and colleagues never claimed their research signaled a causal inference; in fact, they explicitly state it did not.

The expert analysis by David Hewitt illustrates this approach. (At the time, Hewitt, a medical doctor with a master's degree in public health, was the director of occupational health services for the Center for Toxicology and Environmental Health.) He writes, "There is an established and accepted methodology by which scientists and physicians determine cause and effect relationships from chemical exposures" (2005: 2). In his expert opinion (i.e., his "training and experience in epidemiology, toxicology, and occupational and environmental medicine"), "a causal relationship between living near areas of oil exploration in Ecuador and health conditions such as . . . cancer cannot be supported based on an inability to satisfy basic criteria for establishing causation" (10). In order to affirm a link, Hewitt continued,

Cases suspected of having a causal relationship to oil exploration-related exposure must be evaluated in a systematic way to determine the validity of such an association. The chemical of concern must be identified; the health effect must be consistent with known health effects of the chemical [i.e., a chemical's signature toxicological effect]; an exposure pathway must be identified; the exposure must be quantified to determine if the health effect is consistent with known dose-response relationships; and the temporal relationship between exposure and the health effect must be consistent (i.e., the health effect must occur within the expected timeframe after exposure). Finally, alternative explanations or confounders for the health effect must be examined and ruled-out before attributing the health effect to a chemical exposure . . . [such as] infectious disease, non-chemically related skin conditions, nutritional status, habits, past medical history, and others. (9)

This method of determining a causal relationship echoes that advocated in virtually every epidemiology textbook—criteria that emerged from the work of Sir Austin Bradford Hill, an eminent English epidemiologist and statistician for the greater part of the twentieth century. In his research on the transition from association to causality, Hill (1965) emphasized the importance of paying attention to (1) the strength of association (i.e., the relative risk), (2) the consistency, (3) the specificity, (4) the temporality, (5) the biological gradient (or dose-response nexus), (6) the biological plausibility, (7) the coherence, and (8) the consideration of alternate explanations.

Hewitt further discredited Hurtig and San Sebastián's research on another front. Although benzene has long been associated with leukemia, Hewitt noted, that association has been with acute myeloid leukemia. Hurtig and San Sebastián (2004) report on cases of acute lymphoblastic leukemia (ALL), which, Hewitt contended, epidemiological studies had not associated with volatile organic compounds (VOCs) (2005: 4–5). Clearly, however, that was not a foregone conclusion. Reviewing the research on benzene's carcinogenic capacities, the IARC affirmed (seven years later) a link between benzene and ALL in children (2012: 283–84). As the esteemed publication recognized, many more constituents of crude are carcinogenic than otherwise thought.

On its websites dedicated to the lawsuit, Chevron underscored the weaknesses of the plaintiffs' epidemiological claims and stated: "The science is clear—The overwhelming body of credible scientific evidence presented to the Ecuadorian Court from . . . years of judicial inspections demonstrates that the people of the Oriente region face *no significant oil-related health risk from the areas remediated by Texpet.*"[23] The plaintiffs' lawyers had ignored what many international development agencies recognized: "The primary causes of disease in the Ecuadorian Amazon are poverty."[24] Invoking studies conducted by UNICEF, Chevron's lawyers observed that poor sanitation, infrastructure, and diet "result in [the] frequent exposure to infectious parasite- and bacteria-related disease"; and chronic malnutrition, serious vitamin deficiencies, and inadequate housing increase the likelihood of infectious disease.[25]

Of Epidemiology and epidemiology

As controversy around the lawsuit was heating up, the *International Journal of Occupational and Environmental Health* (2005: 217) published a letter to the editor that excoriated Chevron for attempting to undermine "scientific integrity." The letter—signed by sixty-one physicians and epidemiologists across the globe—was written in response to a full-page advertisement that Chevron ran in five national newspapers (*El Comercio, El Universo, Hoy, La Hora,* and *Expreso*) in Ecuador earlier that year (February 10, 2005). Echoed on the company's website, the advertisement stated, "Leading international epidemiologists and medical experts have thoroughly analyzed and discounted the health studies promoted by the [plaintiffs'] lawyers . . . arguing that they are biased and flawed, and often ignore the more plausible causes of the health problems of the Oriente region."[26] Upon quoting the work of six of its scientific consultants, Chevron concludes: "The truth is that there is *no* credible scientific evidence linking health concerns to Texpet's former oil operations."[27]

Hailing from twenty countries and five continents, the sixty-one public-health experts who signed the *IJOEH* letter to the editor were incensed by the way Chevron used consultants to produce purportedly unequivocal assessments of peer-reviewed research and, by that very act, claimed to have trumped its opponents with a superior scientific truth. The letter called the corporate consultants—three with academic positions or affiliations (Boston University; MIT; University of Texas, Houston)—"Texaco's protagonists" and "hired experts" and questioned the ethics of providing remunerated analyses to a corporation in the midst of a lawsuit. As Dr. LaDou, director of the International Center for Occupational Medicine at the University of California, San Francisco, and an *IJOEH* editor, noted, "The scientific community is getting a little bit impatient with these hired guns who are willing to have quote after quote of criticism of the scientific literature appear in corporate-sponsored Web sites, while at the same time ignoring a scientific process that reviews articles and generates scientific truth" (Guterman 2005).

Exonerating his actions, one of these esteemed commissioned experts— Kenneth Rothman, a professor of epidemiology at Boston University; author of two widely used epidemiological textbooks (1988, 2012) and coauthor of a third (2008); founding editor of *Epidemiology*; and vice president for epidemiology research at the Research Triangle Institute Health Solution—said that his analysis "is just scientific criticism" (Guterman 2005) and that Chevron had not pressured him to conform to corporate interests when reviewing the studies by San Sebastián and colleagues.

"Just scientific criticism." The comment underscores a deeper tension (and ethical quandary) haunting the science and art of epidemiology, which (adapting one of Bruno Latour's conventions) can usefully be distilled to Epidemiology and epidemiology.[28] *Epidemiology* with a capital *E* refers to a universal science of detached observation rooted in statistics, while *epidemiology* with a small *e* refers to a locally grounded, observational science directed by context and a broader mission to protect public health. Big-*E* Epidemiology exonerates Rothman's "scientific criticism," because extracting truth from numbers defines analysis as technically exacting, value free, impersonal, and unbiased (Porter 1995). Rothman's conclusion—"As a body of data, these reports collectively contain little material information about the relation between oil development in the Ecuadorian Amazon region and health effects among residents of that region" (Rothman and Arellano 2005: 6)—results from his analysis that errors in methodology and design render the findings useless (2–5). The "truth" was that the numbers (i.e., incidences of cancer) that San Sebastián and colleagues reported were not significant (2–3).

Arrogance and scientific hubris aside, a number of Chevron's commissioned experts raise legitimate concerns. Take the cancer studies by San Sebastián and his team. At the core, they are geographical studies assessing a possible correlation between cancer incidence (where the "incidence" is based on undoubtedly compromised data) and a rudimentary measurement of exposure (i.e., "positive exposure" means long-term oil operations and "negative exposure" means no oil operations in the county in which individuals resided when diagnosed). Both the imprecision of the SOLCA data and the imprecision of how exposure and its effects are defined complicate any direct reading—and understandably so; Hurtig and San Sebastián were working with the data available to them.

Determining the causes of cancer is hardly straightforward, even in the midst of comprehensive records, rigorous science, and research funding (Jain 2013; Michaels 2008; Proctor 1995, 2012; Ross and Amter 2012; Sellers 1997). In more marginalized regions of the world—such as the Ecuadorian Amazon—establishing the etiology of cancer is near impossible. The mere diagnosis of cancer is a high-tech and costly enterprise, not readily available to the inhabitants of the Amazon region. As such, what statistics are available on the incidence of cancer are most probably incomplete, lending analysis based on them to be imperfect. A host of forces (economic, racial, age, gender, and geographic) influences who might gain access to diagnosis and thus be officially counted. Ecuador's only oncology hospital that can diagnose and treat cancer is in Quito, the country's capital, a day's travel (or more) from where rainforest peoples may live.

As the epidemiological studies by San Sebastián and colleagues demonstrate, ecological epidemiology seeking to gain insight into the connection between residence and industrial discharge is difficult to pull off and can never be proven definitively. It is virtually impossible to bridge the gap between scales, where data on cancer incidence are undoubtedly incomplete, where measures of exposure are nonexistent, and where the long-term effects of diffuse hydrocarbons do not register on toxicological topologies. San Sebastián and colleagues are well aware of the weaknesses inherent in their studies' design and methodology—especially weaknesses related to the quality of data and the ability to reliably infer from it in this remote, poverty-stricken, and forgotten region of Ecuador (Hurtig and San Sebastián 2002b: 1025). Similarly, they are well aware that their findings point to an *association* or *concurrence* between oil and health concerns; findings do not signal a *causal* relationship (1025).

In response to their critics, Hurtig and San Sebastián note, "The complexity of cause-effect relationships cannot be reduced to a discussion of methodological issues such as P-values and potential confounders" (2005: 1171). It is easy to dissect studies broadly understood to be statistically weak and dismiss them as invalid. For under the cover of capital-E Epidemiology, researchers can assert only narrow truth-claims. But, Hurtig and San Sebastián argue, this approach is reductive. It denies the importance of context in producing understanding of a public-health concern. Epidemiological knowledge rarely emerges as a set of techniques divorced from social context (2005: 1172). Shredding San Sebastián and colleagues' studies because their findings are deemed less than compelling statistically, while simultaneously ignoring the technohistorical context of the industrial production and discharge of waste in the vicinity of human populations, is to enable a process that cannot be good.

Capital-E Epidemiology claims authority to its truths by dismissing something crucial: circumstance and context. In each study San Sebastián and

Hurtig conducted, the authors describe the larger "facts" about Texaco's former production practices in the Ecuadorian Amazon: that over the course of its twenty-eight years of oil operations, Texaco spilled at least 16.8 million gallons of crude oil from its primary pipeline and discharged 20 billion gallons of toxic wastes into the environment (Hurtig and San Sebastián 2002b: 1021–22). These facts are not disputed; they come from Texpet's production records and are calculated and retained by the Ministry of Energy and Mines (Dirección General de Medio Ambiente 1989; Ministerio de Energía y Minas 1989). By adroitly situating their studies in the political economy of oil in Ecuador and reviewing extensively the epidemiological literature associated with hydrocarbon activity and oil spills around the world, Hurtig and San Sebastián's research is highly suggestive and provocative in its attempt to ascertain a crude-cancer relationship under compromised conditions.

Chevron's experts, it would seem, occluded a nearly thirty-year history of Texaco discharging production wastes into the environment—an environment through which campesinos, indígenas, and myriad ecologies live. This is not a "background environment" upon which to lay the truth of numbers. This environment is a labyrinth ecology in which people abide and of which they are constitutive elements. Unquestionably, assessing health concerns with respect to a chemical's signature effect and dose-response gradient has its place. But the absence of data signaling causation is hardly reason for not thinking broadly about implications to public health. Taken together, Texaco's production records and the research on crude oil's harmful propensities could constitute the bases for an epidemiology that analyzes the uneven distribution of forces affecting public health, in contrast to one focused solely on the biomedical, physiological, and toxicological responses of human bodies. Such an epidemiology would take on the complex conditions of those whose health has been undermined and the historical contexts and sources from which suspected deleterious agents may arise. It would welcome the intricate orbitals of coalescence that constitute complexity.

Epistemic Orbitals

Exploring the science of epidemiology over the past two hundred years, Mervyn Susser and Zena Stein (2009) distinguish discrete eras with corresponding theoretical paradigms: the early nineteenth century was the era of sanitary statistics and its reigning paradigm, miasma; the late nineteenth to early twentieth centuries were the era of infectious disease and its reigning paradigm, germ theory; the latter twentieth century was the era of chronic disease and its paradigm, exposure-outcome risk ratio; and the early twenty-first century is the era of eco-epidemiology, in which a complex multiplicity of scales and dimensions will be the reigning paradigm.

This schema reflects distinct empirical, analytical, and epistemic conceits that have shaped the practice of epidemiology. And the tensions spanning the late twentieth and early twenty-first centuries shine insight into the uses of epidemiology in the lawsuit against Chevron. That is, the plaintiffs' and the defendant's contrasting epidemiological analyses reflect epistemological tensions within the discipline. This friction centers around two concerns. First, what is the locus from which "appropriate causal concepts" arise (Susser and Stein 2009: 318)? Do they emerge from universal laws or are they derived from a "localized ecologism" (the complexity of a multidimensional context) (319)? Second, what is the appropriate relationship between the science of epidemiology and public health?

With the end of World War II, scientific innovation, economic expansion, and the relative aging of the population, industrialized nations witnessed a simultaneous decline in infectious diseases and an increase in the so-called chronic diseases of middle age. Whereas medicine viewed chronic diseases as "degenerative . . . intrinsic failures of an aging organism," epidemiologists developed the perspective that "chronic diseases had environmental and behavioral causes" (Susser and Stein 2009: 166). In the United States, researchers at leading universities focused on identifying individual risk factors—the elements clustered together under the conceptual category "exposure"—that "predisposed individuals to chronic conditions" such as lung cancer, coronary heart disease, and peptic ulcer (171). Risk-factor studies sought to estimate "the effect of an exposure on the disease risk of individuals within a given population" (330).

Early research devised what would become highly successful research designs—cohort studies, case-control studies—to infer a causal relationship between exposure and outcome. The work of Sir Brandon Hill and his student Richard Doll (Doll and Hill 1950, 1956) on smoking and lung cancer still stands as exemplary of both the cohort studies and the case-control method. What set this work apart—allowing understandings of chronic disease etiology to shift from being intrinsic to environmental—was the meticulous care and caution of Doll and Hill's scholarship to consider problems and weaknesses in the study design, introduce and rule out alternative exposure hypotheses, and statistically analyze the significance levels and strength of the observed association, thus estimating relative risk.

Diverging from the notion that *a* specific agent caused a disease (a tenet of germ theory), researchers were attentive to multiple causality. Over the following decades, epidemiologists seized upon the success of the risk-factor paradigm and developed increasingly sophisticated mathematical models for assessing the relationship between exposure and disease. Early risk-factor epidemiologists used conventional statistics (e.g., multivariable regression) to understand the relationship between a single outcome or dependent variable (i.e., a disease) and multiple predictors or independent variables (such as age, gender, occupation, behavior, genetics, etc.). The expanded use of computers and computational statistics fostered the advancement of elaborate stratified and multivariate modeling that sought not only to control for confounding factors (how another variable distorts association between exposure and outcome) but also to measure more complex interactions: how the magnitude of association between X (exposure) and Y (disease) varied across levels of particular variables; how the association of X (exposure) on Y (disease) necessitates a mediating variable(s); how the effect of X (exposure) on Y (disease) depends on the level of a third moderating variable; and how the interdependence and interactions among independent variables (exposure) are associated with distinct Yn (diseases) (Holford 2002; Lewis and Ward 2013).

For several study designs crafted during this era, it proved possible to infer a causal relationship and draw implications for prevention: smoking and lung cancer; smoking and cardiovascular disease; various hormonal drugs in pregnancy and birth defects; low levels of radiation and leukemia; saccharine and bladder cancer; coffee and pancreatic cancer; and various concerns around toxic waste, pesticides, and defoliants. As remarkable as this was, it was achieved with little understanding of either the biological processes that linked the exposure to the disease or the social context that gave rise to them (Susser and Stein 2009: 330). The focus was exclusively on risk factors at the individual level within a population, while the broader "social context was held constant" and excluded from "the frame of the investigation" (330).

By the mid-1980s, the statistical wizardry of the risk-factor paradigm had reached its apogee. Seminal texts (Miettinen 1985; Rothman 1988, 2012; Rothman, Greenland, and Lash 2012)—now in their second and third reprinting—explicated how elegant study design, causal inference, and mathematical equations offered insight into the complex interrelations between variables within a disease system. Sophisticated techniques of calculation and quantification tested complex causality. To this day, the risk-factor paradigm—a paradigm that regards highly the ability to mathematically describe the complex system that is disease—remains the dominant focus of training in epidemiology, especially in the United States (Susser and Stein 2009: 333).

The celebrated texts of Rothman (Rothman 1988, 2012; Rothman, Greenland, and Lash 2012) signal two tendencies. First, they distance themselves from the public-health orientation of the pioneers of the "chronic disease era"—such as Bradford Hill. Second, they herald statistical analysis rather than research design as the discipline's central focus. Rothman opens the first chapter ("Introduction to Epidemiological Thinking") of a text that is now part of the discipline's canon by defining epidemiology as "the study of the distribution and determinants of disease frequency"—a definition distinguishing it as the "core science of public health" yet not invested in public-health ends (Rothman, Greenland, and Lash 2012: 1; see also Rothman 1988). As Susser and Stein note, this incarnation of epidemiology is "untrammeled by the call to address disease in social groups, communities, and other formations of the social structure"; "akin to the physical sciences" it searches instead for "the highest level of abstraction in universal laws" (2009: 312). This makes "the scientific endeavor" distinct from "the public-health advocacy" (331).

As many scholars of human health underscore (Hill 1965; Michaels 2008; Murphy 2006; Proctor 1995; Sellers 1997), since World War II, the dominant epidemiological paradigm of our time has been ensnared in concerns over the "biomedical processes of [the] individual" (Hurtig and San Sebastián 2005: 1171). Discrete physiological metrics developed in conjunction with biomedicine, biostatistics, and toxicology form the predominant means by which causality inference is registered (that exposure X led to effect Y). Ergo, the critique of Chevron's experts: where is the dose-response curve; the biological plausibility of an exposure and its effect; the specificity and temporality of an exposure and its effect; and the relative risk and consistency? For Chevron's experts, research design and biophysiological and computational metrics provide a productive and precise understanding of causality. They are scientific tools for determining truth.

But this form of causality—even as Kenneth Rothman and Sander Greenland (2005) recognize, when not donning a corporate-expert hat—is narrowly constraining. Such was the assessment held by Sir Austin Bradford Hill. In his 1965 presidential address to the Royal Society of Medicine, the famed epidemiologist outlined what have come to be known as the "Hill criteria"—criteria still widely accepted today—as key in determining a causal relationship. Indeed, brandishing these criteria, Chevron asserted, "There is no credible scientific basis for suggesting that the compromised health of people in the Ecuadorian Amazon was linked to petroleum-derived environmental contamination."

But the nuances of Hill's 1965 speech reveal a more complex intent. Upon enumerating the eight "criteria" useful in determining the passage from association to causality, Hill underscored that just because an association provided by a given criterion was weak or absent, that was not grounds for dismissing a cause-effect hypothesis. If strong correlation "exists we may be able to draw conclusions without hesitation; if it is not apparent, we are not thereby necessarily left sitting irresolutely on the fence" (297). It depends on circumstance—on context, on milieu, on the play of valence.

To make his point, Hill noted examples of statistically weak associations being causal nonetheless: the relationship of harboring meningococcus and contracting meningococcal meningitis, or being exposed to rat urine and contracting Weil's disease (1965: 296). And he noted cause-effect relations that were not recognized because they were not deemed "biologically plausible" (298)—as, for example, was the case with ALL. Still, he advised, even when the strength of an association between X agent and Y disease supports a causal hypothesis, caution is in order: "one-to-one relationships are not frequent," he noted, and indeed "multicausation" involving a plethora of unknowns was more likely (297).

What congealed, and became christened, as the Hill criteria were, for Hill, anything but criteria—as in, the *minimal* requirements necessary for adequately asserting a causal relationship. Rather, they were for him "viewpoints"— different angles from which to study an association before concluding that a cause and effect determined the relational link. He wrote, "What I do not believe—and this has been suggested—is that we can usefully lay down some hard-and-fast rules of evidence that must be observed before we accept cause and effect. None of my nine viewpoints can bring indisputable evidence for or against the cause-and-effect hypothesis and none can be required as a *sine qua non*" (1965: 299).

For having pioneered a new quantitative phase in twentieth-century epidemiology, Hill was cautious with numbers. Considering the weight given to statistical tests and calculations, he noted, "Often I suspect we waste a deal of time, we grasp the shadow and lose the substance, we weaken our capacity to interpret data and to take reasonable decisions whatever the value of P. And far too often we deduce 'no difference' from 'no significant difference.' Like fire, the χ^2 test is an excellent servant and a bad master" (1965: 300). The power of and trust in numbers had its place. And that place was subservient to, and in the service of, circumstance.

That Which Exceeds Causation

Invoking the lack of scientific rigor to dismiss a statistically weak ecological study uniquely frames an environmental health concern. It reduces a complex economically and racially inflected technosocial health phenomenon into an epistemological problem that can only be understood through mathematical equations and rigid criteria. And it shifts concerns over the stakes of acting to a narrow epistemological question of what justifies action. Configuring the relationship between oil extraction and disease in this way bestows scientific objectivity on some research while denouncing other research as invalid. Precision becomes a substitute for veracity (cf. Salm 2014: 5), thereby making the act of denying an association seem reasonable and far from unethical. This is what the distinctive orbitals of big-*E* Epidemiology can do.

Hill wrote:

> In passing from association to causation I believe in "real life" we shall have to consider what flows from that decision. On scientific grounds we should do no such thing. The evidence is there to be judged on its merits and the judgment (in that sense) should be utterly independent of what hangs upon it—or who hangs because of it. But in another and more practical sense we may surely ask what is involved in our decision. In occupational medicine our object is usually to take action. If this be operative cause and that be deleterious effect, then we shall wish to intervene to abolish or reduce death or disease. While that is a commendable ambition it almost inevitably leads us to introduce differential standards before we convict. Thus on relatively slight evidence we might decide to restrict the use of a drug for early-morning sickness in pregnant women. If we are wrong in deducing causation from association no great harm will be done. The good lady and the pharmaceutical industry will doubtless survive. . . . In asking for very strong evidence I would, however, repeat emphatically that this does not imply crossing every "t," and swords with every critic, before we act. All scientific work is incomplete—whether it be observational or experimental. All scientific work is liable to be upset or modified by advancing knowledge. That does not confer upon us a freedom to ignore the knowledge we already have, or to postpone the action that it appears to demand at a given time. (1965: 300)

Hill clearly believed there were circumstances that called for restricting substances to avoid potential danger. Indeed, using quantitative techniques to adjudicate the significance of epidemiological studies seeking to gain traction on the extraction-cancer connection, and negating that relationship in its entirety when studies do not live up to quantitative criteria, is to turn a proactive science inert. It also makes epidemiology vulnerable to manipulation and a tool for perpetuating inequity. As Hurtig and San Sebastián suggest, eviscerating statistically weak studies will do more harm than good when industry calls (2005: 1172).

During those early days of testimony at the Lago Agrio superior court in October 2003, Miguel San Sebastián stated under oath before the court, "In general, there is a paucity of knowledge about petroleum's possible impacts on health. Given that exposure to petroleum and its relation with cancer is little studied—with the exception [of the carcinogenic effect of] benzene and polycyclic aromatic hydrocarbons—it is possible that the different chemical components of petroleum could be producing different types of cancer which hitherto are unknown."[29]

In a commentary on Hurtig and San Sebastián's cancer study in 2002, a prominent Canadian epidemiologist, Jack Siemiatycki, noted, "Epidemiological research is sometimes used as a cover of scientific legitimacy in calling for sensible public-health precautions. While this definitely puts epidemiologists 'on the side of the angels,' it also risks compromising the scientific credibility of epidemiology" (2002: 1028). Similarly, epidemiological knowledge may sometimes be used, as Chevron has, as a cover of scientific legitimacy to negate impacts and absolve an association; this, too, undermines scientific credibility. But one hurts and the other does not. As Hill instructs, depending on circumstance, action is often called for even when data are more suggestive than indicative.

Siemiatycki continued:

Epidemiology is an eclectic discipline, using an ever-expanding panoply of methods. In assessing methodological quality, we must make allowances for the resources and local conditions in which the investigators find themselves. To require the same standards of research design everywhere would lead to pockets of the world where there is no information at all on various issues. The study by Hurtig and San Sebastián represents a bold attempt to use imperfect data to derive scientific knowledge; it is useful in highlighting the issue and drawing attention to the limitations of the data. But it does not provide strong evidence in favour of the hypothesis. Nevertheless, given the complexity of disease aetiology, and the need to discover both universal and local facets of disease aetiology, we should encourage the conduct of research such as this. (2002: 1029)

Reflecting on the practice of science, Latour states, "There do not exist true statements that correspond to a state of affairs and false statements that do not, but only continuous or interrupted reference. It is not a question of truthful scientists who have broken away from society and liars who are influenced by the vagaries of passion and politics, but one of the highly connected scientists . . . and sparsely connected scientists [all] limited only to words" (1999: 97). But, surely, some would claim, San Sebastián and colleagues are influenced by their "vagaries of passion." San Sebastián hails from the Basque region, a purported bastion of leftists and vindicators of the underdog. Rothman alluded to this sentiment when he wrote, in analyzing the Hurtig and San Sebastián (2002b) article, "Its unbalanced assessment of the epidemiology makes it appear more of an advocacy exercise than a scientific paper" (Rothman and Arellano 2005: 6). And, surely, others would argue, Kelsh and Chevron's experts, enchanted by the generosity of their corporate benefactor, have swayed their science. With remuneration of multiple hundreds of dollars per hour, pleasing the lord holds the promise of further engagements. Clearly theirs are not the voicing of "truthful scientists" deploying sanitized epidemiological reasoning to uncover the statistical unworthiness of studies influenced by politics.

So how to understand what exactly Latour is saying in relation to the Chevron case? Reading further helps:

> If the traditional picture had the motto "The more disconnected a science the better," science studies says, "The more connected a science, the more accurate it may become." The quality of a science's reference does not come from some *salto mortale* out of discourse and society in order to access things, but depends rather on the extent of its transformations, the safety of its connections, the progressive accumulations of its mediations, the number of interlocutors it engages, its ability to make nonhumans accessible to words, its capacity to interest and to convince others, and its routine institutionalization of those flows. (1999: 97)

For Latour, science is not about distance. Rather, it is all about attachment.

This chapter has sought to evince the orbital clouds that usher forth the combining attachments and bonds that forge epidemiological knowledges. What I call "exposure's orbitals" captures the epistemic realms whereby studies differently determined their modes of analysis, regimes of classification, techniques of calculation, and logics for extrapolation in order to generate a singular truth. But the phrase also captures the orbitals, the valenced bonds and rapports, that formed upon exposure: that is, upon a study's release into the world. San Sebastián's and Kelsh's work differently seeped into and stirred reflection within the discipline of epidemiology, the Ecuadorian litigation, and here. And that is because the denser and more extensive the orbitals suffusing research are, the more secure the science and its capacity to provoke thought.

Following Latour, one might venture that none of the epidemiological studies invoked through the lawsuit against Chevron represent a fixed "truth." Neither the work of San Sebastián and Hurtig nor that of Kelsh and colleagues renders a truth on lived reality in the region of Texaco's former operations. How could the broad sweeps of these ecological studies possibly capture that? Rather, these studies produced knowledge differentially based on the parameters they choose for defining exposure's orbitals. As conveyances, these orbital spaces included the depth of locally and historically contextualized knowledge of the region; the logic of their methodology and the sources of their data; their statistical χ^2 test and p-values; the spheres of literature within which researchers situate their findings; the basis for their assumptions in discussing their results; and the sources of their funding. And as conveyances of connection, exposure's orbitals extended to include other scientists citing this research; other scientists intrigued by the controversy this work generated; lawyers deploying their works; and communities and nongovernmental organizations concerned with the place of epidemiology in contamination disputes.

As detailed in this chapter, the work of San Sebastián and Hurtig reflects that of a densely relayed science, whose intensities and proximities profoundly engaged with both its object of study and the epistemic controversy that its knowledge-making sparked. Kelsh's work reflects that of a sparsely connected science whose mystifying methodology, confined discussion, and fraught funding is scarred with interrupted attachments. And the work of Chevron's experts, in general, reflects a suspended ethico-epistemological bond in which the taut narrowness of their "causal criteria" quivered brittle with isolation. What became wrenchingly apparent in Ecuador was how epidemiology—the science of the incidence and distribution of disease—could so variably embrace or detach from local concerns over health and, consequently, what counts as ethical engagement, wounding, and life worthy of being in Amazonia and beyond.

DEATH

On January 4, 2017, the Frente de Defensa de la Amazonía wrote the following obituary, "Legendary Ecuadorian Nurse Who Hosted Celebrities and Battled Chevron over Pollution Tragically Dies of Cancer."[1]

SAN CARLOS, ECUADOR—Rosa Moreno, the legendary Ecuadorian nurse who hosted major celebrities such as Brad Pitt in her small jungle health clinic while serving as a medical lifeline to people battling Chevron over oil pollution, has herself succumbed to cancer apparently caused by exposure to toxins in her community, the Amazon Defense Coalition (FDA) announced Tuesday.

Alphonso Moreno, Rosa's husband and a community leader in San Carlos, confirmed the death by phone on Tuesday. "My beloved Rosa who took care of her own family and hundreds of people in this community has left us forever," he said.

Moreno, 55, who had an infectious smile and quiet grace that captured the attention of people around the world who visited the devastated area, had three adult children and had been married to Alphonso for over thirty years. Among those who visited her clinic in recent years were actor Brad Pitt; actor and producer Trudie Styler, wife of Sting; human rights activist Bianca Jagger; actor Daryl Hannah; and US Congressman James P. McGovern.

[. . .]

For more than three decades, Moreno was on the front lines of the health catastrophe in Ecuador's Amazon region caused by oil pollution in the area where Chevron discharged benzene-laced waste into rivers and streams relied on by local inhabitants for their drinking water. Moreno lived in a small house near the site of a large oil separation station surrounded by dozens of open-air waste pits gouged out of the jungle floor by Chevron and later abandoned. Many of the Chevron pits, constructed mostly in the 1970s, still contaminate soils and groundwater and have pipes that run oil sludge into nearby waterways.

San Carlos became known as ground zero in the legal battle against Chevron given that several rivers and streams pass through the town and carry the oil contaminants into other areas where Indigenous groups live. Litigation against Chevron, first brought by Moreno and others in 1993 in the United States but later transferred to Ecuador at Chevron's request, resulted in a historic $9.5 billion judgment against the oil giant and captivated the world's attention as one of the most successful corporate accountability campaigns ever.

[. . .]

[F]rom her tiny clinic on a dirt road, Moreno hosted a long line of international celebrities and politicians to sensitize them to the health impacts of oil contamination. The clinic itself was often bereft of medicine and lacked any diagnostic equipment; most farmers were poor homesteaders and had no funds to take the long eight-hour bus ride to Quito over the Andes mountains to receive hospital treatment at the cost of a year's wages.

[. . .]

Moreno was mostly known as a person who tried against all odds to stave off the impending health disaster with her compassionate care of young children. The clinic was a short walk from her house and she was often found there seven days per week. Moreno meticulously kept a handwritten log of people in the clinic who had died, often without receiving proper treatment given the paucity of doctors in the area. The list in recent years had grown to dozens of names—many were young children—even though only 2,000 people lived in the surrounding community. Each name on the list had a date of birth and date of death scrawled in Moreno's distinctive script.

Over time, Moreno became an activist as well. On three occasions, she traveled to the United States to speak to the media and to confront Chevron executives at the company's annual shareholder meetings, which she entered via a shareholder proxy. Her comments, along with those of Indigenous leaders from the area, were generally dismissed by Chevron's CEOs David O'Reilly and John Watson who used various technical arguments to claim the company had no legal responsibility for the pollution.

Moreno and most of those who lived in the affected area have been forced to consume water from contaminated streams and rivers given the almost complete absence of potable water systems. Cancer rates in the area have been confirmed not only by several independent health studies but by independent reports from journalists. One analyst formerly with the Rand Corporation, Dr. Daniel Rourke, estimated based on current evidence that 10,000 people in the affected area of Ecuador would die of cancer if no remediation were to take place.

The FDA (the Spanish acronym for the Frente de Defensa de la Amazonía), the grass-roots organization that represents dozens of affected communities in the lawsuit against Chevron, issued the following statement:

"We believe Rosa Moreno's life was cut short due to Chevron's atrocious, irresponsible and criminal behavior in Ecuador. Rosa was fearless in telling the truth about Chevron's role in poisoning her community in San Carlos. She helped care for sick people and she saw many of her neighbors, friends, and family members die of cancer. Now that she too has died of cancer, Rosa has tragically become yet another victim of Chevron's greed. Her proud legacy will live on forever and will help motivate the communities to ensure that Chevron is held fully accountable for the harm it has caused."

[. . .]

Luis Yanza (a friend, founder of the FDA, Ecuadorian community leader, and winner of the Goldman Prize with Pablo Fajardo in 2008) called Moreno "the gold standard for health care workers" and said she did not deserve her fate.

"This is a tragedy for the people of San Carlos who relied on Rosa to provide medical relief and a warm smile in a time of extreme hardship. Rosa's death underscores that Chevron's pollution in Ecuador remains a loaded gun aimed at the heads of thousands of people. Until that gun is removed, it is inevitable that more deaths will follow."

PART II
SPECTRAL
RADICALS

IN CHEMISTRY, BONDING is the effect of electrons associating in milieu. In organic chemistry, this often occurs through covalent bonding, meaning when two electrons (each from a different atom or molecule) join to share a common orbital. Different conditions and contexts, however, can break covalent bonds. And when that occurs, radicals are formed—atoms or molecules with one or more unpaired electrons. Mostly (though not always), radicals are quite reactive and can serve as transient molecules that generate spiraling chain reactions, or they can perform key intermediary roles in more extensive chemical reactions.

Containing a lone electron, radicals seek out other radicals with whom to bond, or they seek to break other covalent bonds in their desire to satisfy their own incompleteness. Chemistry names this process "propagation." The latter radical reaction reflects a specific mode of enacting valence. It is an operation whereby an unstable entity (i.e., the radical atom or molecule) grasps "part" of a stable molecule in order to make itself stable, but now as an altered entity. There are two things to note. First, this operation makes the for-

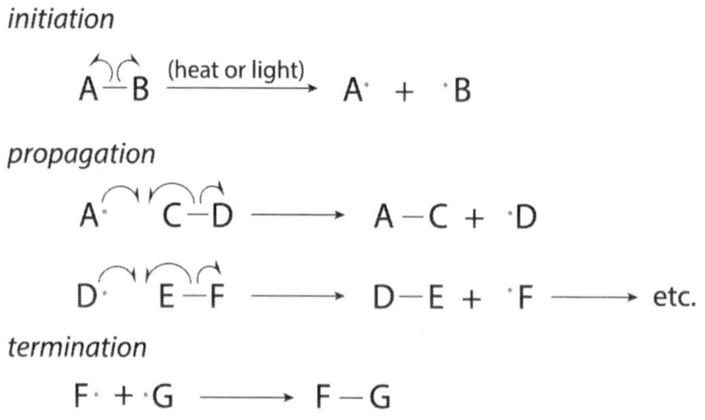

initiation

$$A \!-\! B \xrightarrow{\text{(heat or light)}} A\cdot \ + \ \cdot B$$

propagation

$$A\cdot \quad C \!-\! D \longrightarrow A - C + \cdot D$$

$$D\cdot \quad E \!-\! F \longrightarrow D - E + \cdot F \longrightarrow \text{etc.}$$

termination

$$F\cdot + \cdot G \longrightarrow F - G$$

FIGURE 19   An elementary schema of the radical, the radical being the one with a dot. Source: https://chem.libretexts.org /Courses/Purdue/Purdue_Chem_26100%3A_Organic_Chemistry _I_(Wenthold)/Chapter_05%3A_The_Study_of_Chemical _Reactions/5.5.%09The_Free-Radical_Chain_Reaction.

merly stable molecule, whose part was detached, now unstable, propelling it to seek to cleave a covalent bond and perpetuate a new radical achievement—thus creating a chain-reaction effect. Second, the formerly unstable entity subsumes the appropriated "part" as integral to its new mattering while always being susceptible to future appropriation.

In this sense, the radical magnifies and sets mechanisms in motion that characterize all chemical combinations: not so much a relation as a form of relating—a co-enveloping (or translating, in the Latourian sense)—which carries an entity into the makeup of another, thereby transforming both and their milieu. The radical's particular propensity of movement embodies "modes of expansion, propagation, occupation, contagion" (Deleuze and Guattari 1987: 239). We could think of radicals as mediators in a process they do not control but compel further, regardless. As such, radicals exemplify how an element's capacities change in relation to context. Alone, it is one thing; in different degrees of collectivity, it becomes something else. Indeed, the complex associations that form radicals manifest agencies marked by intensity, pressure, and compulsion that exceed the capacities of the individual elements when added up. That is, they disrupt rational part-whole relations. This is the enigma of the mixt.

In biochemistry, radicals perform functions both essential to and injurious to matter and life-forms—think of free radicals and their association with ozone depletion or cancers. For Deleuze, his "positive ontology" would observe that radical propagation has no moral valence. Rather, the valence is one of passage, a "melodic line of continuous variation" (Deleuze 1978), whereby emergent orbitals of coalescence animate ever new composition, decomposition, and recomposition.

This section explores two very different moments of legal process that transmuted and propagated in ways never anticipated. In one instance, questionable corporate agreements threatened to thwart, as they ricocheted like radicals along the peripheries of this legal saga, the very legal instrument created to sustain the corporate form: the contract. In the second instance, a bewildering cascade of experiential events haunting the judicial inspections transformed into radical iterations of exceptional and singular forms of prehensive knowing.

CATCH

Over the years of the lawsuit, Texaco and then Chevron created websites dedicated specifically to the lawsuit against the corporation. On the most recent website that Chevron has created (theamazonpost.com), one can view a video titled "Chevron to Ecuador: Keep Your Promise, Clean Up the Amazon." The following text reprints a transcript of the audio from this video.[1]

"These pictures of the Ecuadorian Amazon tell an emotional story. And it's easy to blame a big oil company like Texaco for images like this [see figure 20]. There's just one problem. Texaco, now owned by Chevron, already cleaned up its share.

FIGURE 20

Here's the real story. It began in 1964 in Ecuador when Texaco produced oil in partnership with the national oil company, Petroecuador. Almost thirty years later, Texaco, at Ecuador's request, ended oil production in the Amazon, leaving Petroecuador as the sole oil producer. However, before leaving, Texaco spent $40 million and worked with independent environmental experts to clean up its share of well sites—replacing soil, replanting native trees and plants, and investing in community programs.

FIGURE 21

o TexPet Remediation Areas

FIGURE 22

The whole process was overseen and verified by the Ecuadorian government. In total, Texaco cleaned up over 220 impacted areas. And in 1998, the government and local communities approved the cleanup and granted Texaco a release from further liability. Meanwhile, Petroecuador, which owned over 60 percent of the operations, agreed to clean up all of the remaining sites. Now fast-forward almost twenty years. Many of the sites Petroecuador promised to clean up are still there, unremediated. In fact, they started to clean them up but were stopped by the government at the request of the lawyers behind the fraudulent lawsuit against Chevron who feared it would hurt their case and deprive them of photo opportunities to use against Chevron in the media. Since 1990, Petroecuador has even expanded its operations, more than doubling the number of wells and racking up almost two thousand spills in the process. Rather than doing what is right, the Ecuadorian government, together with calculating lawyers and deceitful activist groups, are seeking to blame a big oil company and cash in.

FIGURE 23

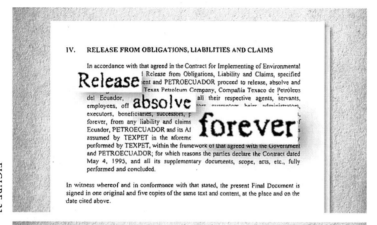

FIGURE 24

It's time for Petroecuador and the Ecuadorian government to stop pointing fingers, live up to their agreements, take responsibility, and clean up the Amazon."

Alchemical Deals

Of Contracts and Their Seepage

ON THE MORNING OF NOVEMBER 11, 2004, Dr. Adolfo Callejas, Chevron's chief Ecuadorian lawyer, stood before the then president of the superior court of justice, Dr. Efraín Novillo. Buttressed by his fellow counsel, he began:

> Your Honor, ChevronTexaco Corporation requested this inspection exclusively to demonstrate the quality of the remediation work of Texaco Petroleum Company—who is not the defendant in this lawsuit—conducted at this well site. . . . The purpose of our inspection tour today is precisely to allow you, Your Honor, to visit and be immersed in these sites where remediation efforts took place—that is, to be at three pits that Texaco remediated, one oil spill that was also cleaned up by Texaco, and, of course, the location of the well head.[1]

It was just past 9:00 A.M., and the judicial inspection of the oil well Sacha-57 and its surroundings was underway. The sweet smell of recently macheted pampas grass lingered as the judge, lawyers, scientists, and technicians gathered around the wellhead. Dr. Callejas felt confident about this inspection. The equatorial sun already drew light perspiration from everyone's foreheads. Directly across from him stood Dr. Mónica Pareja Montesinos, the plaintiffs' then presiding lawyer, and Pablo Fajardo, who at the time was assisting the plaintiffs' legal team. This inspection had not drawn a crowd of spectators; only a smattering of technical assistants and a handful of local residents looked on.

The inspection entourage proceeded along a scrub forested path to a nearby opening spotted with low pasture grass and some native shrubs. Dr. Callejas began again:

> Your Honor, according to the information we have, this site is where once stood an oil-well drilling pit. It encompasses an area of approximately 712 square meters. In 1995, it was identified as containing petroleum and other substances. In June 1996, Texpet remediated this pit as part of the remediation work it needed to complete in accordance with its ownership [percentage] in the consortium's concession. Technicians used the same remediation system here as they would in other places: they washed the soils with a detergent-like product in order to facilitate the better recuperation of hydrocarbons from the earth and left the soils in a state such that they do not pose a risk to human health or the environment.[2]

As would soon be evident, Dr. Callejas assured, the soil and water samples extracted by technical experts would demonstrate that Texaco had complied with its mid-1990s mandate and cleaned up the area.

Dr. Pareja scuffed at the soil of the former pit on which they stood. Her stature was petite, yet her voice was anything but demure. She interjected:

> Your Honor . . . allow me to remind you that we are *not* here to determine, at this moment, whether or not a contract celebrated [*celebrado*] between the Ecuadorian government and Texaco has been fulfilled. We are conducting a judicial inspection as part of a lawsuit filed by local inhabitants who have been affected by Texaco's operations. We are here to determine if this area of Texaco's operations is or is not free of contamination.[3]

Dr. Pareja continued:

> Your Honor, may it please the court, if we are here to engage in a judicial inspection, it is not to determine whether the contract to which Dr. Callejas alludes was complied with or not, given that this contract was not celebrated among the parties who make up this litigation. What we seek to accomplish here is to determine the truth of the claims made in a lawsuit, or the truth of claims made in response to a lawsuit. What we are doing here is determining the existence, in this place and at various sites where Texaco operated, of environmental contamination: contamination of waters, contamination of soils, contamination that has affected the ecosystem.[4]

Over the course of the day, the inspection entourage walked along narrow paths curving through thick grassy weeds and dense tropical brush to six locations previously designated for investigation: the Sacha-57 oil-well site (drilled in 1973 and capped in 1981), three separate waste pits, an area where crude had been spilled, and an area of ossified crude where petroleum had been burned. These were the topographical remnants still evident on the landscape of oil operations that Texas Petroleum (Texpet) had initiated thirty years prior. Technical assistants from both sides of the litigation augured numerous cores, pulling up subterranean soil samples at various depths, and took water samples from underground aquifers and surface waterways. In total, fifty-seven soil and water samples were extracted and transported (through a chain of custody and command) to their respective laboratories for analysis.

By midafternoon, Dr. Pareja furthered her legal arguments as she stood on a swath of ossified crude that covered the ground like thick, undulating asphalt:

> Your Honor, the inspection of Sacha-57 and its surrounding area of influence, including existing rivers, water ways, and marshlands, has been solicited to demonstrate [constar] the environmental effects of the petroleum [activity] performed by Texpet, a subsidiary of Texaco Inc., in its capacity as the technical expert, operator, and executor of the consortium CEPE-Texaco, and, consequently, responsible for the negative consequences of said activity. Texpet . . . had within its responsibility the design, construction, installation, and operations of the infrastructure and equipment necessary to explore for and exploit petroleum. Texpet, in its capacity as a subsidiary . . . , was subject to the parent corporation with respect to technical, administrative, and economic concerns. During its operations in Ecuador, Texpet used methods and procedures, known and approved by Texaco Inc., that had been prohibited in other countries because of their injurious effect on the environment and human health.[5] . . . Because of Texpet's irresponsible actions, which explicitly reflect the economic interests of its parent company, Texaco Inc., and because of *the complete lack of respect* for the [local] inhabitants . . . there exist to this date contaminating elements in the environment that were deposited by said company in the process of exploring for and exploiting oil on Ecuadorian soil.[6]

His agitation rising, Dr. Callejas retorted: "Your Honor, our duty here is to determine if Texaco complied with the remediation and technical standards that were established contractually" in 1995, when the parameters in the remediation contract between Texpet and the Republic of Ecuador were determined.[7]

> No legal or contractual rule gives the plaintiffs the right to challenge the contract of May 4, 1995, . . . or its executing documents, the Acta Final, signed on September 30, 1998. This lawsuit cannot be used to dispute decisions that the Ecuadorian people legitimately and sovereignly made through their leaders; nor can it be used to object to a release that was validly granted in favor of Texpet, through the aforementioned Acta Final, which recognized that Texpet fulfilled its contractual obligations in a responsible manner and in accordance with the environmental legislation and then-current practices, thus ensuring the cleanup of its percentage of sites that were contaminated upon the termination of the 1973 concession.[8]

Dr. Pareja countered:

> Release from responsibility, be it granted by Petroecuador or the Ministry of Energy, does not compromise nor can it compromise the rights of those, like the plaintiffs, who are not part of that agreement. Neither the government of Ecuador nor Petroecuador can . . . release the operator, and its parent company, of the responsibility that it must assume with respect to the population of these provinces, whose right to live in a clean environment is guaranteed by the Constitution.[9]

According to her reasoning, since Texpet was the consortium's operator (it constructed, maintained, and operated all oil facilities 100 percent), then Chevron bore all liability, as the material cause of the damage.

The Spectral Alchemy of Deals

This chapter explores the alchemical work of contract—legal contract—in the lawsuit against Chevron. Contracts, of course, are deals, agreements for setting rights and responsibilities around a particular concern. They aspire to clarity and precision, rigor and comprehensiveness, fixity and finality. In law, "contract" as a legal form, when complied with and fulfilled, especially when configured as a settling agreement, is equivalent to *res judicata* (*seguridad juridical*)—matter judged, issue closed, end of story. This armature gives a contract, as an idealized form, the illusion of resolute stability and absoluteness. In the Ecuadorian litigation, however, a contract's capacity as a legal form to stabilize, contain, and bring finality was recurrently displaced. Contracts seeped, oozed, and overflowed into a cascade of proliferating reactions, sometimes into improbable realms. All too often, contracts became what I call "alchemical deals."

A plurality of contracts precipitated and flowed from one particular contract at the core of this chapter—a couplet agreement signed in 1995 and ratified in 1998 between Texaco (through its representative, Texpet) and the Republic of Ecuador (through its representatives the Ministry of Energy and Mines and Petroecuador). The couplet agreement affirmed that the government of Ecuador would grant Texaco immunity from any future obligations were it to successfully conduct specific remediation work. As a contract, the 1995/1998 couplet agreement detailed comprehensively what was needed to cut ties between the oil conglomerate and the Ecuadorian state. It defined the parameters by which a site was deemed contaminated, effecting the mattering of toxins. It detailed the cleanup actions necessary to mitigate that contamination. And it determined that once the corporation effectively completed these duties, it would be precluded from any future environmental liability and protected against future environmental claims lodged by the state. That is, the matter would be closed.

However, the closing of the matter—res judicata—did not happen. At least not in Ecuador. Rather, the 1995/1998 contract between Texaco and the Ecuadorian government, whose intent was to put to rest any and all future obligations between the parties, transmogrified into an alchemical deal that triggered a chain of reactions—spectral radicals—whose incomplete closure haunted and conjured unanticipated events. The alchemical effect unraveled foibles. It spun criminal indictments. It triggered sting-operation fantasies. It contributed inroads in concocting and uncovering corruption schemes. It instigated international investment disputes. And, most disturbingly, it served as the basis upon which a tribunal of the Permanent Court of Arbitration (PCA) determined that the Republic of Ecuador breached a bilateral investment treaty (BIT). For years, these proliferations would, in the eyes of the Ecuadorian court and, for a while, those of the larger international public, undermine a core anchor in Chevron's claim to immunity. But in time, by the fall of 2018, the legal contract as form would mutate one more time, revealing the corporation to be an alchemist extraordinaire—that is, an exquisite legerdemain executing astounding transformation.

Contracts are curious legal instruments. As the eminent legal scholar Charles Fried observes, the moral and philosophical basis of a contract as a legal form is "the promise" (2015). He writes that "the generating genius" of contract law is "the promise principle"—the principle by which "persons may impose on themselves obligations where none existed before" (134, 1). Together with other principles—the "security of persons" and the "sanctity of property"—Fried notes that the "obligation of contract" constitutes the core of Euro-American "liberal political morality" (7). Contract signifies a "moral invention" that functions as a legal device to coordinate "mutual wills" and "facilitate human collaboration" over time through "self-imposed moral obligation" (134, 8). The deep liberal assumptions propping up this notion of persons with their purported equality, autonomous will, and capacity to freely enter into contract are widely recognized (cf. Pateman and Mills 2007). As Hannah Appel clearly demonstrates within the oil and gas industry, contract is crucial to the "licit life of capitalism" and to how the dense debris of postcolonial racial inequalities can serve as "arbitrage opportunities" (2019: 143) through which hydrocarbon capital both maximizes profit and relinquishes responsibility. Yet there is also more to learn about the function of the contract form.

Annelise Riles's analysis of the "swap" in high finance offers intriguing insights. As she notes, the swap is "an agreement in the present to exchange something at a definite point in the future" (2011: 801). In the financial world, parties mitigate exposure to risk (i.e., default, bankruptcy) inherent in the temporal gap by collateralizing their agreement. And this collateralized contract performs "a remarkable act of substitution": parties "agree to swap the politics of their relationship—the nature of their mutual entanglements, its asymmetries, and intrigues—for a known quantity, the collateral. In the definite but attenuated present of the swap, [Party A] will hold [Party B's] collateral, and that collateral will precisely stand for, be the measure of, the extent to which it can compel [Party B] to act as promised" (802).

The 1995/1998 couplet contract between Texaco and the Republic of Ecuador was similarly a swap where two parties agree in the attenuated present to exchange one thing of value for another at a determinate time in the future: in exchange for environmental cleanup, the government agreed not to pursue any legal action against the corporation. Although not similarly collateralized, the agreement gave the government of Ecuador leverage—a legal release from future liability—that compelled Texaco to fulfill a promise and perform environmental remediation. In order to acquire the state-dangled treat, Texaco needed to complete a specific task. Formalistically, the couplet agreement similarly detailed "a remarkable act of substitution": the parties agreed to swap the murk of their complexly valenced connections for a clear absolute—definitive legal severance.

As Riles (2011) demonstrates, however, the wizardry of collateralization never in practice released the parties of the swap from the politics of their relations. It carried them to new entanglements and relational modalities, despite the document of the swap agreement assuring substitution. Similarly, the Texaco-Ecuador contract abounded in a double life. As form and a material document, it was an absolutist pact of severance, an unwavering legal fiction of release recurrently resurrected. As process, the contract evinced an encumbered, messy material world haunted by compromised efforts to negotiate, comply with, and invoke it as absolute. The contract's presumed ease of fulfillment, closure, and res judicata never quite obtained. Or at least not for two decades.

What happens when a contract, that idealized legal form, bares scant relation with practice? What happens when the dictates of a contract reflect not the behavior of parties (even if attenuated) and instead generate transmogrifying possibilities? The Texaco-Ecuador couplet contract was never a swap between parties who substituted the politics of their relationship for a known quantity. Instead, the very politics of corporate-state relations—"the nature of their mutual entanglements, its asymmetries, and intrigues" (Riles 2011: 802)—fundamentally defined the parameters and terms of the 1995/1998 agreement. And in its fulfillment, the corporate-state contract detailed the material processes through which the very politics of interested and contentious relations became ever more present. Said more precisely, Texaco, and subsequently Chevron, used contract, an elemental, abstract legal form, and the 1995/1998 couplet contract specifically as an alchemist's tool. Though, as is the alchemist's fate, command over processes of consequential and intentional reactions was often elusive. When it suited, the corporation deployed contract formalistically with the aim of closure (res judicata), only to have that recurrently deferred by other reactions beyond its control. And when it suited, the corporation invoked contract to bring anything but closure, engaging instead in unrelenting, protracted, derivative litigation in a multiplicity of jurisdictions and forums.

Amid the melee of unfolding events, contract is revealed to be an essential legal instrument for attenuating liability and mitigating responsibility. It served to extradite the corporation from its viscous entrenchment in the afflicted socioecological conditions that its rainforest operations forged. More disconcerting, the alchemical deal—of which the 1995/1998 couplet contract was at the forefront—transformed a dispute over contamination into a dispute over a contract. And as will be clear toward the end of this book, the alchemical deal partook in transforming a dispute over contamination into a dispute over corruption *and* in transmuting a dispute over contamination into a dispute over investment. As an instrument for sealing res judicata, contract clearly had its place. But contract's more consequential role in the legal saga against Chevron was its capacity to transfigure purpose and prolong litigation solely with the aim to debilitate. Chevron's alchemical acumen proved unparalleled.

The Work of Contracts to Absolve

The Ecuadorian lawsuit against Chevron was the product of a legal battle for jurisdiction in the United States over the course of the 1990s. The lawsuit was first filed in the United States in November 1993 and then refiled in Ecuador in May 2003. During that interim decade (1993 to 2002) in which the trial was being shuffled between the US district court and the court of appeals, much happened in Ecuador. Most important, for my purposes here, was a key contractual event over the span of three years. On May 4, 1995, in what some characterize as a scramble to cover its corporate backside, Texaco entered into what was called the "1995 settlement agreement."[10] Scaffolded atop a collection of legal understandings (the "memorandum of understanding [MOU]," the "scope of environmental work [SOW]," and the "remedial action plan [RAP]"), the 1995 settlement agreement between Texpet, the Ministry of Energy and Mines, and Petroecuador detailed select environmental-remediation actions that Texpet would undertake in exchange for release from further liability. The 1998 *acta final* (final release document) blessed and consecrated Texpet's cleanup efforts and absolved the corporation of its sins.

Far from existing as an isolate, however, the couplet agreement, consisting of the 1995 settlement and 1998 acta final, was linked to and hinged on a slurry of prior contracts between Texaco and the Republic of Ecuador over the previous thirty-plus years. Elsewhere I trace this history in detail (Sawyer n.d.). Below are the major contour lines that determined the legal and geographic configuration of Texaco's oil concession in 1992, on which the 1995/1998 contract was based. As will be evident, the concession with which the corporation began (in 1964) and with which it ended (in 1992) were hardly the same thing. What stands out is that the contract—that purportedly weighty and precise legal instrument—was anything but stable. Over a thirty-year period, the contract, in both form and substance, exhibited a radical shape-shifting capacity.

In February 1964, the government of Ecuador granted Texaco Inc. and the Gulf Oil Corporation a forty-five-year contract to explore for and exploit oil in an area of the northern Ecuadorian Amazon that abutted the border with Colombia. The oil concession was a 1.4 million-hectare oil block (called the Napo Concession).[11] The joint operating agreement (JOA) that governed the Texaco-Gulf 50/50 consortium detailed that both parties shared the rights and obligations of the concession, although Texaco was the sole operator.[12] Spurred by its recent oil discovery in southern Colombia only thirty miles north of the border with Ecuador, the Texaco-Gulf consortium embarked on what would at the time be the first substantial oil operation in the region.[13] In comparing the original 1964 Texaco-Gulf oil concession with the CEPE-Texaco oil concession on which the 1995/1998 agreement is based, clear discrepancies arise. Not only are these oil concessions not the same, but they barely overlap.

In 1965, something anomalous happened. The Texaco-Gulf consortium was conducting extensive geological exploration—seismic lines, aerial photography, and air magnetometry to determine the region's subterranean stratigraphic columns—on land outside the limits of its Ecuadorian oil concession (Sass and Neff 1966; Sawyer 1975: 1146–48). Why would that be? The ins and outs are complicated (Sawyer n.d). The short story is that, unbeknownst to the Ecuadorian government, the Texaco-Gulf consortium had orchestrated to annex part of an abutting oil concession previously granted to Minas y Petróleos. By the end of 1965, the concessionary lands that the consortium controlled ballooned by 667,731 hectares, while the government, ignorant of the true rightsholders, believed it had approved a transfer to a different corporate entity.[14] Through a series of questionable dealings among subsidiaries of subsidiaries, the original 1964 concession of 1,431,000 hectares had swelled in 1965 to 2,098,731 hectares (see figure 26). As it happened, the newly acquired area would prove to be the choicest, most commercially productive hectarage for the consortium in the Amazon region.

In March 1967, Texaco-Gulf drilled its first exploratory well, Lago Agrio-1. The strike proved rich in oil and, by the end of the year, the consortium drilled four more wells to define the extent of the Lago Agrio subterranean reserve (*American Association of Petroleum Geologists Bulletin* [AAPGB] 1968). These 1967 operations were within the confines of the original 1964 Napo Concession

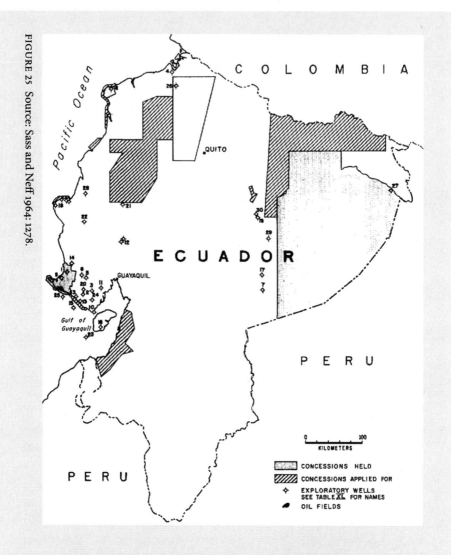

FIGURE 25 Source: Sass and Neff 1964: 1278.

that the Republic of Ecuador had granted to Texaco-Gulf. But geophysical and seismic exploration in the new annex would soon lead to wildcats (or drilling risky oil wells based on negligible geological knowledge) that vastly surpassed those early discoveries. In 1968, Texaco drilled Shushufindi-1 and struck oil in early January 1969. One month later, Texaco discovered the even more abundant Sacha reserve (see the graph in Neff 1970: 1375). One year later, in 1970, Texaco discovered the Auca reserve. These three subterranean reserves, all situated within the surreptitiously annexed region, would prove to be the most productive oil fields with the best-quality crude Texaco would discover in Ecuador.

In late 1969, President José María Velasco Ibarra decreed a renegotiation of the Texaco-Gulf-state agreement. Along with changes in royalties, labor requirements, and the pipeline to the Pacific Ocean, the new contract redrew the boundaries of the original Texaco-Gulf concession, decreasing it from 1.4 million to 500,000 hectares. Despite the mandated return of land, the consortium retained its choicest acreage (Sawyer 1975: 1147), with the consortium—and Texpet as operator—still controlling the annexed region, which was proving to be a more profitable venture (with a total of 1.15 million hectares).

By the end of 1969, Minas had sold or relinquished its former concession to an array of petroleum interests, leaving Ecuador's Amazon patchworked with oil blocks. At the end of his term, Velasco Ibarra signed the 1971 Hydrocarbons Law that went into effect with the projected date for the first Amazonian crude export. The 1971 Hydrocarbons Law substantially altered the

FIGURE 26 Source: *American Association of Petroleum Geologists Bulletin* (AAPGB) 1968: 1406.

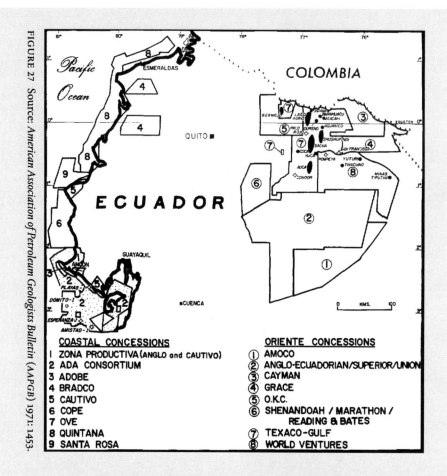

FIGURE 27 Source: *American Association of Petroleum Geologists Bulletin* (AAPGB) 1971: 1453.

COASTAL CONCESSIONS
1 ZONA PRODUCTIVA (ANGLO and CAUTIVO)
2 ADA CONSORTIUM
3 ADOBE
4 BRADCO
5 CAUTIVO
6 COPE
7 OVE
8 QUINTANA
9 SANTA ROSA

ORIENTE CONCESSIONS
① AMOCO
② ANGLO-ECUADORIAN/SUPERIOR/UNION
③ CAYMAN
④ GRACE
⑤ O.K.C.
⑥ SHENANDOAH / MARATHON / READING & BATES
⑦ TEXACO-GULF
⑧ WORLD VENTURES

political-economic relations of petroleum. It affirmed state ownership of subterranean resources, asserted the state's integral role in establishing all hydrocarbon relations and policies, brought an end to an older era of *arrendimiento* (a concessionary policy based on renting land to corporate entities), and set concession areas considerably smaller and government royalties higher than allowed by the previous law. In practice, the new Hydrocarbons Law had little immediate effect; it was not retroactive.

Within a year, however, pre-1971 concessions were no longer exempt. The 1972 military coup d'état ousted Velasco Ibarra before the end of his term and maximized the effect of the Hydrocarbons Law. In June 1972, the junta promulgated Supreme Decree 430 mandating all oil contracts be renegotiated with the newly created state oil company, Corporación Estatal Petrolera

Ecuatoriana (CEPE, legally established on June 23, 1972) (Sawyer 1975: 1148). Among Supreme Decree 430's more controversial clauses were further limitations on the size of oil concessions and the option of CEPE to exercise the right to acquire a percentage interest in any oil operations.[15]

In August, the Texaco-Gulf consortium signed a renegotiated 1973 concession contract with the Republic of Ecuador. The 1973 contract reduced considerably the size of the original and annexed concession area and consolidated them into one single area, henceforth called the Napo Concession.[16]

FIGURE 28 Source: *American Association of Petroleum Geologists Bulletin* (AAPGB) 1984: 1478.

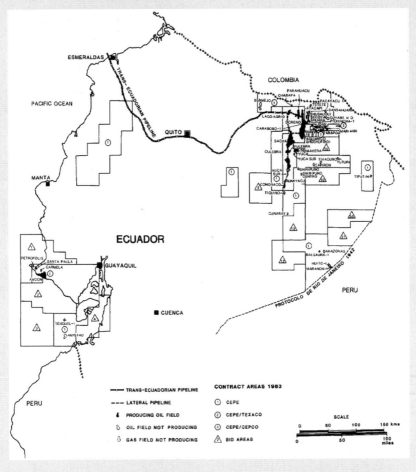

All in all, the former 1.15 million hectares under Texaco's operations were reduced to 491,000 hectares: comprising 170,000 hectares of the Napo Concession (officially granted by the Ecuadorian state in 1964) and 321,000 hectares of the annexed zone (Sawyer 1975: 1148). By that time, Texaco had "discovered 11 oil fields and [had] 99 productive wells producing 220,000 barrels daily" (1149), and the 1973 contract demarcated the 1973 concession such that the consortium retained all of its infrastructure investments and discoveries.

The 1973 contract also stipulated a CEPE participation percentage of 25 percent by acquiring 12.5 percent of the rights and obligations of each of the consortium members, such that the consortium consisted of 25 percent CEPE, 37.5 percent Texpet, and 37.5 percent Gulf. By participation, the parties meant that CEPE would begin to receive 25 percent of the net extraction of oil, as well as to supply 25 percent of the costs of extraction. This agreement was officially enacted in June 1974.[17]

In late 1976, Gulf decided to withdraw its operations from Ecuador. By the end of the year, CEPE acquired Gulf's 37.5 percentage interest in the consortium. These events transformed the ownership interest in the Napo Concession, such that Texpet, the concession operator, had a 37.5 percentage interest in the consortium and CEPE had the majority at 62.5 percent (Martz 1987: 168; Vargas Pazzos 1976).[18]

This history of contractual shape-shifting—a contractual radical—volatilized as Texaco and Chevron professed the legal contract sacrosanct. This imaginary of the legal contract as unwavering and of the consortium's oil concession as singular was the basis upon which another contract (the 1995/1998 couplet agreement) was built—meant to absolve Texaco of future claims. This imaginary conceals the displacement of process inherent in the form.

Remedial Alchemy

Being the "operator" throughout the life of the concession, Texpet designed, built, and ran all the infrastructure and facilities needed to explore for, extract, produce, transport, and export oil in Ecuador. These operations indelibly transformed the northern Ecuadorian rainforest with thousands of miles of seismic grids, 336 oil wells and 20 injection wells in 28 petroleum fields, over 900 open waste pits, 22 processing facilities and numerous pumping stations, an oil refinery, a network of thousands of miles of primary and second-ary pipelines, and the bare-bones infrastructure essential for petroleum oper-ations.[19] In 1992, Texaco's rights to operate the oil concession terminated, the company pulled out of Ecuador, and its operations reverted to Petroecuador. Over its quarter century as operator of the concession, Texaco had extracted nearly two billion barrels of crude.

When Texaco's contractual right to extract petroleum ended in 1992, the Texaco-Petroecuador consortium was obliged to perform an environment audit to assess the impact of Texaco's operations on the concession. Two en-vironmental assessments transpired: one commissioned by Petroecuador and Texpet and a second commissioned solely by Texpet.[20] Both audits detailed a number of negative environmental impacts resulting from former Texpet well sites and production stations. The numerous waste pits that Texaco ex-cavated alongside each oil well posed great concern.

In December 1994, Texpet—keenly aware that a lawsuit had been lodged in the New York federal court against Texaco for environmental contamina-tion in its Ecuadorian operations—signed a MOU with the Ecuadorian min-istry and Petroecuador; the MOU outlined actions to remediate the environ-mental damage (Texaco and Ministry of Energy and Mines 1994). The parties agreed that they would develop, define, and approve a detailed scope of the work needed for environmental restoration and that Texpet would contract a reputable environmental engineering firm to complete this work.[21] In ex-change for the execution of the remediation, the parties would "negotiate the full and complete release of Texpet's obligations for environmental im-pacts arising from the operations of the consortium" (Texaco and Ministry of Energy and Mines 1994: Art. IV).

This brings us to the 1995/1998 couplet contract. Over the following year, a series of negotiations between Texpet, the Ministry of Energy and Mines, and Petroecuador resulted in the May 1995 settlement agreement. The agree-ment stipulated that Texpet had to design a "remediation action plan" (RAP) that detailed the precise environmental cleanup work essential for remedi-ating specific sites. The agreement affirmed that "Texpet agrees to under-take such environmental remedial work in consideration for being released and discharged of all its legal and contractual obligations and liability for environmental impact arising out of the consortium's operations" (Texaco and Ministry of Energy and Mines 1995: 3). Since, following the 1973 conces-sionary contract, Texpet held only a 37.5 percentage interest in the ultimate Petroecuador-Texpet consortium—despite Texaco being 100 percent respon-sible for executing and maintaining all technology and operating facilities in the oil concession—the corporation argued that a remediation agreement must necessarily reflect Texaco's proportional responsibility. Consequently, the 1995 settlement agreement outlined a cleanup and restoration program that reflected this proportional contamination and remediation liability.

In mid-1995, Texpet hired an environmental engineering firm, Woodward-Clyde, to design a RAP following the parameters outlined in the 1995 settlement agreement. The remediation agreement between TexPet and the government of Ecuador mandated that the corporation remediate and close the pits at 135 well sites, remediate oil spills at 27 well sites, and modify how formation waters are discharged at nine production stations (Texaco and Ministry of Energy and Mines 1995: "Annex A, Scope of Environmental Remedial Work," 1–2).

Between October 1995 and September 1998, Woodward-Clyde performed the required environmental remediation outlined in both the SOW and the RAP. Texaco spent $34 million on environmental remediation and restoration and contributed approximately $6 million to community-development, obligations as outlined in the 1995 agreement. Over the three years in which Woodward-Clyde conducted its work, representatives from the Ministry of Energy and Mines and Petroecuador inspected impacted sites and certified that they had been successfully remediated—issuing interim certificates of completion. On September 30, 1998, Texpet, the Ministry of Energy and Mines, and Petroecuador executed the 1998 acta final. This "final release" certified that Texpet had performed all its obligations under the 1995 settlement agreement and released Texpet from future obligations and liability arising from the consortium's operations. The release from liability reads as follows:

> In accordance with that agreed in the Contract for Implementing of Environmental Remedial Work and Release from Obligations, Liability and Claims, specified above, the Government and Petroecuador proceed to release, absolve and discharge Texpet, Texas Petroleum Company, Compañía Texaco de Petróleos del Ecuador, S.A., Texaco Inc. and all their respective agents, servants, employees, officers, attorneys, indemnitors, guarantors, heirs, administrators, executors, beneficiaries, successors, predecessors, principals and subsidiaries, forever, from any liability and claims by the Government of the Republic of Ecuador, Petroecuador, and its Affiliates, for items related to the obligations assumed by Texpet in the aforementioned Contract, which has been fully performed by Texpet, within the framework of that agreed with the Government and Petroecuador; for which reasons the parties declare the Contract dated May

4, 1995, and all its supplementary documents, scope, acts, etc., fully performed and concluded. (Texaco and Ministry of Energy and Mines 1998: §IV, "Release from Obligations, Liabilities and Claims")[22]

This is why Dr. Callejas kept stating that the purpose of the judicial inspection at the Sacha-57 well site was to verify that Texaco had complied with its remedial obligations as specified in the 1995/1998 couplet contract.

Res Judicata

But there was also another reason. On October 21, 2003, the first day of the litigation against Chevron in Ecuador, Dr. Callejas articulated yet another legal defense: "As you know, Your Honor, one of the most important effects of decisions that are handed down in the resolution of legal claims . . . is that of 'res judicata.' [This doctrine holds] that, once it is final, it must be accepted that litigation that has been concluded cannot be argued once again. The complaint to which I am responding constitutes an attack on the immutability of the judicial decisions that approved the aforementioned 1995 and 1998 settlement contracts. This is unacceptable and must be rejected by you, Your Honor."[23]

According to Chevron, the 1995/1998 couplet contract had certified that the corporation had already fulfilled its obligation to address any negative impact that its operations had in the concession area. And in so certifying, that agreement exempted the corporation from any further liability associated with the consortium's operations; consequently, the lawsuit was misplaced. The defense based its argument on res judicata: that which has previously been resolved in a final judgment, either through settlement or judicial decision, shall not be litigated again.

Res judicata is foundational to the workings of law. Indeed, it instantiates finality and fixity and, consequently, legal security—components understood to be essential to an authoritative and just legal system. The capacity to believe that a judgment is binding rests on the knowledge that legal proceedings ultimately reach an indisputable end. Res judicata bars litigation, adjudications, and decisions between the same parties regarding the same matter once all appeals have been exhausted. The Ecuadorian Code of Civil Procedures states, because a judgment in the last instance (nonappealable ruling) is irrevocable, a new legal proceeding involving the same parties (i.e., where there is equivalent subjective identity) and "claiming the same thing, quantity, or fact based on the same cause, reason, or right" (i.e., where there is equivalent objective identity) cannot be brought. Under the Ecuadorian Civil Code, a settlement agreement is a "contract in which the parties extra-judicially terminate a pending litigation or prevent a potential litigation."[24] And, as in many Latin American countries, the code also states that "a settlement agreement has the effect of *res judicata*."[25]

In making its case, Chevron's argument proceeded as follows. First, that the 1995/1998 couplet contract and the Lago Agrio disputes revolve around the same material fact: Texpet's operations allegedly had a negative ecological effect that compromised the environment. Second, that the couplet contract and the Lago Agrio disputes invoked and sought to vindicate the same legal right: the right to a clean environment and public health to be secured through remediation. And third, that the couplet contract and the Lago Agrio disputes involved the same parties: that the 1995 government of Ecuador and the 2003 plaintiffs are the same individuals in that they purport to represent the Ecuadorian community; and that Texpet (and its affiliated entities) and Chevron are the same entity.

To substantiate this argument, Chevron characterized—and in so doing homogenized—the violated rights in both the 1995/1998 couplet agreement and Lago Agrio disputes as being "diffuse rights" as opposed to "individual rights." The bifurcation of rights went as follows: diffuse rights are public, collective, and indivisible, belonging to a grouping or class of individuals connected by circumstance (for example, the right to live in a clean environment). By contrast, "individual rights" are divisible, belonging to a juridical being (for example, the right to compensation for damage to property or person).[26] Creating this dichotomy was crucial—for it allowed Chevron to assert a logical indisputability of "subjective" and "objective" identity between the 1990s agreements and the 2000s lawsuit. This is important, as it will come back to haunt this legal saga—the focus of chapter 7. So, let me explain.

When both the 1995 settlement agreement and 1998 acta final were signed, Chevron argued, the Ecuadorian state was the only entity that could seek restitution for an infringement of diffuse rights.[27] Chevron noted that a cluster of Ecuadorian legal provisions provided private individuals with the right to sue parties for past and future personal injuries. And the 1998 Constitution (Articles 23[6] and 86) gave an individual the right to sue the state for not complying with the broad obligation to protect the environment. But Chevron argued that at the time of the 1995/1998 couplet contract there was no legal mechanism whereby individuals could gather collectively to file a lawsuit based on an alleged violation of a so-called diffuse right. Furthermore,

Chevron maintained that the state was the entity empowered to oversee and protect diffuse rights—such as the right to live in a clean environment.

Chevron propped up this claim by noting that Article 5.2 of the 1995 settlement agreement expressly referenced Article 19.2 of the Ecuadorian Constitution, which provided that all citizens had "the right to live in an environment free of contamination" and that it was "the State's duty to ensure that this right not be violated and to safeguard the preservation of nature."[28] Because the right to a clean environment is a "diffuse right," Chevron argued, and because the Ecuadorian state was the only entity entitled to protect and vindicate that right, then, the corporation argued, "the government was necessarily representing the diffuse rights of its citizens in settling with [Texaco]."[29] That is, according to Chevron, in fulfilling exclusively its right and responsibility to protect a diffuse environmental right, the government of Ecuador acted as a representative of all citizens when it negotiated and signed its agreements with Texaco.[30]

Thus, the argument proceeded as follows: because the Lago Agrio plaintiffs— like the Ecuadorian state in the 1990s—acted on behalf of (as representative of) the affected community, the real party of interest was the community. That is, the negotiation and litigation shared the same *subjective* identity. Furthermore, the plaintiffs in the Lago Agrio lawsuit were a constitutive part of the "community" that the Republic of Ecuador represented during the 1990s remediation negotiations. And because the 2003 lawsuit—like the 1995/1998 couplet agreement—sought restitution for alleged petroleum-induced environmental harm, Chevron argued that the cause, reason, and right were the same. That is, the negotiation and litigation shared the same *objective* identity. Consequently, according to the corporation, the doctrine of res judicata deemed the lawsuit against Chevron null and void. The 1995 settlement and 1998 acta final had already settled the *community*'s diffuse right to a clean environment and had released Texaco from further liability for repairing any environmental harm caused by its oil operations.

These arguments will resurface to consequential effect at the end of this book. Suffice it now to clarify that, for the litigation in Ecuador, Chevron's argument failed. Although the Lago Agrio court recognized the Texaco-ministry settlement, it recognized it as an "administrative contract" between parties.[31] The 1995/1998 couplet contract did not express "the unilateral will of the State" and "much less [was it] signed by the Government in the name of all the Ecuadorians."[32] In the court's judgment, the language of the Texaco-ministry agreement precisely established this fact. Article 5 of the 1995 agreement read as follows: "the government and Petroecuador shall hereby re-lease, acquit and forever discharge Texpet [its parent company, employees, and others] *of all the government's and Petroecuador's claims* against the releasees for environmental impact arising from the operations of the consortium."[33] The Lago Agrio court concluded that "the scope of the release from liabilities and lawsuits is limited to those that could come from the government, from Petroecuador, or from its affiliates."[34] The Lago Agrio plaintiffs were not sig-natories to the couplet contract, and "legal transactions" of this nature could not extend to the "inalienable rights"—the right "of [legal] action and peti-tion guaranteed by the Constitution"—of "third parties."[35] The Lago Agrio court dismissed Chevron's res judicata claim.

The Closure That Never Was

Closure was not, for now, the remediation's outcome. Instead, Texaco's agreed-on remediation triggered the seepage of the lawsuit beyond Chevron's control. And, in all its legality, the 1995/1998 couplet contract foretold the lawsuit's propulsion and viscous swirling into improbable realms.

As part of the 1995 agreement, the RAP outlined the criteria for determin-ing whether remediation was necessary and at what point it was complete. It established the sampling measures, testing procedures, and numeric val-ues that both the government of Ecuador and Texpet agreed were to be used as standards to determine what, if any, remediation would be required at a particular location and also whether remediation, once completed, had been successful.

The RAP specifically noted that the "criteria and guidelines were devel-oped in accordance with the Ecuadorian regulations applicable at the signa-ture date of the contract for the execution of the remedial action work (May 4, 1995) . . . and current practice in tropical forest environments."[36] But in 1995, as Texaco was negotiating its remediation plan, "Ecuador had no nu-merical standards for soil remediation or pit closures" (Alvarez, Mackay, and Hinchee 2006: 12). That is, the state did not have any regulatory standards that determined what constituted a measure of crude contamination. Conse-quently, defining criteria for when cleanup was necessary and whether it had been sufficiently achieved was a task of significant consequence.

First was the matter of deciding when and how remediation was needed. Crude oil, one might remember, is a brew composed of thousands of hydro-carbons, and total petroleum hydrocarbons (TPH) is the umbrella concept and measure used to capture this chemical complexity. Although hardly

TABLE 3   RAP Criteria for Remediation Work in Soils and Muds

TPH Value (ppm)	Action Performed
< 5,000	None—material to be left at the site
5,000–20,000	Stabilization with dehydrating agent and material used for filling
20,000–50,000	Bioremediation and material used for filling
> 50,000	Recovery of crude oil and material used for filling

Adapted from Woodward-Clyde International 2000. See also Cabrera Vega 2008: "Appendix H: History and Inventory of Waste Pits Opened for Texpet Operations in Ecuadorian Amazonia," 8.

straightforward, as chapter 1 demonstrates, TPH is the measurable amount of petroleum-based hydrocarbons in a physical matrix and is an indicator of petroleum deposits in an environment. With respect to the Texpet-ministry remediation contract, both parties had to agree on a numerical TPH value at which point remedial action should be taken and on how action might change with varying TPH values.

In line with contemporary protocol elsewhere, Texpet and the ministry agreed to a tiered regime of remediation. First, they agreed on a cleanup standard of greater than 5,000 mg/kg. Were soils in a former Texpet site deemed to have a TPH value that was below 5,000 parts per million (ppm), then no action would be taken; were the TPH level higher, remediation would occur relative to the TPH concentration, as set out in table 3.

After various adjustments, the SOW pursuant to the 1995 agreement required that 133 well sites be assessed for necessary remediation, taking into consideration all pits and spills at each site. Woodward-Clyde identified 225 pits at the 133 well sites. Of those, they characterized 76 pits as having TPH levels below 5,000 ppm and thus exempt from remediation. As more former waste pits were discovered during the process of remediation, the number rose; ultimately, the firm remediated 162 pits and 6 spills in Texaco's former concession area.[37]

As expected, the plaintiffs' lawyers denounced the corporation for only remediating a small portion (162 out of 900-odd) of the total number of pits excavated and used by Texpet over the course of its operations. However, the judicial inspections would soon reveal that more was awry with the Texpet-ministry remediation agreements. As the inspections proceeded, an in-

FIGURE 29 Oil well Sacha-57 (pit #1) during Texaco's remedial action plan (RAP). Taken by Woodward-Clyde (the environmental remediation contractor) on May 5, 1996, the photo depicts heavy crude embedded at the bottom of the pit after liquids and viscous hydrocarbon contents were removed. Source: *Maria Aguinda Salazar y Otros v. ChevronTexaco Corp.*, Case No. 002-2003-P-CSJNL (2011-63-1), p. 11,849, Provincial Court of Justice of Sucumbíos, Nueva Loja, Ecuador.

creasing number of technical reports indicated that many of the pits that Texaco remediated between 1995 and 1998 exhibited TPH levels well above the 5,000 ppm TPH level. Chevron's own samples indicated a number of pur-portedly remediated waste pits with TPH levels in the multiple thousands and tens of thousands (e.g., Sacha-51 at 63,000 ppm TPH and Lago Agrio-16 at 170,095 ppm TPH). Even the scientific report that Chevron submitted to the court for Sacha-57—not a particularly remarkable site—had TPH levels over 8,000 ppm (the TPH readings for the plaintiffs ranged between 2,418 ppm TPH and 262,581 ppm TPH). As the judicial inspections continued, test results well above the 5,000 ppm TPH level proved not to be sampling anomalies.

Advocates for the plaintiffs quickly seized on the TPH readings with levels greater than 5,000 ppm as evidence that the remediation was a sham. How-ever, this was not a case of a simple coverup. Rather, high hydrocarbon read-ings unearthed during the judicial inspections at alleged Texpet-remediated sites underscored a duplicity written into the contract. Along with establish-ing "action levels" (the TPH value needed to initiate remediation), the Texpet-ministry agreement established the process for authenticating whether remediation had occurred successfully or not. The agreement stipulated that a specific laboratory test—a soil leachate assay called the toxicity characteristic leaching procedure (TCLP)—was to be used to verify remediation action and that a reading of less than 1,000 mg/L from that procedure would be the measure indicating whether proper remediation had been achieved.

Between 1995 and early 1997, Texpet's contractor performed remediation work on 154 pits (approximately 90 percent of the total number of pits slated for cleanup) and the ministry and the contractor assessed that work based on the soil leachate assay. In March 1997, however, the ministry contractually modified this assessment method; for the remaining eight pits still to be remediated, the company had to supplement the TPH-TCLP soil leachate with analyses demonstrating that the soils at post-cleanup sites contained at a maximum a TPH level of 5,000 mg/kg. Ten years later, the significance of changing the contract terms (laboratory tests and methods) for assessing whether remediation work was sufficiently successful would prove consequential. As became clear, the contractual change was not incidental—seeking to enable more successful remediation (as Chevron has claimed). It spurred allegations of criminal intent. But I get ahead of myself.

According to the US Environmental Protection Agency (EPA), the TCLP was a test "developed to simulate the leaching of constituents into ground water under [the acidic] conditions found in municipal solid waste landfills" (EPA 2006: chap. 2). Under set laboratory conditions, the test evaluates the degree to which metals, volatile and semivolatile organic compounds, and pesticides might leach from waste sites and the degree to which that waste might be classified as hazardous under the Resource Conservation and Recovery Act (RCRA; 40 CFR Part 261). Although, as the EPA notes, "the TCLP is the most commonly used leachate test for estimating the actual leaching potential of wastes," it is not appropriate for "all situations or conditions and for all types of wastes" (EPA 2006: chap. 2). Indeed, the EPA specifically clarifies that "the TCLP might not be appropriate for analyzing oily wastes. Oil phases can be difficult to separate (e.g., it might be impossible to separate solids from oil), oily material can obstruct the filter (often resulting in an underestimation of constituents in the leachate), and oily materials can yield both oil and aqueous leachate which must be analyzed separately" (EPA 2006: chap. 2).[38]

During the litigation against Chevron, the legal team for the plaintiffs claimed that between 1990 and 1994 "not one of the 50 U.S. states used the TCLP method for determining acceptable levels of TPH in soil" as part of cleanup efforts after petroleum mishaps and spills.[39] Experts for the plaintiffs explained the test this way:

> The TCLP test only measures the amount of contamination that leaches out of the soil after [the soil] is mixed with acidic water for a short period... the equivalent of pouring water over coffee grounds and then measuring the amount of caffeine in the water, rather than the total amount of caffeine in the grounds. The TCLP test thus returns a rough estimate of contamination that might be released in a single rainstorm, but it doesn't come close to estimating the cumulative environmental threat posed by the toxic source, especially when dealing with toxins like TPH and heavy metals.[40]

As the plaintiffs' lawyers argued, "Texaco's use of the TCLP test allowed Chevron to claim only a tiny fraction of the contamination actually existing in purportedly remediated soils. Indeed, one could have poured crude oil onto the ground overnight and a soil sample from that ground might not fail the TCLP test."[41] According to the plaintiffs' lawyers, the Texpet-ministry agreement "ensure[d] that its remediation obligations under the contract could be satisfied even if no remediation occurred. The result: Texaco's widespread contamination would remain in place and, in exchange, Texaco would receive a 'full release' of liability from the Government of Ecuador."[42]

As the litigation against Chevron in the 2000s proceeded, questions arose as to the appropriateness of the TCLP test for measuring successful remediation. Upon review it became clear that the TPH-TCLP leachate standard negotiated in the Texpet-ministry 1995 agreement established an infallible setup. That is, it was impossible to fail. It would be chemically impossible for a sample (whether taken from a purportedly remediated or unremediated pit) to reach, let alone surpass, the 1,000 mg/L TPH level as measured by the TCLP test. Indeed, Raoult's law—a theorem named after the late nineteenth-century chemist François-Marie Raoult, which formalizes the principles of vapor pressure for components of a solution and which is used to calculate the dissolvable concentration of a chemical in water—predicts that "the dissolvable concentration [of TPH in water] never exceeds the 10 mg/L" (O'Reilly, Magaw, and Rixey 2001b: 8). That is to say, although the "dissolved concentration of the compound increases with TPH concentrations at low levels . . . the maximum dissolvable concentration remains limited to the value predicted by Raoult's Law"—10 mg/L (8). In 2001, when he published this analysis, O'Reilly was employed by Chevron (Chevron Research and Technology Co.) and served on a soil and groundwater task force for the American Petroleum Institute. In

2009, O'Reilly worked as an expert consulting for Chevron on the Ecuadorian lawsuit and is among the corporate scientists I cite in chapter 1.

Between 1995 and 1997, Texpet touted its remediation program. Remediated sites produced readings for the TCLP test well below the 1,000 mg/L TPH standard of acceptability; in fact, many read at 5 mg/L TPH (Woodward-Clyde International 2000: 3–15). By early 1997, however, state scientists monitoring the remediation publicly voiced irregularities. Not only was the 1,000 mg/L TPH-TCLP reading one hundred times higher than the maximum solubility reading that a measure of crude oil could render, but also the 1,000 mg/L TPH-TCLP soil leachate level was multiple times more permissible than the 5,000 mg/kg TPH level for soil. In March 1997, the ministry amended the 1995 agreement, changing the tests and criteria used to verify whether remediation had been adequately conducted. From that point forward, both the 1,000 mg/L TPH-TCLP soil leachate assay and a 5,000 mg/kg TPH soil standard were used to declare a site sufficiently remediated.

The politics of corporate-state relations—what Riles calls "their mutual entanglements[,] . . . asymmetries, and intrigues" (Riles 2011: 802)—hardly volatilized in the swap between remediation and immunity. Rather those asymmetries and dependencies shaped the parameters of the 1995/1998 couplet contract. Not only were there incongruities in the very architecture of the agreement, but also its fulfillment provided the material processes through which the politics of interested and unequal relations became ever more enacted. As stipulated in the 1995/1998 couplet contract, Texaco's compliance in remediation and subsequent immunity hinged on the truth of a scientific standard that it could not fail.

When Contracts Ooze into Improbable Realms

SPECTRAL RADICAL TAKE I:
FRAUDULENT REMEDIATION SPINNING CRIMINAL INDICTMENTS

The incommensurability of methods in the 1995 agreement opened the possibility that more was awry. In August 2008, Ecuador's prosecutor general indicted a Chevron vice president for legal affairs in Latin America, Ricardo Reis Veiga, and a Chevron Ecuadorian counsel, Rodrigo Pérez Pallares (along with seven ministry and Petroecuador officials), for *falsedad ideológica*. Both Reis Veiga and Pérez Pallares had worked for Texaco. The entanglements of politics and contract are intriguing here.

The suggestion of improprieties first emerged in 1996. In May, the head of Petroproducción (Petroecuador's production company that had assumed full operations of former Texaco facilities in 1992) noted in a memo to his superior, the president of Petroecuador, that Texaco's remediation amounted to "covering [some] pits that had never been inspected or had their water and soil analysis certified; terminating remediation work despite water and soil levels not being fulfilled; using completely inadequate technology to treat crude ([such as] open air incineration); dumping [waste] water into the surroundings; . . . and sealing pits without treating the sediments."[43] In September 1996, the deputy secretary for environmental protection informed the minister of energy and mines of a number of irregularities associated with Texaco's remediation. Two stand out. First, the initial inventory of the number of pits associated with former Texaco oil wells was incomplete, according to official records; over two hundred excavated waste pits had not been documented. Second, normative methods, techniques, and standards for assessing contamination (in particular, those pertaining to the TCLP) were implemented with no reason other than to accommodate the interest of the corporation.[44]

Between 1997 and 2001, the Contraloría General del Estado (Office of the Comptroller General, akin to the General Accounting Office in the United States) conducted onsite audits of Texpet's remediation. These audits culminated in a 2002 report by Ecuador's then comptroller general, Dr. Genaro Peña Ugalde. The contraloría is an autonomous technical and advisory body that oversees the use of public resources, and it conducts internal and external audits to determine administrative, civil, and criminal responsibilities in the use of resources. The contraloría's report concluded that "legal and technical omissions and deficiencies affect the [Texaco-ministry] contract, and the extent of the activities and remediation plan . . . such that the result is a very limited remediation with insufficient technical parameters" (Contraloría General del Estado 2002: 70). In October 2003, the comptroller general submitted a complaint to Ecuador's prosecutor general, asking him to investigate possible criminal activity associated with Texaco's remediation. In May 2004, the prosecutor general (Mariana Yepez Andrade) initiated a preliminary investigation into the alleged falsification of the contract and certification of the remediation. But two years later, a different prosecutor general (Cecilia Armas) dismissed the criminal complaint, citing that there was insufficient proof of criminal wrongdoing.[45] It would appear the case was closed.

In March 2008, however, further evidence garnered from the lawsuit against Chevron prompted Ecuador's then prosecutor general, Washington

Pesántez, to give notice of an imminent reinvestigation into possible fraud. With the statutes of limitation expiring on September 28 (ten years to the day after the 1998 acta final was signed), the prosecutor general ordered a criminal investigation that August of nine high-ranking individuals for fraudulently conspiring to certify that Texaco had properly completed a thorough environmental cleanup, when in fact it had not (Fiscalía General del Estado 2008). According to Articles 338 and 339 of the Ecuadorian Criminal Code, if a conviction rules that an agreement or contract was procured through fraud—by way of "ideological falsification/falsehood"—then said contract is declared null and void *ab initio*.[46]

Scientific reports from the judicial inspections served in part as new evidence warranting the 2008 reopening of what had otherwise been the dismissed claims of fraud tainting the Texaco-ministry contract and release from liability. In addition to the years of laboratory analyses that emerged from the comptroller general's investigation between 1997 to 2004, and then further expert reports from 2008 to 2010, there were reports from five years of judicial inspections between 2004 and 2008. Combining the data that Chevron, the plaintiffs, and court experts' technical teams gathered, fifty-seven of the pits from which they extracted samples had also been remediated by Texaco and received a certificate of compliance and completion. Yet, of those, twenty-seven exhibited TPH levels higher than 5,000 ppm—the level indicating that remediation should begin, according to the 1995/1998 couplet contract. The RAP also indicated a collection of pits that did not need any remediation due to their condition at the time—eleven had been previously closed and four exhibited no impact. Of the eleven previously closed pits, the judicial inspections determined that seven exhibited TPH levels greater than 5,000 ppm, with one pit (Lago Agrio-16) that Chevron sampled having a TPH of 170,095 ppm. And of the four that seemingly had no impact, one (Sacha-51) sampled by Chevron had a TPH level of 63,000 ppm. Simply to note, the allegedly remediated pit at Lago Agrio-6 that according to one LAP sample had a TPH level of 299,430 ppm meant that 29.9 percent of the total weight of soil in that particular sample was crude oil.

While visiting the San Francisco Bay Area in April 2008, Pablo Fajardo noted that "Chevron has a serious problem." Fajardo, who in 2005 became the plaintiffs' chief lawyer in the litigation against the corporation, had traveled to California to receive a Goldman Environmental Prize—an award touted as the environment's "Nobel." He continued: "There is extensive evidence that a fraud was committed and a significant portion of it comes from Chevron itself."[47] It would seem that res judicata was hardly secured.

TABLE 4   **Pits Analyzed in Judicial Inspections
That Had Also Been Part of the RAP**

Determinations of 1995/1998 Couplet Contract	Determinations of Judicial Inspections
NUMBER OF PITS CHEVRON CLAIMED WERE REMEDIATED	NUMBER OF PITS EXCEEDING 5,000-PPM TPH
57 pits remediated	27 > 5,000 TPH (8 from Chevron)
	Highest levels:
	LAP measurement: Lago Agrio #6—299,430 TPH
	Court measurement: Parahuaca #3—206,512 TPH
NUMBER OF PITS CHEVRON CLAIMED NOT NEEDING ACTION	NUMBER OF PITS EXCEEDING 5,000-PPM TPH
11 pits previously closed	7 > 5,000 TPH
	Highest levels:
	Chevron measurement: Lago Agrio #16—170,095 TPH
4 pits "no impact detected"	1 > 5,000 TPH
	Chevron measurement: Sacha #51—63,000 TPH

Source: Lago Agrio plaintiffs' court submission compiled from data presented in Cabrera Vega 2008: appendixes H1, H2, H3; all data verified in the Lago Agrio court record.

The prosecutorial investigation carried through for nearly two years, generating additional site inspections and reports. On April 29, 2010, then prosecutor general Alfredo Alvear Enríquez filed a "prosecutorial opinion" in the First Criminal Chamber of the National Court of Justice against nine individuals for having created fraudulent public documents.[48] The named included Ricardo Reis Veiga, Rodrigo Pérez Pallares, and seven other high-ranking officials.[49] The prosecutorial opinion contended that the two Chevron lawyers (who formerly had worked for Texaco) and high-level officials committed the crime of falsedad ideológica when negotiating and signing the 1995 agreement, the certifying documents after each remediation, and the 1998 acta final. Citing German legal theory, Enríquez noted that the crime of falsedad ideológica refers to the act of agents creating a document and "representing it as true" when it contains "facts, or statements of consent, or knowledge" that the authors know to be false.[50] That is, the infraction is committed during the "elaboration of the instrument" when agents insert or record false facts, such that the "document appears genuine" despite holding "mendacious" and "untrue content."[51] This could include "hiding" or "suppressing essential parts or events . . . or presenting a fragmented view of facts."[52]

The prosecutor general registered two areas in which this crime was knowingly committed. First, it occurred in the scripting and signing of documents certifying that Texaco performed appropriate and effective remediation; this included fifty-one certificates and the 1998 acta final. Second, falsedad ideológica transpired when the parties inserted "technical and legal deficiencies" into the language of the 1995 agreement and the RAP that virtually guaranteed that remediation would be "improperly executed," having "consequences for the biodiversity of the Oriente region and the health of inhabitants."[53]

As the opinion noted (drawing from sixty-five binders of over six thousand pages), the government had appointed an interinstitutional committee to observe and to monitor Texaco's remediation.[54] On numerous occasions, committee members underscored the ineffectiveness of using the TCLP analysis to assess cleanup efforts as well as the vast extent of environmental damage caused by Texaco's operations not addressed by the RAP. Between 1995 and 1997, none of their observations were considered. Prosecutor General Alvear Enríquez concluded:

> [The named individuals] committed a clear act of falsehood, by issuing the mentioned final certificate, stating as true and real facts which are false, consisting of affirming that the environmental-remediation work has been fully and totally executed by Texpet, when they knew perfectly well that this was not true. I reiterate, there were observations that indicate that Texpet had failed to meet its obligations, a fact that was subsequently verified during this investigation. . . . . But that is not all, the [1998 acta final] arrives at the serious and astonishing stipulation of releasing Texpet and all its dependents of their responsibilities, thus depriving the state of the legitimate right, such as demanding that a true environmental remediation be carried out in order to guarantee its inhabitants the right to live in a healthy environment that is free of contamination and to enjoy

a decent life. This demonstrates the willful misconduct that fits perfectly with the codified crime addressed in this investigation.[55]

In early 2011, preliminary hearings took place in the First Criminal Chamber of the National Court of Justice against Reis Veiga, Pérez Pallares, and the other officials. By midyear, in June 2011, the court declared the criminal prosecutions against the men null.[56] In September 2013, the prosecutor general opened new "criminal investigations into the individuals who signed the 1995 settlement agreement and related documentation."[57] These investigatory "proceedings were and remain confidential under Ecuadorian law."[58] It appears, however, that they involve a collection of individuals who have engaged in questionable activity in relation to the Lago Agrio litigation beyond the 1995/1998 couplet contract.[59] You will learn about one of these individuals, Diego Borja, shortly.[60]

SPECTRAL RADICAL TAKE 2: US CONTRACT DISPUTE ARBITRATION

Less than a month after the Lago Agrio legal proceedings began in October 2003, Chevron filed a claim with the American Arbitration Association (AAA), a New York–based forum for arbitrating and mediating commercial disputes. The claim contended that the 1995/1998 couplet agreement contractually obliged the Republic of Ecuador and Petroecuador to indemnify the corporation for any judgment and all defense costs that might be incurred in the Lago Agrio litigation. In October 2004, the Republic of Ecuador filed a petition in the New York State Supreme Court to stay Chevron's AAA arbitration proceedings. The case was subsequently transferred to the US District Court for the Southern District of New York, spinning and triggering over the subsequent years a number of parallel and intersecting legal actions through which Chevron sought to fight the litigation in Ecuador.[61] Catalyzing radical reactions, corporate actions that started in the mid-1990s as a cluster of remediation agreements between corporate lawyers, ministry officials, and Petroecuador executives to secure closure and finality around Texaco's oil operations in Ecuador fantastically transfigured into an ensnared chain of derivative US legal actions, the first being Chevron's claim before the AAA.

The basis of Chevron's claim before the AAA was the original contractual agreement consolidating Texpet's and the Gulf Ecuadorian subsidiary's duties and obligations in Ecuador—the 1965 JOA, which outlined the terms of their 50/50 partnership. Like most contracts, it contained an arbitration clause and determined that disputes between the two parties be submitted to the AAA in New York. Similarly, the 1965 JOA contained an indemnification clause stating that Gulf would indemnify Texpet, as the concession's operator, for liability to a third party that might arise from its exploration and drilling activities. Chevron claimed that when Petroecuador assumed Gulf's position in the consortium in late 1976, the Republic of Ecuador assumed Gulf's duties under the 1965 JOA. The Republic of Ecuador repudiated this position, maintaining that neither the state nor Petroecuador signed on to the JOA.

During the legal proceedings in the New York district court in which the Ecuadorian state sought to suspend the AAA arbitration, Chevron entered a counterclaim as further justification for why the Republic of Ecuador must indemnify the corporation for any judgment and all legal expenses emerging from the Lago Agrio trial. Tracing its strategy in the Lago Agrio litigation, Chevron declared that when Ecuadorian officials released Texaco from liability in signing the 1998 acta final, it did so not just with respect to claims by the state but also claims by the Lago Agrio plaintiffs. Chevron claimed:

> The Republic and Petroecuador are in breach of their obligations under the 1995 Settlement and 1998 Final Release of Claims, and the Republic . . . and Petroecuador are obligated to intervene in the Lago Agrio litigation and inform the Ecuadorian court that they owned and released all rights to the environmental remediation or restoration by Texpet in the concession area, and to indemnify and hold harmless Texpet and Chevron-Texaco for any and all fees, costs and expenses relating to the Ecuadorian lawsuit, including any final judgment that may be rendered against ChevronTexaco in Ecuador.[62]

After extensive discovery and multiple motions over three years, in June 2007, Judge Sand of the New York district court granted the Republic of Ecuador's request to stay arbitration proceedings on the grounds that the Republic of Ecuador was not contractually bound by the 1965 JOA, which neither it nor Petroecuador signed. Judge Sand did not rule on Chevron's counterclaims regarding the 1995 settlement and 1998 acta final.[63] Upon appeal at the US Court of Appeals for the Second Circuit, the three-judge panel upheld Sand's ruling in toto.[64] And when Chevron appealed to the ultimate US judicial realm, the Supreme Court denied certiorari—meaning, they refused to review the case—in June 2009.[65] Chevron withdrew its petition to urge Judge Sand to rule on its counterclaims *and* withdrew its claim before the AAA in July 2007. Chevron's counterclaims invoking the 1995/1998 couplet contract would, however, reappear two years later, in September 2009, as will soon become apparent.

SPECTRAL RADICAL TAKE 3: A STING OPERATION GONE AWRY

The summer of 2009 was eventful. In June, video recordings of an alleged judicial bribery scheme appeared at Chevron's California headquarters. Significant portions of the multi-hour tapes show two individuals—Diego Borja and Wayne Hansen—talking with the then president of the Lago Agrio court, Dr. Juan Nuñez, who was presiding over the lawsuit against Chevron. In August 2009, Chevron posted the tapes on its corporate website.[66] Chevron claimed the videos presented evidence that the Ecuadorian court was corrupt. The tapes, allegedly, were made by two civic-minded citizens—one Ecuadorian, the other from the United States—acting as whistleblowers to reveal how fraudulent the litigation against Chevron was. Armed with hidden video recorders in a pen and watch, Borja and Hansen secretly taped meetings with Nuñez in which they posed as the heads of an environmental engineering and remediation firm. Together Borja and Hansen sought to catch the judge accepting their promise of a cut of the remediation costs if their firm were granted the remediation contract essential for cleanup. Borja and Hansen were after two things: first, that Judge Nuñez declare that Chevron would be made liable and, second, that Judge Nuñez accept a bribe for granting the Borja-Hansen "firm" a contract to clean up the contamination.

I was not alone in being perplexed as I watched these videos in late August 2009 on the home page of Chevron's website.[67] Many were dumbfounded. Granted, the image and sound quality of the tapes is poor—dizzyingly inaudible in parts. But between Nuñez's patience, Hansen's painful broken Spanish, and Borja's attempts to assuage misunderstanding, there was no bribery to be seen or heard. In fact, if anything, the tapes revealed that Nuñez was guardedly gracious, that Hansen was often fumblingly inept, and that Borja's efforts to finesse fell short. The Hansen-Borja tag team was less than convincing, and the tapes do not demonstrate corruption.

Delving deeper—which, of course, the plaintiffs' legal team did—served to further undermine the tapes' credibility. And it generated questions about Chevron's involvement in their making. An investigation indicated that Wayne Hansen had several run-ins with the US law, the most serious being in 1987, when he was convicted of conspiring to traffic 275,000 pounds of marijuana into the United States. Far from being an environmental engineer, as he claimed in the tapes, Hansen was, according to an investigation, described by others as a "con-man" and "hustler" with no steady employment, who tended to swindle associates and scoff at the law.[68] This raised doubts about Hansen's character and his purported civic-mindedness. What, one might ask, would inspire a convicted felon who served nearly three years in a federal penitentiary to become civic-minded, let alone concerned, about a lawsuit against a corporation in a distant land?

Similarly, Diego Borja (an Ecuadorian citizen) hardly appeared to be the disinterested "good Samaritan" that Chevron claimed. Rather than a concerned citizen shedding light on the corruption of Ecuador's judicial system, Borja proved densely associated with the corporation. And this association suggested that Borja's motivations were perhaps less than altruistic or transparent. Between 2004 and 2009, Borja worked as a contractor for Chevron. His role was to oversee and facilitate the safe transit of water and soil samples obtained during the judicial inspections to a laboratory for chemical analysis. Borja's name and signature appear on court documents attesting to the chain of custody that secured the passage of samples from the field to the lab. Chevron paid Borja for his services through his company, Interintelg, SA, although on court documents he is said to be a representative of Severn Trent Labs, a US laboratory that Chevron engaged to analyze their samples' chemical content. Further imbricating association, Interintelg was located in a Quito office building, Edificio Borja-Paez, in a swanky part of town that also housed the offices of Chevron's Ecuadorian legal team, Callejas y Asociados, Estudio Juridico. Borja's uncle, who partly owned the building, had worked for Texaco for thirty years.

Orbitals of intrigue intensified, however, when a collection of extended phone and Skype conversations between Borja and a longtime friend, Santiago Escobar, saw the light of day. The conversations began on October 1, 2009, soon after Chevron had posted the Borja-Hansen sting videos on its website. By that time, Borja was comfortably ensconced in San Ramon, California, near Chevron's headquarters. It would appear Borja was the recipient of what is best called a corporate private-protection program—a move Chevron would repeat within a few years for another company favorite. Before Chevron released the videotapes of the purported bribe scheme to Ecuadorian authorities, the corporation moved Borja, his wife, and child out of Ecuador to California and settled them in a home with a swimming pool abutting a golf course in a gated community. According to court documents, Chevron covered all of Borja's living costs, provided him with an SUV, and paid him a generous stipend, as well as all legal fees for a pending asylum case. Bravado exudes from Borja's voice in the audio conversation with his friend when he laughingly goads, "Crime does pay!"[69]

But Borja's voice also intimated an undertone of anxiety. By turns cocky and paranoid, testy and ambivalent, Borja's demeanor and maneuvering shifted dramatically over the course of one day (October 1, 2009).[70] For instance, he boasted of his accomplishments (having within days achieved what Chevron sought to do for years). He worried about how long the corporation would fête him. He threatened to release incriminating information if the corporation "tricked" him. His insecurities heightened (this was yet early in his post-sting Chevron relationship), he stewed over how events might play out. Deals of this sort have been known to go awry. What stands out is that Borja repeatedly makes clear that if Chevron chose not to play nice, he would go public with information that would make the company buckle. The damning evidence that he alludes to—from tampering with scientific data (especially from Texaco's remediated sites) to corporate officials' involvement in masterminding the sting operation—would make the corporation liable under the US Foreign Corrupt Practices Act.

Between August 2009 and December 2011, Chevron paid Borja over $2.2 million in cash and services.[71] Over $900,000 went to pay legal fees.[72] And these were not merely for asylum lawyers. Rather, Borja and the two sets of audiovisual tapes (the bribery scheme with Judge Nuñez and the inchoate disclosures with Escobar) in which he implicated himself catalyzed a number of court proceedings in the US District Court for the Northern District of California.[73] As will become more apparent in time, the Republic of Ecuador *and* the Lago Agrio plaintiffs (LAP) had a great interest in Borja. Soon the Republic of Ecuador would reopen a confidential investigation into unnamed individuals associated with the Lago Agrio litigation (one, most presumably, being Borja). The Republic of Ecuador found itself in an international arbitration. The LAP found themselves being sued in Chevron's Racketeer Influenced and Corrupt Organizations Act (RICO) case. Borja was someone who could potentially implicate the corporation in wrongdoing. Seeking to mitigate that possibility, Chevron handsomely paid (and potentially continues to pay) a high-profile criminal defense team to fight hard in the court and fend off the wolves.[74]

This brings me to this chapter's final spectral radical reaction: how, amid the Borja-Hansen intrigue, the 1995/1998 couplet contract crossed continents and oceans to haunt The Hague.

SPECTRAL RADICAL TAKE 4: INTERNATIONAL
CONTRACT DISPUTE ARBITRATION

In September 2009, Chevron filed an arbitration claim against the Republic of Ecuador with the PCA in The Hague.[75] Arbitrated under the rules of the United Nations Commission on International Trade Law (UNCITRAL), the claim alleged that Ecuador breached the BIT that was signed between the Republic of Ecuador and the United States in 1997. The claim was based on Chevron's counterclaim before the AAA in the early 2000s. Recall that although Chevron's claim before the AAA primarily concerned the 1965 JOA, in the course of the proceedings, Chevron subsequently claimed that Ecuador also breached the 1995 settlement and 1998 acta final. Chevron, however, withdrew this latter claim before Judge Sand ruled in favor of the republic in 2007—slipping its legal argument concerning the couplet contract up its sleeve for future opportunities. That time and place proved to be the PCA.

I will pause here, as chapter 7 takes up the PCA ruling in more depth. For now, I want you to dwell in a space where radicals—those chemical wonders that breed chain reactions that, left to their own devices, continually unfold—effectuate entities with relations integrally implied. Both this and the next chapter tap "radical" as a method tool for paying attention to the proliferating possibilities that accompany a form. Here I underscore how the contract, as form, compelled a perverse proliferation that sanctioned the displacement of liability under the law. In the next chapter, I consider how the judicial inspections themselves gave witness to their own proliferating phenomena—some so ubiquitous that Chevron sought to normalize, some hardly attended to, and some slated to be dismissed.

CLANDESTINE

In April 2015, Amazon Watch, a San Francisco Bay Area–based environmental and Indigenous rights organization, received in the mail digital tapes from a Chevron whistleblower. The tapes document teams of soil scientists and technicians, some working for and others contracted by the corporation, encountering soils laced with crude oil and voicing their frustration in their search for uncontaminated samples. By contrast, the voice and side comments by the interviewer and videographer are gentle and attentive.[1]

The following text provides transcripts and descriptions of a snippet of the tapes that Amazon Watch obtained.

This audio transcript is from a video interview with a Kichwa man at his home near the oil well Guanta-6. The gentleman had lived in his finca for more than thirty years.[2]

"The area was free, unencumbered. We were the only ones here. We were the first to homestead. And then more and more folks came. When we came, Guanta #6 did not exist, nor did other wells. Three years after we arrived, they drilled the Dureno #1 oil well. Then after that they came to make this wellhead here. The little one. They asked my permission to build a road. 'This will be good for you.' . . . But it hasn't been that way. Instead, there has been contamination. It hasn't been so good. Well, clearly, in part there are good things that come with the road. But the work, it contaminates. The petroleum. It has killed three of my children. Three daughters have died. They ran around and got completely covered with crude. All up their legs. And that killed them. Men would come and say, 'Yes, we are going to help; we are going to help. We will make this right.' But up to this point, they have not arrived."

[Off-camera interviewer:] "Your daughters have died? How many?"

[The man's wife responds:] "Four."

"How old were they?"

"The little infant was two months."

"And the doctors didn't tell you what she died of?"

"No doctors come here. Another child was three years old, another five years old, another one and a half years old."

"Were they sick?"

"Yes, they were contaminated. To this day, we ourselves are contaminated. You know, when blood, pure blood comes out of your throat. Sometimes black pus comes out. Also when you're overcome with dizziness, and headaches, and your vision is not clear and you can't see. You can see ten meters, no more."

"And you think it's because of the contamination?"

"Yes, precisely, it's because of the contamination."

"What other problems do you have from contamination?"

"Well, the lands are contaminated. The pastures sometimes dry out and die. And the plantains, they dry out and die. They don't fruit. Or they fruit but the fruit dries and falls. And the yuca rots. Maize, the same. With this, we are screwed. And chonta, too. There used to be many over here, and they all died. And then they burn the crude and a lot of smoke and soot falls, and this contaminates, too. . . . Yes, they made two pits with this well. They are there to this day. One is covered with dirt. The other is open. . . . When they come to clean the pits and wellhead, all the wastewaters flow into the stream, and by way of the stream, that crude contamination reaches us. And the uncovered pit also overflows when it rains and contaminates the stream. This is our only little stream. We drink from there and bathe in there. And what we are drinking and bathing in is this dirty stream. We don't even know all the illnesses that we have or those that will come. And you see over there? Damn, it was pure crude. And it's never been cleaned. It's always been like that since they drilled the well. With the rain it gets worse. The oil rises. . . . Yes, when it rains, it contaminates more and more. . . . A cow and two horses died recently. They were in the swamp and they drank that crude-coated water. . . . The stream we use passes alongside the pit. Because the creek is contaminated, we dug a water well. But there was a spill recently [a rupture in the pipeline that passes through] . . . the pasture and that contaminated the water well, too. And now, where are we supposed to get our water? We catch rainwater but it's not enough."

This audio transcript is from a video of three soil scientists extracting samples from the oil well Shushufindi-8 on January 15, 2005.[3]

"Holy shit!"

"Wow, that's definitely a spill."

"That's crude."

"Something crudo going."

FIGURE 30

"Maybe we just start getting sediments here."

"That might be the way to go. Maybe dig down with a machete and take something at depth. . . . I mean, it looks . . ."

"It looks bad."

"Yeah. It looks bad."

"How far back does it go?"

"We didn't go real far back."

"This is only one of the few places where the canopy breaks. So whether it's the oil that killed it off. Or whether it's a pit that is changing. There is something going on here."

[The team takes soil samples.]

"You know this doesn't smell so bad. It's not horrible. It's super organic-y."

"Yeah."

"But when I drove in my knife, it came up with a little bit of petroleum sheen."

"Yeah, there's definitely a sheen coming through. Not as much as I would have expected."

"Yeah, I couldn't smell it."

"That it's weathered makes sense. It's definitely not fresh."

"Courtney [who is doing the mapping], you got this, right?"

"'Area of interest'—is that what we called it originally? Right? Did you change the name? . . . We're supposed to find two pits. If we can confirm that this is a pit, I think we can call ourselves conquering heroes."

"Alright, man, just don't break your arm patting yourself on the back." [Gentle laughter.]

This audio transcript is from a video of Shushufindi-21 on March 3, 2005. The footage opens with technical teams, flanked at the head and rear by military soldiers, dispersing to explore the site and locate where to take clean samples. A generator-propelled drill bores deep and extracts soil samples to be examined by petroleum experts. Conversation ensues between Dave (an expert consultant) and René (a Chevron representative) as they examine long core samples resting on a table.

FIGURE 31

R: "Nada?"

D: "No."

R: "Damn, this is a pretty good spot."

[It is clear as the video moves in closely that the soil samples are pocked with crude oil.]

[. . . The footage cuts to later (13:16 minutes). Different samples are on the examining table.]

R: "Wow, you found it here. Damn. At about 1.9 [meters]."

[Dave leans over to smell it.]

D: "OK. It's here. Good news. Petroleum!" [Laughter.]

R: "No. No! Check it again." [Laughter.]

D: "You want to smell it? I think it is."

R [Leans over, takes a piece in his hands. Lifts it to his nose. Looks at Dave, and says straight-faced]: "No!" [And then breaks into a big smile.]

D: "No?"

R: "Nah, it is. It is. It is."

D: "Cause, ya know, I don't know what this funk is."

R: "Well, we might as well stop them now. Stop 'em" [referring to the drilling crew]. Just that we're done here. We're trying to find a clean core, and obviously we didn't go out far enough. [Inaudible.] We've now got a headache. Let's core this."

D: "Well, we have another sample coming."

R: "We'll sample this. Just take the worst of it. And we'll just do it for TPH. We won't beat it to death. . . . Nice job, Dave. One simple task."

D: "Who picked the spot, René?"

R: "Don't find petroleum."

D: "Who picked the spot, René?"

R: "Don't find petroleum."

D: "You told me where to drill, René."

R: "Uhh? My fault? My fault? I'm the customer. I'm always right. [Chuckle from Dave.] The customer is always right."

D: "Whose fault?"

R: "Well, that would have been yours, Dave, because you kept finding oil in places where it shouldn't have been."

D: "I try. That's why you hire consultants."

R: "Shoot the messengers."

D: "We can find it where it's not supposed to be. And then it's your fault, not our fault for putting it there. See? It's a brilliantly conceived strategy."

R: "As expected, this is a shithole. [Inaudible.] This pit with all its historical discharges [inaudible] . . . it would be a point of embarrassment trying to find a clean point five hundred meters to the west."

[The footage cuts to another technical team taking auger samples near the African palm plantation. The soil samples look and smell of crude oil. The people whose voices follow are unidentified.]

"How do you think these palms are growing?"

"They could use a little petroleum. A little organic carbon."

"It's become less plastic. It's sticky. The composition may change real soon. See it's different."

[Courtney, the woman on the team approaches:] "Found some Texas tea, ay?"

"It's not that bad, but it's dirty."

"Is this the pit?"

"It must be."

"So it wasn't, uh . . . [long pause as she turns to the videographer] georeferenced wrong?"

"You never escape."

["Georeference" appeared to be a code for remediation. The boring continues, finding crude oil at a depth of five meters. Chevron's efforts to find sites suitable for sampling in the judicial inspections repeatedly failed to contain oil's cunning, attesting instead to the recompositional mattering of hydrocarbons in human and nonhuman lives.]

4 Radical Inspections

Of Sensorium as Toxic Proposition

SEEP BACK INTO THE MIDST of the contentious, increasingly overwrought process of judicial groundproofing. This was not the sort of judicial process to which those in the Anglophone world have become inured—where common law stages a confrontation between legal contestants before a judge safeguarded in a wood-paneled court: an adversarial method of law. No. Seep into a legal spectacle of another sort—that of the civil law tradition as it took shape over the seven-plus years of litigating the lawsuit against Chevron in Ecuador. Note the curious shape, the intriguing form, that one rendering of the inquisitorial method of law could take.

Ooze inside the unfolding of the judicial inspections, multiday excursions to alleged contaminated sites. In attendance are the Ecuadorian judge and his clerk, teams of lawyers, scientific and technical crews, local dwellers, a scattering of press and curious individuals. Ultimately, fifty-four inspections consumed the parties constituting this lawsuit over the course of five years.[1]

At each inspection site, the parties sought to materialize, enroll, and argue over the connections between hydrocarbons, formation waters, excavated earthworks, signed papers, and sensate bodies—that is, over the consequential relations between crude-oil extraction and compromised life capacities. Have the industrial wastes decanted into the environment sickened local residents and the ecological system on which they and other life-forms depend? Have the deposits of industrial activity undermined health and well-being?

The complexity of the situation gets broken down. Is there crude-oil waste in the environment? Is that crude toxic? Is there a correlation between crude contamination and cancer? Both parties appeal to science to found their case. Thousands of soil and water samples are extracted from their milieu. Experts examine samples in situ. Assessing their texture, their color, their smell. Attending to their enmeshment in hydraulic flows, soil matrices, industrial infrastructure, practices of human habitation. Samples follow chains of custody, some more and some less transparently, to national and international laboratories where their chemical constituents are analyzed, standardized, and quantified. Reports emerge months later, multiple versions from each inspection site. Stacked together, the reports are tomes produced by each of the parties involved in the suit. Reports detail local topographies; forest disruption and regeneration cycles; local soil stratigraphy and permeability; extraction infrastructures; and most importantly, the complex chemistries of the hydrocarbon brew that constitutes crude.

Extracted from context, the warm, slick, frictionless feel and pungent scent of hydrocarbons vaporize. As chapter 1 showed, chemical analyses distinctly slice and dice the structure and arrangement of the atoms constituting crude and impute contradictory meaning to the resulting fractionated molecular assemblages. With crude extracted from context, the corroded skin, the shallow breath, the sunken eyes of ailing bodies are made but hauntings. As chapter 2 showed, epidemiological studies distinctly assert the aggregate plausibility or implausibility of a mathematical correlation between oil extraction and cancers. With lives transformed into statistical abstractions, the taut scar, the fifth miscarriage, a grandmother's bloody vomit are made to volatize and dissipate. In a constellation of laboratory equipment and quantitative protocol, assays, and statistical modeling, hydrocarbons and their health effects rematerialize, imbued with specific meanings and possibilities. Differently fractionated molecular assemblages and competing statistical regressions carry a singular ontological politic.

Despite being powerful in their suggestion, however, the scientific results of either party were not truly conclusive. Five years of judicial inspections could not deny the scientific uncertainty clouding the relationship between crude operations and damage to life-forms. The thousands of chemical analyses, over one hundred of hydraulic and degradations reports, and a collection of epidemiological studies provided no indisputable scientific proof and could neither definitively determine nor deny any causal relation.

FIGURE 32 Earth adjacent to a waste pit at the oil well Sacha-53.
Photo by Chris Toala Olivares.

And what of the texture, the hue, and the waft of crude that scientists inspected upon extracting soil and water samples from the forest floors? What
of the people who spoke of themselves and their loved ones becoming ill from
hydrocarbons entering their being? Had they slipped from sensual encounter, recoded as stark data points reaching for the purported cold certainty of
scientific statements, of objective and calculated technique?

No. Half a decade of judicial inspections made clear that a lack of certainty should not be confused with an absence of understanding, of insight,
prudence, acumen, a fluency with complex conditions.

Listen to crude's distinctive resonance. The inspections were active events—
what emerged was always shifting, their effect always contingent. A lot was
going on. The movement of security, both state and private. The armored
cars. Chevron's catered lunches with cold soda under the equatorial sun. The
clanking of augurs. The squish of rubber boots. The crying infant. The bark-
ing dogs. The smell of perspiration. The rustling of brush. The clicking of
recording equipment. Local residents' homemade signs. The shuffle of pa-

pers. The schoolkids holding a banner. The small plantain grove that shriveled after a spill. The middle-aged farmer with pictures of his deceased father and niece. The young woman whose skin ailment caused a lawyer to recoil. The horse whose insides had desiccated after drinking poisoned waters. The inspections rested on deepening knowledge and not solely that provided by chemical analyses and scientific reports. Rather, other understandings were absorbed, shared, and transmitted. Heightened ways of seeing, sensing, assimilating, and subsuming—all directed by the judicial inspections' experiential hold.

Slip along the crude-laced surfaces. Contamination abound in wild profusion, were one sufficiently attentive. The most quotidian of practices—walking, waiting, and watching; or sensing, sniffing, and smelling; or touching, feeling, and tingling—triggered experiential sensibilities. Petroleum and its very materiality emits, secretes, deposits its presence and begs notice. It oozes under foot. It leaks from embankments. It saturates soils. It congeals thickly on the surface of waste pits. It drowns decaying organic matter. It rests hidden below surface sediments. It glistens iridescent on streams. It howls with torrential force when alight. It causes pipes to vibrate, their hum a constant to the pulse of the night frogs. Its aroma tightens the throat. It draws welts on sensitive skin. It embeds in the cracked soles of callused feet. It provokes the stomach to contract, and then bloat, if ingested. For creatures—human and nonhuman alike—the texture, sight, smell, and sound of crude came as proximate, not abstract. Sensations experienced universally, not just by a select few. And they surfaced recurrently. They were not the effect of a one-off, isolated event.

This chapter regathers midflow, returning to the judicial inspections. Seeking to name what remained only partially spoken, yet experienced. Viscerally. In sensoria. Where hydrocarbon residue traced a multiplicity of unannounced encounters. Where without warning an oil-laced labyrinth registered across surfaces. Where a silent patina spun untold stories of uncertainties still to come. Consider the possibility that five years of ground-proofing alleged contaminated sites constituted "radical inspections." Borrowing from chemistry's notion of the radical, the inspections propagated and proliferated derivative potentials for apprehension. And while not the normative toxic-tort story, these derivative potentials for pause and recognition proved as consequential to the Ecuadorian litigation as the extraction and analysis of water and soil samples, or the arguments around epidemiology, or debates around contracts.

Said otherwise, this chapter explores how the judicial inspections unfastened science's hold on the relationship between crude oil and life. It suggests that the inspections obliged understanding beyond codifiable and quantifiable data points, beyond laboratory assays and reports, beyond statistical calculation, beyond the framing of contracts and liability. Without doubt, the inspections provided the material basis for scientific arguments around toxicity and crude. Similarly, they serve as a reference point for the epidemiological studies. And they serve as the stage on which to make legal arguments over signed agreements. Yet the effect of the inspections was simultaneously vastly more spectacular and subtle.

Dozens upon dozens of judicial inspections, individually and collectively, cascaded as radicalizing sensoria—a gloss that seeks to trigger a conceptual grammar sensitive enough to hold the material, kinesthetic, and emotive energetics of transformative processes that, while palpable, were not in the moment fully registered. They exceeded intellectual control.

Conventionally speaking, a sensorium refers to the sensory system that receives and coordinates stimuli conveyed from a body's sensory receptors. Thinking with valence and Gilles Deleuze, a sensorium extends to signal the immersive spheres of connection whereby entities attune to, move through, and are moved—by virtue of their own unique compositional capacities— in en-worlding encounters. Bodies from atoms to humans, and everything in between and beyond, imbibe—and thereby are transformed by—their worlds via the sensorial faculties singular to them. A radicalizing sensorium cares about the cascading ways entities are receptive, even susceptible, to the impress of others and, simultaneously, the cascading ways they initiate, transmit, and withhold impressions. Such prehensions of immanence register, animate, and subsume transformations in beingness, implying that an "entity" is never singular, never contained. Entities do not exist and then enter into a sensorium. Rather, they exist by virtue of their very participation in sensoria (cf. Stengers 2012; Strathern 1999a, 2005). As Stengers notes, "Such a coming together is the first and last word of existence" (2012: 7). Valence is that relational, compositional capacity; radical gives it a mode of specificity.

The judicial inspections unfolded through iterative practices.[2] And iteration mattered. The same configuration of characters—the judge, lawyers, scientists, technicians, villagers, rural inhabitants, the press—touring (at times slogging through) former Texaco sites on a set itinerary: stops at drilled wellheads, waste pits, gas flares, pipelines, separation tanks, rivulets, streams, water sources. As such, the inspections were choreographed in that, at each site, the lawyers advanced legal arguments and challenged through legal questions, scientific crews gathered technical samples and assessed their provenance, and rainforest residents made their presence felt by observing, holding placards, and giving testimony.

It would be a mistake, however, to see these iterations as repetitions. No more than living among these sites from day to day, no more than the flow of life from moment to moment, no more than the combining achievement of an atomic radical, is repetitive. Scientific assays assigned that which was repetitive—that which in being made consistent could be captured and inscribed by unique instruments and protocol—as an instance of the real. But bodies in movement always have a somewhat unpredictable trajectory, as movement unfolds responsive to relative circumstances, circumstances they also have a part in generating. More provocatively, the judicial inspections could be considered a refrain—if one understands, as do Sasha Engelmann and Derek McCormack (leaning on Deleuze and Guattari 1987), "refrains . . . [to be] patternings of materiality that express the tendency to return while never doing so in quite the same way" (2017: 254).

As refrains, the inspections were practices of attunement for those unfamiliar with the texture and cadence of rainforest life in the refuse left by extraction. That is, they entrained an increasingly skilled capacity to note relations among entities—waters, soils, plants, animals, humans, waste pits, wellheads, pipelines—and crude. They instilled a way of becoming attentive to how hydrocarbons gathered around, adhered to, precipitated from, saturated, repelled, pocked, and debilitated. They exposed variations, always of a circumstance, never identical, yet which made the force of elements palpable. The inspections conditioned an encounter with non-Lavoisierian knowing, an encounter with sensoria whose complex ecologies impressed a slow, accruing prehension of contamination and its effects. And this is what is meant by "radicalizing sensorium."

This chapter expands on three movements that figured the judicial inspections as radicalizing sensoria: one, geoengineered; a second, aquatically agitated; and a third, vernacularly embodied. Resonating through eons-old earth depths, a geoengineered sensorium collapsed and distended the spatial-temporal formation of waste pits. An aquatically surfaced sensorium revealed, through molecular, atomic, and subatomic forces, how water evinces hydrocarbons. And a vernacularly embodied sensorium spoke through testimonies of the untenable human predicaments of abiding with crude.

FIGURE 33 The massive gas flares popularly called "el dragón" at the Drago Norte-1 station. Photo by author.

Literally and figuratively seeping out from under foot, a cascade of stirrings eluded the logic of science. The judicial inspections as a unique judicial assembly invited space for taking notice. Taking notice came on multiply through sensing, through attuning, through experiencing. Momentarily inhabiting Texaco's form operations, experiencing the ubiquity of crude's presence, and listening to testimonies by local peoples triggered realities to surge forth and imprint, make an impression. These geoengineered, aquatically agitated, and vernacularly embodied sensoria animated the possibilities that mattering—what matters—runs beyond the constraints posed by science. Thinking of the judicial inspections as a radical form extends a specificity to what valence-imbued mattering does—for radicals are about passages and propagations that always enfold a modicum of change. Those participating in the inspections (the judge, lawyers, etc.) became enmeshed, though differentially, within these sensoria. They traced material networks of proximate dangers, unearthing a consideration that evidence could be other than what is abstracted and made reproducible. That proximity was substantial, sensual, and radical; it seeped into and altered experience

FIGURE 34 Horizontal flares burning over a partially dirt-covered, crude-oil waste pit. Photo © Amazon Watch. Used with kind permission.

such that what existed was different from what existed before. That affection surfaced an iterative, embodied, and sensorial understanding of harm; and that understanding was far from singular, far from precise, and far from detached.

Lurking on the periphery, even when in plain sight, radical sensoria transformed. Incrementally. Never linearly. Often transitorily. And that is because radical valence is not about origins, progressions, and ends. Rather, it is about orbitals and intensities that destabilize dominant or prior forms. Extending Deleuze and Félix Guattari, radicals generate "modes of expansion, propagation, occupation, contagion, peopling" (1987: 239). They are about bodies moving and being moved, affecting and being affected: the affects and affections of a substance (Deleuze 1988: 27, 45) never singular or fixed. The judicial inspections tacitly encouraged this attunement. Iterative practices impressed again, and yet again, the lively substrate through which crude and matter reacted to the point of producing different forms of recognition for the court. As will become evident in chapter 5, together these sensoria provoked alternative forms of meaningful deliberation.

Color is what first catches the eye upon approaching open, excavated waste pits. Against the tropical greens of the equatorial forest, red soils rich with iron form a striking contrast. From the crest of the pit's dikelike embankment, the packed-earth clay transforms in color yet again, as it does in texture and smell, too. Sloping down the contours into the pit, blacks and browns stain the rusty red soils. Half a meter or less from the embankment's height, a pool glistened with heavy crude. Soon a dissonant acrid smell hits the nostrils, tightens the face, turns the throat raw. Ever so slowly, heat wafts up, barely noticed.

Geoengineered Sensorium

Enfold within the vertiginous changes of geological time.

Two miles below the rainforest floor, crude oil resides in subterranean reserves. The Euro-American geologic narrative from whence crude comes tantalizes.[3] Stretching beyond human time frames and temporalities, what is given as solid and inert shifts in the changing composition of a yielding earthly mantle. Crude's presence below the rainforest floor signals vestiges of a deep planetary history. A world of vast, shifting oceans teaming with primitive life and unrecognizable and transitory continent-masses variously covered in bogs, forests, seas, and ice. Many millions of years ago—in the era of "ancient life" (the Paleozoic, from the Greek *palaeo* [παλαιός], "old, ancient," and *zōē* [ζωή], "life")—what we know as the Upper Amazon was thoroughly submerged, resting beneath waters off the western shore of the then-largest terrestrial landmass, Gondwana.[4] During the latter third of the Paleozoic, or the Carboniferous period, dramatic shifts compelled Gondwana to converge with another landmass to form the mega-continent Pangaea.

Geologists hold that in the Carboniferous period, colossal tectonic shifts occurred (roughly 350 to 150 million years ago), thereby forming most of the earth's hydrocarbon deposits. In what would become the Upper Amazon, massive land and ocean plates shifted and collided over hundreds of millions of years, creating in the process vast fluctuating sea inlets and swamplands. Over the millennia, these seas and wetlands capped with warm waters were incubators of microscopic life; they teamed with a dense and bewildering array of minuscule sea creatures—diatoms, foraminifera, plankton. Suspended organic matter drifted into murky sediments as ancient oceans distended into swamp, early lands swallowed seas into lagoons, and age-old rivers shifted and churned to a slow.

The geological record suggests that microscopic blooms of these simple-celled plants and animals were so dense that their cascading detritus outpaced their decay on the sea, swamp, or lagoon floor. Semidecomposed zooplankton, spores, and pollens accumulated into coagulating biotic sludge covered by accruing and shifting silts and muds. Buried in legions of sediments, these former marine organisms were captured in the perfect anaerobic conditions. The weight and pressure of the sediments above and the earth's internal heat from below precipitated a chemical reaction to transform the organic matter into kerogen, a dense insoluble organic compound: source rock.

In what would become the Upper Amazon, ancient sedimentary rock served as the perfect source rock. As vertical miles of sedimentation amassed atop source rock over the subsequent millions of years, the pressure, compaction, and heat from the earth's core cracked the hydrocarbon bonds composing kerogen. The resulting smaller hydrocarbon molecules were light enough to migrate upward through the porousness of sedimentary formations. Beginning in the early Cretaceous period (140 million years ago), tectonic force thrust a deep ocean plate (known as the Nazca Plate) under the South American Plate. In its wake, the thrusting bowed the ocean floor of what over geologic time would become the Upper Amazon into undulating patterns. These sea-floor folds and faults forged anticlines, or traps, of impermeable rock. There, the migrating hydrocarbons nestled, unchanged for millennia, in a capillary sedimentary formation capped by an impervious shield. This is what is called reservoir rock. All crude oil found in the Ecuadorian Amazon resides within the Hollin and Napo Cretaceous formations.[5]

Stumble through the techno-temporal insolence of petrocapital.

Far from being underground lakes, then, an oil reserve is a geoformation where hydrocarbon molecules inhabit the microcrevasses of porous reservoir rock. Reaching those hydrocarbons, as you have learned, is an arduous, capital-intensive, and risky affair. And Texaco's particular engineering of that endeavor in Ecuador was fraught with the imperial bluster that accompanies all arrogant indifference to how the mandate of progress destroys and defiles. Once seismic exploration (which in itself is disruptive of forest life) delineated zones holding the possibility of discovery, Texaco flew in heavy equipment—bulldozers to clear and level forest floors; cranes to build a platform and erect a derrick; chokes and drill bits to control the drilling; stacked pipes, stories high, to compose the drill-string; tons of cement to hold the well casings fast; excavators to dig large pits; and the basic infrastructure to house around-the-clock work crews. Then came Texaco's network of roads that snaked between oil facilities through the ever-dwindling forest.

For weeks or months, depending on the challenges posed, the drilling rig bored through miles of rock. The process was complex, but here is a sketch. Drill bits moving roughly two hundred rotations per minute ate away at clays and rocks, opening space for two concentric pipes to penetrate the earth. The outer pipe casing was cemented to the deepening hole to prevent its wall from collapsing and resist the pressure that different formations affected. Synthetic muds poured down the well. The muds lubricated and cooled the bit, stabilized drilling force, and helped seal the borehole for the outer pipe casing. The mud flowed back to the surface along with earth cuttings and terrestrial fluids through the space between the two pipes.

Geophysics dictates that fluids move from high pressure to low pressure. As diamond bits cut through lithostrata thousands of feet deep, geophysical propulsion forced earthly liquids from microfissures. If the bit encountered reservoir rock, subterranean pressure and heat propelled a crude-fluid-mineral assemblage from the rock formation to the borehole, where it ascended through miles of interlocking pipes. As long as a pressure differential existed, conventional (or unassisted) extraction ensued. All the wells that Texaco drilled flowed this way, at least initially.

Petroleum, of course, is never alone, in isolation. It belches from the earth's depths conjoined with other material forms. The hydrocarbon brew that we call crude oil surfaces as a complex concoction of liquid hydrocarbons, gaseous hydrocarbons, and densely salinated formation waters, earth cuttings, and mineral particles, along with synthetic muds and solvents.

As noted earlier, at each of Texaco's 330-odd wells, the company excavated rectangular-shaped pits to hold the drilling and extraction wastes. Depending on the complexity of subterranean stratigraphy and the problems encountered during drilling, Texaco quarried between two and five pits alongside each well. Each was approximately three meters deep, with the largest extending the expanse of a soccer field, some filling the size of an Olympic pool, and many encompassing a tennis court or smaller. All were rough earthen craters, opened, unlined, and unprotected. It is estimated that Texaco dug more than nine hundred such pits.

Texaco's former installations pulsed of eroding industrial ruins and the despoils of ancient rock formations never meant to be unearthed. A geo-engineered sensorium reeked of human and nonhuman radical compulsions.

The embodiments of smug petrocapital practices and deliberate material affordances, industrial waste pits festered. During the drilling process, pits were receptacles for drilling muds, cuttings, subterranean liquids, anticorrosive chemicals, and often cements. Most wells were productive, in which case adjacent waste pits held the effects of drilling plus quantities of discovered crude itself. To determine the productivity of a well, oil engineers let the crude jut forth into pits for intervals of an hour at a time so they could calculate the consistency and constancy of the flow. These "flow tests," as they are called, continued through the life of a well, belching forth crude into nearby pits during routine well maintenance. And such are the lively substances composing waste pits.

At the time of the judicial inspections, Texaco's former waste pits were not all in the same state of being. Several, rimmed with ocher earth, were richly viscous and black in appearance. At others, vibrant green foliage crowded the perimeter and honed attention on the pit's otherworldly glutinous brew. Tossing a rock into the center of these pits effected a slow-motion gulp. Other pits were covered over with dirt, either because they had been part of Tex-

FIGURE 35 Crude seeping up through saturated soils upon the impression of a footstep crossing a covered waste pit–cum–soccer pitch near Guanta-8. Photo by author.

aco's 1995–98 remediation efforts or because the company had buried them earlier. Plant life struggled to grow atop remediated pits, and at times it incongruously thrived near unremediated ones. Indeed, a strange semblance of forest regeneration consumed some pits. Over the decades, certain pioneering species had adapted to the presence of these pits, and as they grew, the leaves they dropped formed a thick layer of humus on the pits' surface. This being the rainforest, shallow-rooted, rhizomatic vegetation sprouted from this ersatz topsoil only to further extend its reach, releasing more decomposing plant matter. In some cases, the matted humus was so thick that it could hold the weight of a human body. Sometimes kids played on the undulating movement rippling from below as if jumping on a giant waterbed.

OOZING

It appeared to be a soccer pitch. Actually, it was, with goalposts at either end. One would never guess it had been a waste pit transformed into a playing field. That is, until it rained. The morning deluge had saturated the soil, giving it a slight spongy density. And as one walked across the grass, a glisten tainted the rainwater that quietly surfaced and gathered around each step. Trapped by meters-thick dirt, the crude in the former waste pit rose through soil microcavities and seeped to the surface. Only then did it make sense why a local farmer had dug a small trench around the soccer field's perimeter. Rainwater shimmered with a faint iridescence as it drained into the trench. The farmer's hope was to capture this crude-laced water and redirect it away from his home.

An Aquatically Dispersed Sensorium

Texaco's waste pits imprint on the senses. Their expanse. Their ubiquity. Their smell. Their heat. Their daring exhibition of black stained on red clay soils. Their skill at leaking contents that never tire of oozing. Their fevered clamminess as thick blankets of heavy crude absorb and radiate the equatorial sun. Their conjuring, as their oily film surface reflects the clouds. Their persistence, even as creeping vegetation seeks incompletely to engulf them. Their stubborn capacity to make their presence known. Their ceaseless warrant for explanation.

FIGURE 36 Clouds reflected in a crude-oil waste pit encroached upon by vegetation. Photo by author.

It was around these pits that lawyers from each party advanced legal arguments for and against the consequences of Texaco's former operations. And it was from these pits and environs that scientific experts extracted water and soil samples, surveyed hydraulic flows, calculated soil permeability, and assessed the bioavailability of crude for the court. But scientific reason and method were not the only index of contamination.

In concert with a geoengineered sensorium, extending from the waste pit complex, seeped subtler ways that crude hydrocarbons made themselves known—in ways less spectacular yet with ubiquitous effect. The gentle swelling of pools of crude from leaking gaskets. The trickle of crude-peppered formation waters from goosenecked pipes. The turbid suspension of once-buried crude oil released from sediment upon being disturbed. The oozing of crude out of soil embankments. The shimmering puddles percolating up from crude-saturated soil with each movement of one's feet. The slip of crude underfoot. The stubborn encrusting of crude in horses' hooves. The encasing of crude on vegetation after inundation. The languid repelling of crude on plastic boots. The diaphanous film of crude oil in a water well. The rainbowed swirls of oil on flowing streams.

On their own, hydrocarbons would only scantly be able to evince themselves in such a flourishing array. Without water, most hydrocarbons do not move much aboveground. Miles within the earth's crust, imposing pressure and heat compel crude oil's passage, although even then always in the company of formation waters. In Amazonian headwaters, where forest-generated mists, fogs, and rains inundate the region with rivulets, streams, and rivers, crude oil and water further interlace. Lazily, torrentially, water enfolds and, crucially, enables and compels crude to spread, smear, seep, leach, and ooze.

Add a few drops of crude oil to water. The drops may break into smaller beads on impact. Before long, however, the droplets combine to form a larger whole. This molecular movement of oil globules coalescing into larger wholes gets glossed in popular chemistry as oil's "hydrophobic" character. Like other nonpolar organic compounds, oil is said to be immiscible or insoluble in water. And it does appear that oil resists association with H_2O. But hydrocarbons as hydrophobic—fearful of water? Chemistry offers scant insight into a molecule's mode of anxiety; however, it does teach that we cannot presuppose a molecule's force of existence outside the relational valence (the combining capacity) compelling configuration and movement. And in this case, the valence of water molecules—their intra-atomic and intermolecular combining capacities—does matter.

For such a small molecule, H_2O confounds. Its capacities border on magical: its solid phase is lighter than its liquid phase; its boiling and freezing points are remarkably distant (meaning, it takes dramatically more energy to heat or crystallize liquid water molecules than ever could be anticipated); and it holds remarkable tensile strength and creeping powers. Scientists theorize that these capacities emerge from H_2O's atomic composition. The tendential location of electrons and their subsequent magnetic effects have consequences for the architecture of water molecules as collectivity. The model of H_2O's chemistry—as well as its uncertainties—is instructive, so indulge me.

Most atoms bond by virtue of sharing, displacing, delegating, appropriating, and so forth, electrons among each other. H_2O chooses sharing. Hydrogen has one electron to share. Oxygen has a potential of six, offering one electron to bond with each hydrogen molecule, leaving two unbonded electron pairs. Because electrons have a negative charge, the unbonded pairs and the bonded pairs mutually repel each, pushing each other equally apart.

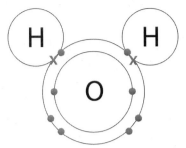

FIGURE 37 Lewis structure rendition of H_2O.

Therefore, the bonded hydrogen atoms hover at one end of the oxygen atom. Because oxygen as an atom manifests more electronegativity (i.e., is electron desirous) than hydrogen, the electrons forming the hydrogen-oxygen bonds cloud closer to the oxygen atom in their swirling orbitals. This gives the hydrogen molecules a slight positive charge, while the two pairs of unbonded electrons give the oxygen molecule a slight negative charge. Such a configuration makes H_2O a polar molecule. And this polar configuration gives H_2O in collectivity another way of forming bonds: that is, hydrogen bonds form *between* H_2O molecules. The slightly positive hydrogen atoms of one molecule bond with the slightly negative oxygen atom of another H_2O molecule. And these bonds are tenacious.

A few things of note. Liquid water is considered molecularly dense, made up of small, tightly packed, and highly agitated molecules in constantly shifting, interlocking composition. This means that hydrogen bonds are transitory. But because they perpetually form anew as quickly as they break, hydrogen bonds give water an impressive coherence. Were this not the case, the great speed at which densely packed water molecules move would generate so much energy that liquid water would boil and evaporate into gas.

In contrast, hydrocarbon molecules, regardless of their atomic configuration and structure, are nonpolar; they sustain no charged dimension. They are neither repelled nor attracted to one another. When brought together, hydrocarbon molecules easily, unconditionally, find relation and mingle, attracted by slight and temporary electrostatic energies generated by the movement of their electrons.[6] But they do not actively seek each other out.

Thus, when drops of oil come in contact with water, hydrophobia is not what is at issue, or, if it is, oil drops are immobile around that fear. A better way to think about the encounter is via valence. Because water molecules are so drawn to each other by virtue of their polar forces, they inadvertently move

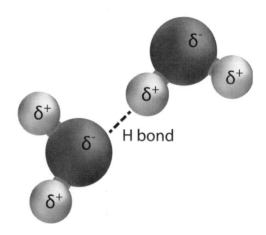

H bond

FIGURE 38 Schematic of hydrogen bonding between water molecules.

the oil drops out of the way and squeeze them together. That is, H_2O molecules' strong drive to form affinities with each other obliges them to reduce their surface contact with oil drops and consequently herd hydrocarbons into a collectivity. At play is not water's essence or oil's phobia, but the dynamic relations whereby polar aquatic fluids importantly partake in evincing oil.

Industrial contamination is more complex than events in a petri dish. Remember, crude oil's chemistry is intricate and changes over space and time. Extremely light hydrocarbons largely volatilize upon exposure to oxygen. Others slowly degrade once exposed to light. Some do slightly dissolve in water and are metabolized by aquatic bacteria. Others become suspended and float on aquatic surfaces, forming a vibrant, animated sheen. Some heavier compounds settle and sink, lodging themselves in sediments, attaching to particles and soil bacteria. Still others congeal into a dense colloid, suspended on the surface of molecularly heavier liquid.[7] At each point, polarly skilled H_2O is agentive in oil's mattering.

In the Upper Amazon, it is water with its astounding capacities that largely compels the movement of hydrocarbons. Those capacities are both intriguing and not fully understood. Take, for example, the fact that hydrocarbons float on water in a film or colloid form. In part, this occurs because the molecular structure of hydrocarbons is lighter (less densely packed) than that of water. Clearly, water's densely packed molecular structure and its polarity usher away and isolate hydrocarbons into collectivities. But water's atomic structure, arrangement, and intermolecular attraction also grant it a remarkable tensile strength, affording it a skinlike surface. Because the hydrogen bonds among the H_2O molecules that form water's outer surface are particularly strong, water holds a remarkable surface tension that actu-

FIGURE 39 Crude oil oozing along the earth. Photo by author.

ally supports (within reason) matter on its surface. Surface tension and tensile strength are what enable water molecules to clump in drops rather than spread out in a translucent smear.

Water's supple surface skin can carry light hydrocarbons, allowing their iridescent sheen to drift. And it can bear the crude suspended on the surface of waste pits. When pits overflow, water's surface tension further extends and displays crude as it runs down a slope. When rains erode a pit's fortifying berm, water channels the crude oil suspended on the surface, a tumescent pulsing through the breach. As formation waters drain through outlet pipes, heavier hydrocarbons spin turbulently in the mix. Carried by the flow, these crude compounds settle into the silts and intermingle in the sediments of rivulets and streams. And when an animal, or child, or scientist disturbs those sediments, it is once more water that evinces crude, magnified now in swirling patterns and textured light. As water swells through the landscape, it carries crude oil, largely hinged to the movements of its host. Water can carry a spill across a landscape, into waterways, through living spaces, into animal troughs. And when the waters recede into soils, some coalesced crude adheres to surfaces gleaming on leaves and humus, some hydrocarbons submerge into sediments nestling into temporary obscurity, and other hydrocarbons seep with water back into the earth.

Along with seeping, water has the capacity to creep, to move in a capillary fashion through the interstices of porous matter. This property comes from water both wanting to cohere with its own (cohesion) and wanting to adhere to other molecules that attract it (adhesion). Adhesion will incline H_2O molecules along an outer edge to move upward like scout molecules. Cohesion inclines other H_2O molecules to follow suit. When scout molecules attach to another substance, other H_2O molecules cling to them, forming an ever-elongating meniscus. This is water's capacity to creep, to climb—as it does up paper, or roots and stalks, or through former pits. These two propensities toward cohesion and adhesion move much in our worlds.

In the process of capillary propulsion, water carries what it has dissolved (chemicals, minerals, nutrients) as well as what it surrounds and embraces (oil). This is the case for lipids in our bodies and for hydrocarbons in the rainforest, as water propels crude oil through rock fissures, between soil particles, and into plant structures. Capillary action spurs water to convey the residue of buried heavy crude as it seeps through pit walls. Capillary momentum facilitates crude oil to seep up through soil-filled pits and puddle on surfaces. Capillary momentum transports crude oil with water through the roots and up the stalks of plants.

Succumbing to the rainforest's swelter, H_2O molecules evaporate, transpire, condense, and coalesce into rising mists and clouds, only to return to the earth as rain. And in the fall back to earth, H_2O molecules gather atmospheric hydrocarbons. The gas flares that burn twenty-four hours a day alongside separation stations and some operating wells emit fine hydrocarbon vapors. Today all such flares emit vertically from two-story-ish pipes. But on occasion, Texaco directed gas flares to shoot horizontally across waste pits. Once sparked, the incendiary gas flares would combust the thick layer of crude floating on top and burn for days, inciting riotous black clouds of hydrocarbon ash. Rain precipitated those hydrocarbons—from fine vapors to dense ash—back to earth, further spreading hydrocarbon effects.

Commanding movement and senses, hydrogen-carbon/hydrogen-oxygen assemblages unfolded a radical sensorium in the midst of the judicial inspections. Mediated by water, the effects of hydrocarbons could be sensed in the compulsions of human and nonhuman materials, objects, and devices. Molecular and atomic agitation stirred processes of conjoining, becoming-with, triggering access to an otherwise invisible elemental abiding. Molecular curiosity with its appulse (its energetic bidding variously toward and away)—its valence—made matter animate.

Immersed in a water-crude complex, hydrocarbon particles darkened sediments devoid of life. Pitted corrugated tin roofs. Shriveled plants and crops. Left clothes speckled gray, schoolwork stained, water's gossamer luster to be skimmed away. Festered into bodily sores. Ate at raw, sensitive skin. Caused dogs and cows and horses to lose hair. Coated plants with a dusting of soot. Weighted the wings of insects that could never again fly. Suffocated amphibians by clogging their glands. Ran scorching through corroded pipes. Made diaphanous water-saturated soils. These chemical-material gatherings were a radical elemental sensorium in which the properties and capacities of hydro-petro-matter shaped organic and inorganic forms.

This shifting water-oil assemblage populated by molecular obligations incited sense-awareness. Molecular relating obliged distinct crude compounds to coalesce, occasioning water to be the medium for evincing the presence of hydrocarbons. With ingressions, entities multiplied—a plurality of water-oil matrixes shifted forms, faintly or dramatically, such that the anteriority, purity, or singularity of each (water/oil), and that with which they touched, proved immaterial. A water-crude molecular sensorium invited orders and forms to leach and, simultaneously, enabled extensiveness—a becoming of continuity. An aquatically dispersed sensorium amplified realities that slipped from the calculations and claims of scientific detection.

Such molecular-material assemblages placed in question trust that a natural order could be deciphered through the details of fact. That chemical analyses and epidemiological studies had succeeded in describing a world of stable entities in functional interaction (despite controversy between legal parties) spoke less to an order of truths than to those *aspects* of an order of truth that distinct scientific disciplines recognized as their own. Theirs was a colorless, odorless, mute, and bifurcated nature (Stengers 2011), devoid of any sensorium. And the unstable sensorial expressions (human and nonhuman) of chemical compositions and molecular movements—they could not be captured by science's "intolerant rule of abstraction" (136).

The waste pit was fifty meters to the east of the wellhead. This time, locals had widened the forest path for the inspection teams to gain access. Smaller than the first two, this pit didn't appear on aerial photographs. Situated at a slightly lower elevation, over the decades it had become engulfed with densely growing marshland vegetation such that it only hinted at its original size. Pablo stepped out onto the mat of floating vegetation. It undulated like a giant boa under his weight. Reaching a long pole toward the center, technical assistants scooped up thick coagulated crude that dripped like slow molasses. Rhizomatic roots permeated the leaf-litter-soil now dense with years of decay resting atop the pit. And from it, verdant foliage grew, engulfing, but only partially, the noxious brew below.

A Vernacularly Embodied Sensorium

Percolate through the openings of the judicial inspections.

The manifest Other of imperial exploits and national progress—those colonos and indígenas living among oil extraction—affirmed their presence. Haltingly. The weight of disdain for people whose very being hinted at regression loomed thick. As the judicial entourage trekked through scrub forest, tracing the conduits connecting drilled wellheads to waste pits to flaring gas pipes, locals occasionally led detours to forgotten, covered waste pits. They guided the collective down slippery gullies to obscured goosenecked overflow pipes where industrial waste trickled into rivulets and wallowed in slow-moving streams. They escorted the judge and parties to low-lying swamplands where shimmering crude kept the secret of a berm collapsed two decades and a half a kilometer away. And they steered the inspection group to note settlement arrangements and the compromised water sources used for drinking, cooking, cleaning, and bathing.

Long before the roads were paved, Texaco used to pour heavy crude scavenged from the tops of waste pits onto the roads to form faux asphalt. At the inspection of Sacha Sur, one resident, José Segundo Córdova Encalada spoke about his experience on these roads:

> We used to walk barefoot, my family and I, because we are not wealthy, on these roads coated in crude. And around midday the humidity and

FIGURE 40 Campesinos guiding observers to swamplands whose waters reek of crude oil. Photo © Amazon Watch. Used with kind permission.

heat would make the crude-oil road smoke and steam. The body would catch those breathings [*aspiraciones*]. That was pure contamination. I am not an expert on oil, but many men from our community began to suffer, from their mid-bodies down, and many women from cancers of their reproductive organs. I think it was caused by the fluid [melting petroleum]. Walking on the steaming crude made you feel like your body was on fire.[8]

Interspersed with tracing oil operation's geoengineered infrastructures, local residents gave sworn testimony before the judge about the effects of hydrocarbon activity in their lives.[9] Testimonies, told by men and some women living near former Texaco facilities, were often emotional, sometimes cathartic, events—at times their words were wrenched with anguish from decades of racial and economic derision, neglect, and disavowal. At other moments, those who testified were self-possessed and angry. Local narratives relayed case after case of the human suffering of their own being and those they loved and on whom they depended. The mere act of testifying in the presence of the palpable animus of some Chevron lawyers was an act of defiance. And to the extent that giving testimony forced corporate representatives to be confronted by local peoples' presence and hear their agony, doing so also exacted a bond. A bond less to instantiate a relationship and instead to expose the harmful effects of the corporation's historic form of relating: how Texaco's oil operations seeped into being-ness and dislodged any stability. The toddler who died vomiting oil wastes. The otherworldly flesh erupting from within. The horse whose stomach had been eaten away. The twins whose bones never ossified. The chickens who drowned in a waste pit. The deep, cutting scar from a cancer surgery or an inconclusive biopsy. The amputated limb. As a radical sensorium, the narratives bore into the makeup of those present and similarly transformed their being, if only momentarily.

Hugo Ureña, Sacha Sur judicial inspection, March 8, 2006:
"I moved here thirty-four years ago. We came from the coast. Texaco had found petroleum to the north in Lago Agrio, but not yet here. My father worked at the station [Sacha Sur separation and pumping station]. He died of cancer in 1995. Later my aunt died of cancer, too. More recently, less than a year ago, my niece died of leukemia at age seventeen."[10]

FIGURE 41 Campesinas holding photos of their deceased twins. Photo © Amazon Watch. Used with kind permission.

Gerardo Plutarco Gaibor, Aguarico Station judicial inspection, November 9, 2009:
"I settled here with my wife and two daughters back in 1979. Murky, dark waters ran through this stream. And it was salty. I had a little house downstream from here on the left bank. The girls would go down to bathe and play in that water, and they fell sick. They contracted typhoid fever and fungal infections. We didn't know the water was contaminated, and we would bathe in it, too, and the illness re-surfaced."[11]

Guamán Romero, Aguarico-2 judicial inspection, June 12, 2008:
"First I cultivated coffee. Then I planted pastures for cattle. But it so happened that the pastures, they all died. The grass desiccated and died because of the problem with petroleum. It was all pure crude. And, here, the pigs died. They were full of daring and life, and when you least expected it, they fell dead."[12]

Lilia Perpetua Mora Verdesoto, Sacha Norte judicial inspection, April 26, 2006:
"A number of years ago, we dug this well to get water for household consumption. My husband bored it. But out came dirty, contaminated water. It has a terrible smell and has a layer of oil on the surface. So, I haven't been able to make juice. That is why we began to collect water for cooking from the river. But the contaminated water that is in our well goes directly into the river and into the very stream where we get our drinking water."[13]

Gustavo Ledesma Riera, Shushufindi-4 judicial inspection, July 25, 2005:
"It all started when I bought this farm. Since we had no drinking water—because we couldn't use the water from the streams—I did what I normally would do and dug a well eleven meters deep to get good water to use. . . . When water began to seep into the well, we left it for a day to see how much would surface. But as water filled the well, something appeared on top of it, some kind of oil. Since we needed the water badly, we started to clean it, skim off the oil, and used it up until ten months ago, when we stopped using it because all the workers who came here, their children would fall ill, and even the workers themselves got sick, and they didn't know why. . . . This water is useless. The water from the surrounding streams is useless."[14]

The court registered this local witnessing. Close to two hundred humble forest dwellers offered testimonies, to which Judge Zambrano would give considerable thought in his 2011 ruling. Embodied vernacular voicings. Viscerally precisioned. Rashes spread across infants' tender skin. Scars carved across thyroid flesh. Swaddling blankets stained with hydrocarbon parti-

FIGURE 42 Infant bearing the telltale scars of being bathed in contaminated waters. Photo by author.

cles. Blistered hands bandaged after rescuing the chicken that fell into a pit. Choked tears of a husband's loss. Intoning silenced, ailing worlds, the body did not explain. It was what testified, pronouncing the multiplicity of relations between corporeal conditions and their durative surroundings. Tracings incarnate—the living contours ensnaring vital conditions and oil installations through time.

José Holger García Vargas, Sacha Sur judicial inspection, March 8, 2006:
"My wife is confined, prostrate, in bed. She is dying. And she and I are suffering because of contamination. We used to use the water from the river because we had no other source for water. We used the water from the river to bathe and wash, for cleaning clothes and food. And that is why we have this festering fungus on our bodies."[15]

Aura Fanny Melo Melo, Shushufindi-13 judicial inspection, July 28, 2005:
"That water has always been like that. When it rains, more crude oil comes out of the ground. And the weedy vegetation that leads toward the stream gets tainted; it becomes coated in crude. Even my daughter, one time when she was searching for fish, put her foot in the stream and her foot started to burn and now it can't be cured. . . . Look at her feet."[16]

Gerardo Plutarco Gaibor, Aguarico Station judicial inspection, November 9, 2005:
"Let me show you the imprint of disease on my body. . . . My skin is contaminated. It has been years. I still have the disease [shown in various photographs] on my skin. I'm not cured yet. This has been going on for approximately twenty years. I still have infections on my skin, Your Honor, I'm not lying, you can see for yourself. That water that you see there, it comes out from where they would inspect the pipe; a great deal of salinated water would spill out."[17]

Miguel Zumba, Sacha-13 judicial inspection, November 10, 2004:
"Our entire family suffers from headaches and stomachaches because our drinking water is contaminated."[18]

Amada Francisca Armijos Ajila, Cononaco-6 judicial inspection, November 16, 2006:
"When we arrived in 1982, the river was not contaminated. Within a few years, however, that changed. But we weren't fully aware of this. And we kept using the river water until my husband became sick. . . . He has been deceased since March 22, 2002, when he died of cancer. . . . We couldn't use the water from the river to wash, to cook. . . . My youngest daughter is also constantly sick; she suffers a lot and has difficulty learning."[19]

Far from repetition, the testimonies during the judicial inspections formed refrains, softly accruing, gently overlapping, densely enfolding, dwelling and submerging within ecologies that agitated and reassembled from site to site. They formed recursive crucibles in which the established criteria for who constitutes a legitimate interlocutor and knowledge producer were jostled from place. Layered upon decades of popular epidemiology (what Callon, Lascoumes, and Barthe [2009: 78] have called "research in the wild"), rural residents through their testimonies linked what had been unconnected—diverse bodily dis-ease (skin rashes, miscarriages, tumors, cancer, deaths)—in an enchained refrain. Gathered, individual malaise folded into a symptomology, collective patterns signaling the probable presence of hazards contaminating the landscape in which people lived. In a region wrought with racial prejudice, testimonies were a reprise, a taking up yet again, that pushed against conventional regimes of representation and truth. Testimonies rendered a vernacular radical sensorium that authoritatively evinced the effects of crude otherwise.

Máximo Celso, Lago Agrio North judicial inspection, January 26, 2005:
"There was a huge spill here—from here to over there [an area of roughly one hectare]. The petroleum flowed that way and they burned it. They set

it on fire. That was what they did to clean up; that was their practice. And what a pestilence [it created]. After five days, there was such a stench from the animals that had died in that area. We found dead deer, you know, and small animals."[20]

Máximo Celso, Lago Agrio North judicial inspection, January 26, 2005:
"I had a pig farm here. I lost all of my pigs from the moment Texaco started to dump the water in this area, 120 pigs, of which 30 were female [and] mature enough to procreate. . . . At that time, I cultivated coffee. And in the middle of the coffee plantation, I had a banana grove that enclosed a pig farm. From the moment [they started dumping], I lost all my animals. There was continuous formation water. The pathway for dumping the formation water crossed through here. They told us it was healthy, that it was even good for drinking. And trusting that, I didn't remove my animals from that place, because I believed what they had told me. But when the animals, when the sows farrowed, along with the stillbirths and placenta, their bodies ejected their uteruses. I asked a doctor about this and he told me this was the effect of a very serious contamination problem."[21]

Simón José Robles, Lago Agrio North judicial inspection, January 26, 2005:
"When Texaco made its pits, those enormous excavations, they would fill them with crude, and they burned them. They would blaze for days in entirety and the smoke would billow. So much that if we were to put laundry out to dry at night those clothes would be black the next morning."[22]

Carlos Quevedo Quevedo, Sacha North judicial inspection, December 8, 2005:
"I've lived in this area since 1970. And often a dirty liquid with a bit of oil would be running through here. It would drain into the streams. And in relation to crops, the plants would become damaged. The fruits were left totally contaminated as a consequence. Let me give you an example, the papaya, the papaya trees would bear fruit and reproduce. But were one to eat the fruit, it had a horrible smell, and it would give you a headache and stomachache afterwards."[23]

Privileging the carnal, a radical, embodied sensorium spoke a vernacular— a form of knowing and meaning-making—conveying that what we discern always has a beyond, dimensions that do not fit and cannot be deciphered within normative categories. The vernacular sensorium communicated by way of an immediacy that, despite everything, was not a language of accuracy and fact. Whereas science—the chemistry of crude and epidemiology—

sought to reduce ambiguity, the vernacular sensorium *was* an irreducible, embodied complexity whose very condition of existence was precarity and indeterminacy. The judicial inspections forced those participating—the judge, lawyers, the press, observers—to partake of that precarity and indeterminacy, again, if only momentarily. A realm of amorphous provenance where the compromised conditions of life are fractured with unclarity. Where complexity inheres in existence and extends the vulnerability of the ailing, bereaving humiliation of watching self and others suffer without clear recourse. Where compounding effects cascade, disappearing into incalculability.

Within canonical logic, with its mind/body, science/culture, and fact/value binaries, the manifest Other of (and obstacle to) imperial exploits and national progress was an index of the "body" (the space of irrationality, emotion, backwardness, and primitive sensate capacities) in opposition to the "mind" (the seat of rationality, knowledge, cognition, and language) of experts and lawyers. In general, the logic sustains a skewed valorization; it discounts practices of sense-making that cannot be recognized as such, that do not confine themselves to the strictures of rational knowledge. Rather than hone, concentrate, and reduce the world to determinate facts, those who lived amid the residue of eroding industrial infrastructure magnified by their very presence the deeply compromised condition of being human, being animal, and even being plant in the wake of Texaco's operations. What does it mean, what are the lived consequences, what toll must the body bear when living in the vicinity of pits filled with crude wastes, washing in streams glistening with oil, scooping up well water acrid with petroleum, living along rivers that are dead?

Emergildo Criollo, Central Guanta Station judicial inspection, March 26, 2009: "When the company arrived, the water became totally contaminated. Although we are not technicians, we know what contamination is. Various illness emerged. A disease of the skin, cancer, And here today, I hear the words of Texaco's lawyer saying these waste pits are not contaminated. Yes, they are. How did my two children die? They drank water from the Aguarico River. The spills and formation waters from oil well Lago Agrio 1 passes by the Teteye River and then on to the Aguarico. And contamination from many wells does the same. My children drank from the Aguarico River. That river for me is completely contaminated. Texaco is culpable for the death of my two children. I also lost an aunt from cancer. And now they [Chevron's lawyers] are trying to diminish the contamination. But the soils, the air, the water was already contaminated by the 1970s.

Texaco is a criminal. Isn't killing people a crime? The company killed my children, That, for me, is a crime. But they don't kill directly, it's done by contamination."[24]

Gerardo Plutarco Gaibor, Aguarico Station judicial inspection, November 9, 2005:
"I have lived in this area since 1979. And when we moved here we brought a few head of cattle. Some of them drank that water and had miscarriages. Some would drink that water, and their insides would be eaten away and they died. The cattle died. All the animals would die, Your Honor, once they ingested that water."[25]

Máximo Celso, Lago Agrio North judicial inspections, January 26, 2005:
"At that well there, when they would burn the oil, that was an inferno! You had to stay several meters away because the heat and fumes were unbearable. And the smoke was deadly. After two or three hours, it would rain petroleum. In that pit, because [the company] didn't want to dump [formation] water directly over there out the other side [facing the road], they changed the direction of flow so that [the wastes] would all go into the stream that's behind the station. They created a pit there that is now buried, and we made a huge fuss, because they were contaminating an area that was used for cultivation." [Question from the judge:] "And did this happen when Texaco was here?" [Answer:] "Precisely when Texaco was here. I even remember the names of the men who directed these operations." [Question:] "Did Texaco divert the water that emerged with petroleum in this direction?" [Answer:] "They diverted the water along with petroleum."[26]

Carlos Cruz Calderón, Yuca-2B judicial inspections, November 14, 2006:
"Over here in this area, there was an oil pit where they dumped a huge quantity of barrels of oil. Many heads of cattle fell into that pit, along with pigs and chickens. Everything disappeared there, because it really was a pool [excavated pit] with an immense amount of petroleum."[27]

Luis Vicente Albán, Yuca-2B judicial inspections, November 14, 2006:
"At that time in 1980, when I arrived here with my cattle, I took possession of the land and started to work on the pastures, as you can see. I didn't know anything and had no idea that oil was bad. And I started to fall sick and my animals started to get sicker and sicker. Many of them have died. About 30 head of my cattle have died. The contamination is still there—do you notice that—and it continues to flow way down toward the village."[28]

Whereas for science, accuracy is the achievement of finite conclusions, for a vernacularly embodied radicalizing sensorium, accuracy resided in the inability to fully capture the unfathomable. The testimonies provoked a registering of that unfathomability *and* forced a witnessing of that which is untenable and yet abided within. Testimonies intoned in refrain what got foreclosed when truth was proofed only through science. Clear conclusion escaped the clutch of toxicology and epidemiology in being confronted by the contradictory conditions pervading lived worlds. Ambiguity surged forth where a language of prehension encountered a language of proof, where sensate experience could not translate into the language of precision, where the enfolding of entities into the experience of the self could not be decoded and recoded. Rasping and chafing, abrasive contact induced uncertainty and unease. And this provoked doubt, within judicial reason, about the language of precision and proof as well as second thoughts about its efficacy.

Miguel Zumba, Sacha-13 judicial inspection, November 10, 2004:
"When I bought this farm here, I excavated wells to get water. . . and found that this entire area was contaminated with oil. First, I dug a deep well, about six or seven meters from the house, and at about three meters deep or less we encountered water contaminated with crude. It smelled like rotting mud, but it had petroleum in it. And the deeper you went, there was still petroleum and water. We dug that well down to roughly twelve meters, and we still found oil contamination that deep. We dug another well a little bit further from there and we also found contamination at about five meters below the surface. So, then we dug another well about sixty meters from here, you can still see traces of it there, and there, too, we also found water contaminated with oil. As a consequence, we had to gather and carry water from our neighbors. . . . We boiled that water so we could use it. Of course, the water was not clear. So we used bleach and some other things, in order that the color of the water disappeared, but it had a stench, a bizarre smell, and indeed, that is the water we have used during all the time we have lived here."[29]

Hugo Ureña, Sacha Sur judicial inspection, March 8, 2006:
"My property is the size of forty hectares, most of which is covered with pasture, though I have very few animals now because they either died and/or had miscarriages. We asked the company to acknowledge this and compensate us for our losses. So, the company investigated with some experts from the Department of Agriculture and they found petroleum in the livers, kidneys, and intestines of the animals."[30]

Aura Fanny Melo Melo, Shushufindi-13 judicial inspection, July 28, 2005:
"I have had cattle here but my cattle die, always. Always. I gave some pictures to Mr. Padilla, who was the president of the community, but I have one here and I'll give it to you. I can't find the others. Some of my cattle were as skinny as this when they died. A goat was practically losing its skin when it died. A piece of its snout fell off. But I didn't take pictures of that because I didn't know to. And the fish, the fish don't develop right."[31]

Testimonies gave witness to the words, emotions, and affections of what living amid an un-bargained-for landscape meant. But it also presenced a radicalizing sensorium not solely verbal in its mode, whereby contamination inflected the bodies and movements of those (human and nonhuman) inhabiting its space. Some gave oral testimony. Others conveyed merely through their bearing. Others conveyed through their loss. All gave witness through the body—for the body and its configurations were a source and repository of ineffable knowing, the possibility of discerning otherwise. Appealing to Stengers, "the body [was] therefore not what explain[ed] but what testif[ied]" (2011: 69). Judicial refrains and corporeal iterations evinced a reality seldom gathered in courts of law. For this testifying and witnessing was of the milieu, of the middle, enfolded always in a mixt. And this alternative judicial assembly opened up realms of human being-ness usually unrecognized through official legal channels and enactments of law—and in the presence of a corporation that had largely shown disdain.

Radicalized Inspections

In 1964, under (as we have learned) somewhat dubious conditions, Texaco embarked on its Ecuadorian operations with a modernist zeal. Texaco was the engine of modernity, bringing civilization, development, progress, and prosperity. And like all petrocapital ventures of the time, it held a moral compass of certitude, rectitude, and betterment that simultaneously dripped with arrogant racism. By the end of the judicial inspections in 2009, it was clear that this modernist project had failed the Amazon region. An estimated 70 percent of the population lived in poverty. The proposition that petro-techno-capital would bring stability, control, and certainty was illusory. The opposite reigned. The public character of the judicial inspections intensified how palpable and urgent the instability and uncertainty were. Vast expanses of ignorance haunted questions about the effects of extraction.

FIGURE 43 Indigenous leader Emergildo Criollo (A'i Kofan) guiding Indigenous
leaders Flor Tangoy (Siona) and Nemonte Nenquimo (Waorani) and her daughter
to a waste pit thirty-two kilometers from Lago Agrio. Photo by Mitch Anderson /
Amazon Frontlines. Used with kind permission.

The judicial inspections were a collective spectacle recurrently confronted
with how entities in turbulent variability succumbed to and accommodated
the presence of weathered, crude worlds. As such, the inspections as process schooled a particular attention that could not help but be guided by
radicalizing sensoria. Water ubiquitously lured crude—inducing it to skim
surfaces, coat textures, be held in suspension, submerge into hiding, absorb
into bodies. The deceptiveness of oil on streams; the wiliness of oil on skin;
the waxiness of oil on leaves; the slickness of oil in soil; the cunning of oil in
sediments. Emanating through fissures and membranes, transforming tactil-
ity and traction, provoking with exhibition and concealment, a crude-water
complex posed and reposed the question of what it means to live in, move
with, and be intruded upon by the elemental force of crude and its wastes. In
many ways, an aquatically dispersed sensorium mediated the traffic between
that of the geoengineered and vernacularly embodied. Reverberating in molecular agitation, a crude-water complex beckoned awareness for how life-
forms and matter, conditioned by the geophysics of crude and its extractive
machinery, experience, exist, corrode, and perish amid industrial ruin.

One would be hard-pressed to avoid the pain. The judicial inspections posed an imperative: what is the response to this unmeasurable yet enduringly immanent loss? For those affected by Texaco's former operations, crude's valence was a relentless companion in the "tyrannical bitterness of everyday lives" (Foucault 1983: xiv). Testimonies heard and received by the judge, legal parties, and all present extended a pause; it allowed those affected to honor their experience of illness, suffering, and sorrow as well as strength, solidarity, and community. And it allowed those who listened to be affected, for the valenced encounter to possibly transform. What was emitted was not precision and linearity, but rather compassionate consideration and an opening into the very setting in motion of radicalizing sensoria. Radicalizing sensoria offered the inspections a different way of being susceptible, vulnerable, and responsive to the loss. That immersive experience—abrasion with that which is untenable and yet abided in—unsettled. It triggered discomfort and obliged inspection participants to grapple with the unnerving fact that passage for Amazonian life-forms was tenuous. That tenuousness indexed the transmogrifying risk of crude's valence—the appulse of its combining capacity.

Judge Zambrano paid due attention to the reams of scientific reports. But in its indeterminacy, the metrics of science offered direction but uneven traction. Radicalizing sensoria, however, moved through the intensities, proximities, and articulations that escaped scientific notice and that, despite all, consistently endured. Science—that which conventionally announces itself as foundation, authorizing a position and making a claim—Zambrano transformed into a constraint, a constraint to be respected, but upon which he confers an altered meaning: scientific truths are not the same as legal truths. Geoengineered, aquatically dispersed, and vernacularly embodied sensoria allowed the varied valences of harm to register. Attuned to how the properties of one become subsumed into the constitution of the other, the radical inspection summoned, echoing Alfred North Whitehead, one question: "What does it make matter?" (in Stengers 2011: 19).

Many years ago, Emergildo Criollo, an A'i Kofan elder, told me the story of KuanKuan. Emergildo lives in Dureno, an Indigenous community along the Aguarico River. The community has long been affected by Texaco oil operations, and Emergildo personally has suffered tremendous losses. Two of his children died in his arms as young toddlers, vomiting blood caused by crude contamination. For decades, Emergildo collaboratively organized Indigenous A'i Kofan and others in the lawsuit against Texaco; more recently, he collaboratively founded the Alianza Ceibo, a regional Indigenous rights alliance. On October 21, 2018, I asked Emergildo if he could tell me the story of KuanKuan again the way he would tell it to his grandchildren. Below is an abridged version of his longer tale.

FIGURE 44 Emergildo Criollo recounting the story of KuanKuan. Photo by author.

"This is the story of the proper protocol for knowing the underworld," Emergildo said. And then he laughed uproariously.

"Let me tell you the story of KuanKuan, the owners of, the ones who tend, all the forest animals.

"Once upon a time, there was a person in the community who simply was not able to hunt animals. But he had a wife. And his wife wanted to eat huangana[1] [white-lipped peccary] meat. Despite practicing his technique and training with his blowgun, the husband couldn't kill a thing. And the poor woman! When people in the community had meat and threw away the innards of the huangana, she would gather them up, prepare them, and eat only guts. The couple lived this way for a very long time.

"But one day, the wife decided she no longer wanted to live with her husband; he was worthless, she said. But the husband did not want to separate from his wife. So, he said, 'OK, I've decided, I am going to commit myself, even if it kills me, to capturing huangana.' And that day he went into the forest in pursuit of huangana. For three nights he slept in the forest. In time, he became attentive to their signs, identifying where the huangana had passed by or where they had slept.

"On the fourth day, he was very close to them and he realized that he was hearing voices. They were saying things like 'Perfect. We are going to sleep here,' or 'Here we can get food.' Or the little ones would say, 'Mamita, I'm thirsty,' or 'Mama, I'm tired.' And he thought to himself, fascinated, 'They are persons, personalities. I'm going to follow them to wherever they take me.'

"The next day, the huangana climbed a tall mountain. They climbed along intricate winding paths, when, suddenly, they disappeared into a cave. As the man approached the mouth of the cave, he, too, entered trepidatiously. To one side of the path was a huge precipice. Swoosh!! To the other side was another huge precipice. Swoosh!! And darkness sealed in around him. He couldn't see and walked with his arms out in front of him to try to feel his way. Then over in the distance, everything was bright and clear, as if under open sky, though still underground. On the slope below was a river and on the plateau above was a collection of large houses filled with sleeping KuanKuan persons. They were immense. In this underground world everything was in reverse: the KuanKuan slept by day and worked in the forest by night.

"For months, the man lived with the KuanKuan and had many adventures. . . . [Here, Emergildo recounted the man's wild adventures with the KuanKuan over the course of almost twenty minutes.] In time, the KuanKuan asked the man, 'Don't you want to return to your home?' 'Yes, I will be returning.' 'Well then, what would you like to take with you?' Contemplating briefly, the man said, 'I would like to take back huangana, for my wife. She wants very badly to eat huangana meat.' The KuanKuan replied, 'We will gift you huangana to her fill. The man dreamt of the fat huangana the KuanKuan would give him and how excited his wife would be. But on his departure day, the KuanKuan presented him with one, scrawny, recently born huangana. The man exclaimed, 'Why would I take this emaciated newborn huangana to my wife?!' The KuanKuan ignored him. 'Beware,' they warned, 'as you pass through the long, narrow cave entrance, hold on tight to the vines hanging along the precipice so that the huangana don't make you fall.' Perplexed, the man bid farewell and carried the scraggy huangana back along the journey out of the cave. As he approached the dark narrow path to the cave entrance, he began to run, and suddenly thousands of huangana stampeded behind him. 'Oh!' he remembered, 'grab a vine!' He held on tight. Countless huangana pushed past him and threw him off balance. But grasping the vine, he did not fall, nor did he loosen his grip around the baby huangana. In fact, something else occurred. Protecting his baby huangana, a bond formed, a way of becoming accustomed and familiar, between the man and the huangana, of protection and care. And the huangana, the thousands of huangana, followed the path to the community. From that day forward, the man's wife was happy and so pleased that huangana roamed their forest.

"Now, understand that the KuanKuan lives in the underworld. They live both as rocks and as persons. They turn back and forth from one to the other. They are rock-beings and the owners of, the ones who are responsible for, thousands of animals who thump through the forest. For this reason, it is said that the petroleum that the *petroleros* extract from the earth is the blood of KuanKuan. KuanKuan live there in the subterranean world. By drilling into that world, the company is extracting the blood of KuanKuan. And, doing so, they are killing the KuanKuan, rock-being-persons who help in countless ways. Bleeding the life out of the KuanKuan will ruin and kill them, and the petroleum will end, too, because there is no more blood. When the KuanKuan are no longer, there will no longer be more blood. And so it is. So the history of the KuanKuan is told."

Anthropologists who have conducted extensive fieldwork with the A'i Kofan suggest that including petroleum into the story of the KuanKuan is an effort to enfold recent anti-oil environmental campaigns within Indigenous tellings (Cepek 2016; Krøijer 2019). It is not, they suggest, conveying petroleum's ontological status as blood for the A'i Kofan. Sidestepping the debates on Indigenous ontologies, it is worth noting the interconnective effect of concerted Indigenous/environmentalist opposition to oil operations in Ecuador (of which the A'i Kofan have been a part) beginning in the late 1980s and early 1990s. One epigraph ubiquitous on city walls at the time throughout Quito, Puyo, and Lago Agrio captured this relation. It reads, *"Fluye el petróleo, sangre la selva"* (as petroleum flows, the forest bleeds). Evoked in this epigraph was, clearly, a connection less dramatic than that posed by the Awa, an Indigenous group in northern Colombia, who were connected to Ecuador via collaborating environmental groups and whose international campaign forced an oil company to abandon its operations. For the Awa, crude oil was and is the blood of the earth, and they threatened mass suicide if oil activities proceeded. But the tale of KuanKuan, along with similar ones from surrounding Indigenous groups, interjected the possibility of a vital and vibrant underworld that stirred inspiration among a burgeoning Indigenous / ecological antihydrocarbon coalition. What holds salience in the melding of Indigenous and ecological narratives is not the truth of oil's ontological status, but its critique of the modernist story that names crude as a raw energy resource unencumbered by relations and awaiting liberation through capitalist extraction. Intriguingly, petroleum in the A'i Kofan tale is no longer alienated (and alienable) from a dense geohistory and is instead a constituent part of a subterranean world neither dead, nor inanimate, nor incomprehensibly scaled geologically, but rather fabulously lively, reactive, and complexly bonded with human consequence.

PART III
DELOCALIZED
STABILITIES

IN CHEMICAL MODELS, electrons—residing in distinct and variously shaped orbitals—are deemed the subatomic particles essential for the forming of bonds. Simple atomic melding occurs via covalent (where two atoms share electrons) or ionic bonding (where one atom donates an electron to another atom to form a bond). However, more complicated models for bonding exist; a ubiquitous one in organic chemistry is "delocalized stability." Crucial to the biochemistry of all life-forms, delocalized stability refers to a reactive association where electrons disassociate from their origin-atom orbital and consort in a new orbital immanent of the transformed conditions of the emergent compound. The new orbitals span the entire molecular composition and impute their valence with a stability, intensity, and durability that far exceed the capacities of constitutive elements. Benzene is the iconic exemplar of this composition.

Being the fundamental aromatic compound, benzene is both highly toxic and a key structure of life—presenting in a number of amino acids. As a chemical formula, benzene = C_6H_6, and in a Lavoisierian chemical world, the specific elements that make up benzene give rise to its capacities. But benzene is a fiercely wily character and the sides of the equation do not add up.

It took nearly a century for chemists to work out the chemical structure and thermodynamic capacities of benzene. In 1865, Friedrich August Kekulé reasoned (actually, he allegedly dreamt) that benzene had a ringed structure of six carbon molecules (to which one hydrogen atom bonded), and he hypothesized that the ring structure was composed of alternating carbon-carbon single and carbon-carbon double bonds. To accommodate the fact that double bonds and single bonds as a rule have unequal lengths and yet in benzene all the bonds are equal, Kekulé theorized that benzene was a "resonance structure"—a structure in which the bonds oscillated between single and double from one split second to the next.

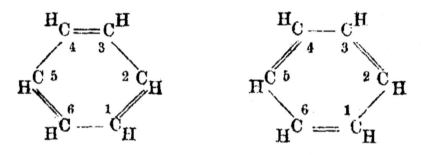

FIGURE 45

Although Kekulé's theory was productive theoretically, it still could not account for the unique structure or properties of benzene. The theory of alternating single-double bonds only accounted in an ad hoc manner for bond lengths, as well as bond angles (the angle formed by three carbon atoms and one hydrogen atom) being equal. Furthermore, benzene exhibited remarkable thermodynamic fortitude and stability.

Linus Pauling's work in the 1930s on the nature of chemical bonds allowed quantum theory to make sense of the benzene structure by reconsidering the way in which the electrons in the second energy level of each carbon atom were forming bonds. The second energy level is composed of two sublevels—s and p. Remember, carbon's electron configuration is $1s^2\ 2s^2\ 2p^2$. And in forming the benzene ring the carbon atoms transmogrify this configuration. Three electrons hybridize, and a fourth jumps a subenergy level and is formed unpaired.

According to models, the three hybridized orbitals of each carbon atom arrange themselves as a planar equilateral triangle and form covalent sigma

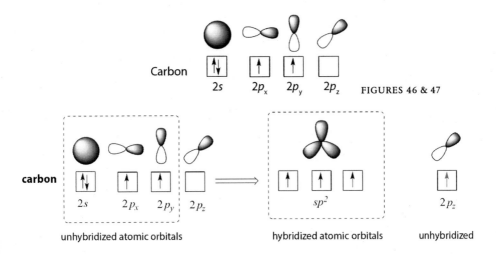

Carbon

$2s$ $2p_x$ $2p_y$ $2p_z$

carbon

$2s$ $2p_x$ $2p_y$ $2p_z$

unhybridized atomic orbitals

sp^2

hybridized atomic orbitals

$2p_z$

unhybridized

bonds with two other carbon atoms and one hydrogen. And the electron in the unhybridized orbital dissociates from its atomic orbital and abides in, as it takes part in forming, six molecular orbitals. This means that the unhybridized electrons of six carbon atoms form bonds over the expanse of the molecular structure and not within the overlapping orbitals between atoms. That is, the electrons in unhybridized orbitals sever their allegiance to a singular atom and commit their obligation to the molecule as a whole. These expansive bonds are called pi (ϖ) bonds and express a delocalized stability.

Figure 48 provides a simplified rendition of this history of molecular understanding. Although the model shows only one delocalized pi system above and below the atomic nuclei, quantum theory actually denotes that three (not one) differently shaped pi systems yield above and below. An electron can abide anywhere in the system while never passing through the middle.

The effect is that benzene as a molecular species is dramatically more stable and potent than would otherwise be thought from a compound evincing carbon double bonds (alkenes). In fact, once established, very little will destabilize the structure of benzene. When placed in reactive conditions with other compounds, benzene does not react as anticipated by alkenes. Usually, adding something to a compound alters its quantum capacities. Benzene, however, is resistant to addition reactions. Rather, unless under very high temperature and pressure, benzene remains stable and conserves its form. Instead, benzene cleverly forms what are called "substitution reactions" with particular atoms or functional groups. By substituting a hydrogen atom with a new substituent, benzene preserves its system of molecular pi orbitals and

C_6H_6

Benzene
Molecular formula

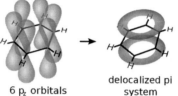

Kekulé Structures
(Isomers)

Planar Hexagon
Bond Length 140 pm

Sigma Bonds
sp^2 Hybridized orbitals

6 p_z orbitals

delocalized pi
system

Benzene ring
Simplified depiction

FIGURE 48

preserves its aromatic properties. Industrial chemistry's capacity to manipulate benzene's affection for substitution reactions (thus fabricating a multiplicity of derivative compounds) has enabled, in good part, petrochemical production and our obsession with synthetics.

Each chapter in this section deploys benzene's capacities in order to make sense of the dramatically different processes that stabilized the 2011 Ecuadorian and 2014 US court rulings. Chapter 5 suggests that Judge Zambrano reached his decision not through individual and specifically localized evidence that he then brought together through a linear and additive logic. Rather, his rendering occurred through grappling with a recursively dispersed, accruing, and encumbered phenomenal affect or valence. The 2011 Ecuadorian ruling did not situate wrongdoing in specific, isolated events but rather delocalized transgression, unfurling it across and enfolding it within a complex and less than fully known condition. Examining Chevron's Racketeer Influenced and Corrupt Organizations Act (RICO) lawsuit in the United States, chapter 6 borrows insights from benzene's aromatic capacities in order to elucidate how Chevron's fraud worlding was sealed. By and large, a good portion of Chevron's evidence of fraud (problematic testimony, blurry documents and receipts, spliced video clips, partial diary entries) was circumstantial evidence, forming weak links in a racketeering ring. Consequently, to solidify its claim, Chevron unleashed expert analyses that, when dislodged from their original meaning and context, formed cohesive bonds that fortified a comprehensive impression of conspiracy, fraud, and corruption among the LAP legal team. Significantly, the stability that Chevron sealed through this delocalized evidence hinged on misattributions, partial understandings, and misplaced graspings of the science of contamination. Chapter 6 poses three instances where this was the case and where Chevron's fraud claims would have been substantially destabilized were the US district court to have admitted to trial a clear understanding of the complexity of crude contamination and its place within the Ecuadorian litigation.

CEO

On May 28, 2015, Humberto Piaguaje (a Siekopai leader) spoke at Chevron's annual shareholder meeting at the Chevron Corporation headquarters in San Ramon, California. Below are my impressions of the meeting.

With his characteristic composure, Humberto Piaguaje stood at the microphone addressing the CEO of Chevron. The corporate auditorium was full, and in attendance were shareholders, corporate dignitaries, a handful of journalists, and a couple dozen stockholder activists. Over the previous fourteen years, Humberto or other leaders had traveled from the Ecuadorian Amazon to Chevron's corporate headquarters to voice their disquiet.

To attend these meetings is to witness a spectacle: the corporation performing itself. From the moment one approaches the Chevron corporate compound—a walled campus that extends for ninety-two acres—security is impressive. A conservative estimate would venture that, at this particular annual meeting, for every nonemployee attendee there are four security persons—be they from the city police, hired private security, or internal corporate security—monitoring identification and stockholder status, parking, personal effects, and personal location upon entry onto the corporate compound. A pen and paper are the only personal items permitted on the premises. No bags, no electronics.

These meetings are tightly controlled. On that day, the CEO, John Watson, sat on the dais buttressed by his board. Watson provided opening words, board members spoke through financial charts and graphs, and a corporate video—exquisite in its composition—illustrated corporate achievements and expansive prospects. The secretary of the board presented stockholder proposals— a majority by religious groups, environmentalists, and socially responsible investment groups. None pass, although over the years they have garnered more and more votes. Then came the question-and-answer period when individual stockholders or their proxies may pose questions to the CEO. Humberto approached the microphone:

> Good Morning, Mr. Watson, your board of directors, and your stockholders. My name is Humberto Piaguaje, leader of the Siekopai Nationality and Executive Coordinator of the Union of Communities Affected by Texaco, today Chevron, in the Ecuadorian Amazon. I'm here, once again, to talk about the world's largest environmental tragedy that occurred between 1964 and 1992. As a result of your oil operations, there is a festering open sore that has yet to be cured, and no one wants to take responsibility for its mending and redress. . . .
>
> Is it good? Is it a thing of beauty? Do you like hearing the cry of people and then turn a deaf ear when they demand justice for a dignified life in the present, for future generations of our people, and for the vibrant life of nature? Do you think it's good to make money when the cost is paid with human lives, environmental damage, and without responding to the consequences you have caused?. . .

Humberto bore the wisdom of an perceptive Indigenous leader. How to render, in his allotted three minutes, the vast chasm of experience between the humble forest peoples he represents and the CEO of a corporation whose shareholders' common stock value crests above $200 billion? How to register for those present—individuals seeking largely to maximize their financial portfolios—a self-recognition of the ravages and pain of their corporate disdain?

> Let me be clear—this is your responsibility—and not because I say so, but because a twenty-year legal process says so. . . .
>
> You are the criminals—you came, you contaminated, you lost in the courts, and you ran from the law—just like any other thief. And now, after poisoning us for thirty years, destroying our homes, killing our people, you claim to be the victims? . . . Is it your corporate ethic to characterize the communities where you operate as delinquents, extortionists, and fraudsters, and prosecute them, when what we solely seek is social, environmental, and cultural justice? . . .
>
> There is no fraud here. If [the Ecuadorian litigation were] a fraud, I wouldn't be here, and I would be shamed for even speaking of this con-

cern. But here I am, showing my face. I know how my people have suffered and are suffering different types of cancer, leukemia, and other sicknesses that affect human health. I've seen how we lost our rich biodiversity—this is what we are fighting for. . . .

We know very well the political and economic power that Chevron has, just as we know the magnitude of environmental damage and death to human life caused by your company. You can piece together a slick video full of lies, but that will not change anything.

Of all people, you cannot speak of fraud. New evidence shows that the only fraud that has occurred is the result of your own work.

Today we are more united than ever, with growing global support. We will pursue you to the ends of the earth if needed to prosecute you and seize your assets. We will not stop until we have achieved justice—until you, Mr. Watson, provide the remedy that you owe us.

FIGURE 49 Humberto Piaguaje and Leila Salazar, executive director of Amazon Watch, outside Chevron's corporate headquarters in San Ramon, California. Photo by author.

Plurivalent Rendering

Of Prehension Becoming Precaution

5

ON FEBRUARY 14, 2011, Judge Nicolás Zambrano Lozada delivered his judgment.[1] With a price tag topping $9 billion, it was the largest liability that a court had imposed in environmental-contamination litigation. The Provincial Court of Justice of Sucumbíos ruled that the Chevron Corporation was responsible for paying $8,646,160,000—funds to be used to remediate the environment and cover the cost of setting up systems to address the poor health and cultural erosion that its oil operations had begotten. As mandated by Ecuadorian law, an additional 10 percent of the amount, plus legal expenses, was owed to the entity representing the offended party. If the corporation did not publicly apologize for its actions in print within fifteen days of the judgment, its damages liability doubled.

Predictably, Chevron did not apologize and instead appealed the decision to the three-judge Appeal Division of the provincial court. In its January 2012 opinion, assessing all of Chevron's points of appeal, the appeal division noted that while indeed Chevron lawyers had caught a handful of transcription errors (transposed numbers in sample results), these "mistakes" did "not bias" Zambrano's decision "or induce it to error."[2]

> The judge in his ruling did not assess each sample and its results separately, as if they were to have described isolated facts, but instead it is the

collection of information coming from various sources that undoubtedly has created in the trial judge the conviction of the existence of damage, allowing him at the same time to have a minimal margin of error in applying the interpretative method of sound discretion to assess scientific evidence. . . . Moreover, the method of interpretation—the intrapersonal form or mechanism, the psyche—is not subject to the strict limits of any concrete, express legal rule. . . . [The method of interpretation] is a question of a mental operation that brings to bear human elements like experience, logical rules, and even some knowledge of the ranges of human psychology in the assessment [of a case]; this principle is supported by legal doctrine and its most distinguished exponents.[3]

Citing a learned supreme court justice, the appeal division noted that this method

configures an intermediate category between legal evidence and freedom of conviction. Without the excessive rigidity of the former and without the excessive uncertainty of the latter, it makes for a felicitous formula, ever praised by scholarly writings, of regulating the intellectual activity of the Judge faced with the evidence. The rules of sound judgment are above all rules of proper human understanding.[4]

Four years later, in August 2016, the US Court of Appeals for the Second Circuit upheld the lower court's ruling in Chevron's Racketeer Influenced and Corrupt Organizations Act (RICO) countersuit, which found Zambrano's ruling was procured through fraud. The US appellate court found that Zambrano's legal logic—of which the above is the description of only a snippet— could not register, in fact was indecipherable, amid the torrent of facts that Chevron's corruption worlding unleashed. In its ruling, the US court quoted parts of the Ecuadorian appeal division's decision (invoking a different translation of the text):

In sum, as stated by the Appeal Division in the above passages, "The [trial] judge in his judgment [did] not assess[] each sample and its results separately, as if they described isolated facts"; rather he made a "discretion[ary] . . . assess[ment of the] scientific evidence." His "method of interpretation" was "the interpersonal, psychic form or mechanism," which "is not subject to strict limits in any concrete, express legal rule." His "judgment establishe[d] amounts different from those established or stated by the parties in defense of their interests." . . . The record in the present case reveals a parade of corrupt actions by the [Lago Agrio plaintiffs'] LAPs legal team, including coercion, fraud, and bribery, culminating in the promise to Judge Zambrano of $500,000 from a judgment in favor of the LAPs. The Appeal Division's Opinion provides no basis for an inference that the Lago Agrio Judgment was not the result of those corrupt acts, given its description of Judge Zambrano as having reached his decision without "assess[ing]" discrete "facts," without following "concrete, express legal rule[s]," and without "consider[ing]" the "economic opinions or parameters that appear from the trial." And given that the Appeal Division in its opinion (a) sets out no findings or damages assessments or calculations of its own, (b) approves Judge Zambrano's approach as "sound," "appropriate," and presenting "no[] . . . reasons to modify what was ordered in the lower court's judgment," and (c) "ratifie[s]" Judge Zambrano's award "in all its parts," we conclude that Chevron's $8.646 billion judgment debt, as approved by the Appeal Division, is clearly traceable to the LAPs legal team's corrupt conduct.[5]

Zambrano's ruling was illegible in the US court, as was the appeal division's review of it. In the US court, whether in snide corridor comments, vague allusions during court testimony, Chevron's legal filings, or judicial opinions, Zambrano's ruling was recurrently depicted as incoherent and obtuse. As will become clearer in chapter 6, Zambrano himself was derided for his seeming lack of sound legal reasoning and his judgment was the object of ridicule—sentiments indirectly extended to Ecuador's entire judicial system.

This chapter takes Zambrano's ruling seriously and shows how legible it was. It deploys the notion of "delocalized stabilities" as a conceptual tool for parsing what, unlike most US legal rulings, is a nonlinear and nonadditive narrative. Zambrano's legal decision coalesces in clouds around central concerns, but it does not rest on, punctuate, and secure specifically localized evidence as "aha occasions" that fix stable elements, allowing for an argument to link like a collection of covalent bonds. Rather, Zambrano's text is best understood as a delocalized stability engaging key issues and grappling with the indeterminacy of the evidence at stake. His "method of interpretation" determined fault not due exclusively to isolated and singular "facts" but through the recursively dispersed, less than systematic, encumbered intensity of the valence of crude. His "method" unfurled and enfolded a multiplicity of transgressive orbitals, delocalizing wrongdoing uniquely from the certitude of evidence and dispersing it through an uncertain cloudscape of a less than fully known condition.

To decipher Zambrano's method, this chapter holds in tension different dimensions of his ruling (although he did not engage them in this order): extant Ecuadorian law of the time; corporate geoengineering practices and knowledge; evidence unearthed during the judicial inspections; and an emergent legal logic codified in Ecuadorian legal decisions around environmental harm. In gathering these dimensions, I ask what intensities, proximities, and coalescing—what chemistries, methods, and ethics—produced for Zambrano compelling associations between hydrocarbons' activities and a landscape of human suffering and biotic degradation.

Zambrano's ruling is bewildering, at first glance, for those unaccustomed to the then-textual form of Ecuadorian legal judgments. Inscribed within the confines of sixteenth-century Spanish jurisprudence, the entirety of the ruling constitutes one single paragraph of single-spaced type stretching across 188 pages. There are no gaps, no open spaces, no indentations, no table of contents. Rather, left- and right-adjusted monochromatic text fills the page and is stifling in its visual uniformity. A remnant of medieval judicial formalism, the *sentencia* or *fallo*, as it is known in Spanish, methodically and precisely occupies every legitimate space on a page so as to ward off surreptitious tampering.

Admittedly, Zambrano's decision appears visually distinct from a ruling rendered in a US federal court—where judicial decisions are made to be eminently readable, prefaced by a summary of the decision and a detailed table of contents that displays the issues examined and the decision's logic with tiers of nested subheadings. And, admittedly, Zambrano's ruling reads more circuitously than the skillfully written text (often on numbered lines), terse analysis, and argumentation produced by many US federal court judges. But then again, admittedly, US federal court judges are buttressed by a small cadre of clerks, while Zambrano had himself, the court's legal secretary, a personally paid typist, and judicial colleagues with whom he might discuss the case.

Expend a bit more effort, delve into the sea of words composing the circuitous sentences of a 188-page paragraph of formal legalese, and a method emerges. Indeed, the very form of the sentencia gives presence to a method that coalesces through dispersion. A thorough Zambrano takes his time to carefully address and reason through the multiplicity of dimensions raised and at stake in this seven-year litigation. A legal saga that in Ecuador alone generated a 216,692-page case file, produced over 100 expert reports, and enlisted fifty-four site-visit inspections.[6] Zambrano's ruling is divided into fifteen sections, each alerted with the number written in all caps (i.e., "PRIMERO," "SEGUNDO," etc.) and focused on a distinct concern. Sections 1 through

Juicio No. 2003-0002

JUEZ PONENTE: AB. NICOLAS ZAMBRANO LOZADA
CORTE PROVINCIAL DE JUSTICIA SUCUMBIOS. - SALA UNICA
DE LA CORTE PROVINCIAL DE JUSTICIA DE SUCUMBIOS.
Nueva Loja, lunes 14 de febrero del 2011, las 08h37. **VISTOS**.- En
relación a la causa signada con el No. 002-2003 que por daños
ambientales sigue María Aguinda y otros, en contra de la compañía
Chevron Corporation, atendiendo su estado procesal se dispone.-
1).- Téngase por incorporado al expediente los anexos y escritos
presentado a las 16H24 de 03 de febrero del 2.011 por el doctor
Adolfo Callejas Ribadeneira, Procurador Judicial de Chevron
Corporation, en atención al mismo se dispone negar su solicitud de
revocatoria de providencia de fecha 02 de Febrero del 2011 las
17H14, en virtud de que no se le está impidiendo el derecho que le
asiste de presentar peticiones que se encuentren amparadas en la
ley y el derecho.- 2).- En lo principal, María Aguinda, Ángel
Piaguaje, y otros, amparados en el contenido de los artículos 2241 y
2256 de la anterior codificación del Código Civil (en adelante CC),
actualmente artículos 2214 y 2229 respectivamente, según la
nueva Codificación publicada en Registro Oficial del 24 de junio de
2005, para fundamentar la obligación de reparar el daño; en el
artículo 169 de la OIT para fundamentar el derecho a compensación
de los pueblos indígenas; y en cuanto al derecho a reclamar las
reparaciones derivadas de una afectación ambiental, en el número 6
del artículo 23 y en el artículo 86 de la Constitución de 1998, así
como en el artículo 2260 de la anterior codificación del Código Civil,
actualmente artículo 2236, que dice "Por regla general se concede
acción popular en todos los casos de daño contingente que por
imprudencia o negligencia de alguno amenace a personas
indeterminadas. Pero si el daño amenazare solamente a personas
determinadas, sólo alguna de éstas podrá intentar la acción", y en el
41 de la Ley de Gestión Ambiental – en adelante LGA (fs. 78 y 79),
comparecen desde fojas 73 a 80 demandando la eliminación o
remoción de elementos contaminantes y la reparación de daños
ambientales, en contra de CHEVRON TEXACO CORPORATION, que
cambió su nombre a CHEVRON CORPORATION, conforme lo indica
y demuestra mediante documento presentado por su Procurador
Común, adjunto al escrito presentado el 23 de agosto del 2005, a
las 08h05; esta demanda en sus considerandos Primero al Sexto
resume los antecedentes (donde alega que el detalle de las obras
realizadas por Chevron está comprendido en el anexo A de la
demanda), métodos contaminantes empleados por Texaco, los
daños y la población afectada, la responsabilidad de Texaco, los
fundamentos de derecho descritos anteriormente, y expone las

1

FIGURE 50   Rendition of page 1 of the February 14, 2011, ruling of the
Provincial Court of Justice of Sucumbíos.

5 (pages 1–60) address distinct motions that Chevron filed to have the case dismissed from the court. Sections 6 through 12 (pages 60–176) focus on determinations of harm and the legal basis for liability. And sections 13 through 15 (pages 176–88) delineate the measures for addressing harm and procedures for executing them.

This chapter focuses on the core of Zambrano's ruling (sections 6-15) that grapples with the controversy over environmental contamination and health harm, and it details his method finding Chevron liable. In highlighting these concerns and the legal logic of his adjudication, I do not mean to discount the importance of the first third of his ruling. Chevron's challenges to the court were significant and relentless. The corporation asserted that the Ecuadorian provincial court lacked competence; that the provincial court had no jurisdiction over Chevron; that the provincial court improperly pierced the corporate veil; that the provincial court improperly applied law retroactively; that the plaintiffs lacked a connection to Chevron; that litigation procedure did not follow the Ecuadorian Civil Code; that the court was tardy and delayed; that the court improperly canceled judicial inspections; that plaintiffs' experts were unqualified or unacceptable; that the plaintiffs' experts and legal team were mired in fraud and misconduct (Chevron would file a total of twenty-six lawsuits against these experts in the United States); and that plaintiffs' signatures had been counterfeit and the power of attorney was invalid. In addressing Chevron's concerns, Zambrano used a different "method." Each motion received a precise and judicious dismissal. Where he came down was absolute—and, by use of analogy, here Zambrano's ruling resembles the covalent sigma bonds (not the delocalized pi bonds) of a benzene ring. This is not the case with the remainder of the ruling.

One caveat. There is a danger that my analysis makes Zambrano's decision seem more self-evident and straightforward than it actually was. It is nearly impossible to overstate the complexity of this case. De-composing the ruling as I do risks reducing that complexity. In retrospect, it seems it would have been nearly impossible to rule in favor of Chevron, given the layers and layers of culpability that accreted over the course of the litigation. But as the litigation came to a close, and a judicial opinion was imminent, uncertainty abounded. And invariably, the corporations' officious vitriol, procedural brashness, and obsequious disdain made writing the ruling all the more demanding.

Zambrano's ruling extended, I suggest, a mode of agency—a jurisprudence—that embraced uncertainty and redeployed it to expand the possibilities for alternative ways of abiding in a messy world. This is the generative possibility

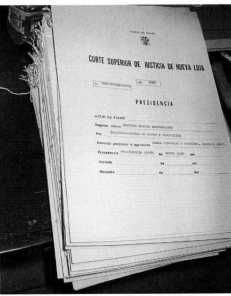

FIGURE 51 The secretary of the Provincial Court of Justice of Sucumbíos, Llcd. Gloria Cabadiana, recording court documents. Photo by author. FIGURE 52 A stack of the 100-page *cuerpos* that made up the over 200,000-page case file of the lawsuit against Chevron. Photo by author.

of delocalized stabilities, where legitimacy does not hinge on the bonding of convention and expertise, but rather unfurls and coalesces across multiplying and dispersed orbitals of evidentiary knowing. And in this sense, Zambrano's decision—and the seven-year trial on which it was based—constituted a unique legal assemblage for debating socio-eco-techno-controversies and what counts as injury, life, and ethical engagement.[7] The legal ruling and litigation resonated with what Strathern calls the "proliferation of the social" (1999b: 174) into presumably pure scientific and technical domains—a proliferation that melded and transmuted the orbitals between science, technics, and politics. And it begged the question of how might uncertainty and imperfect knowledge engender responsibility, obligation, and efficacy?

Extant Ecuadorian Law

In addressing these concerns, I want to start with a substrate within which Zambrano's method resides: did the Republic of Ecuador have laws at the time that Texaco's operations contravened? One of Chevron's repeated defenses during the Ecuadorian litigation was that Texaco broke no law. The corporation claimed that since hydrocarbon activities were not regulated— that the Ecuadorian state had not established legal parameters, standards, or maximum tolerable limits for discharging crude or its wastes—during the period in which the company operated, then Texaco had not violated any law. Indeed, Chevron lawyers never tired of restating and retooling this very argument.

In section 6 of his decision, Zambrano takes his time detailing, contrary to Chevron's assertions, a number of national laws and corporate-state agreements on the books that Texaco's operations violated. Together the national laws delineated a "very reasonable legal system for environmental liabilities" (Zambrano's judgment, 66)—outlining liabilities for harm caused, urging caution to avoid liabilities, and establishing a "positive legal mandate to adopt appropriate measures to protect the flora and fauna" (67).[8]

First, the 1964 concessionary contract between the Republic of Ecuador and Texaco and Gulf Oil (Official Registry No. 186, February 21, 1964) granted the Texas Petroleum Company, because it possessed the "necessary technical and economic resources to carry out an efficient exploration in the hydrocarbon fields" (61), authority to explore for and exploit petroleum in the Amazon. Making no reference to the concession's checkered early history, Zambrano quotes select contractual clauses to underscore the company's obligations. Early on, the contract clarifies, "The Concessionaire has the right, for purposes of this contract, to use the lands comprised within the areas that are the subject of the first and second clauses, as well as the waters, timber, and other construction materials that may be there, to destine them to the exploration, production, and development of their concession, without depriving the villages of the flow of water they require for their domestic needs and irrigation, or impairing in any way navigation, or depriving the waters of their potable and pure qualities, or hindering fishing" (clause 10). And subsequently, the contract states that Texpet is required "to operate the concession employing adequate and efficient machinery" (clause 32-G, cited in Zambrano's judgment, 62). The implication here being that Texpet, as the consortium's operator, could use the waters in the concession, provided that their potability and purity were not diminished.

The 1971 Ley de Hidrocarburos [Hydrocarbons Law] (Official Registry No. 322, dated October 1, 1971) expressly stated that operators of a concession have "the obligation to 'adopt the necessary measures for the protection of flora and fauna and other natural resources'" (Article 71) and "prevent contamination of the waters, the atmosphere, and the land" (Articles 29-S and 29-T, cited in Zambrano's judgment, 63–64). Such provisions were codified in every subsequent major piece of legislation passed regarding petroleum in Ecuador.[9] What stands out as remarkable for Zambrano is that these environmental norms were in effect before Texpet produced its first export barrel of petroleum in 1972. Prior to Texpet's operations, the northern Ecuadorian Amazon had not seen industrial activity; thus, Zambrano maintained, the prior "purity of its waters" (64) was a given. Underscoring this obligation, the 1974 Reglamento de la Exploración y Explotación de Hidrocarburos (Hydrocarbon Exploration and Production Regulations, Supreme Decree 1185, Official Registry No. 530, dated April 9, 1974) declared that the operator of a concession must "take all appropriate measures and precautions when performing its activities to prevent harm or danger to persons, property, and natural resources and to sites of archeological, religious, or tourist interest" (Article 41, cited in Zambrano's judgment, 61).[10] Furthermore, the new concessionary contract entered into in 1973 between the Republic of Ecuador and Texpet and the Gulf Oil Consortium stated, "The Contractors shall adopt all convenient measures to protect the flora, fauna, and other natural resources, and shall also avoid contaminating the water, atmosphere, and land, under the control of the pertinent organizations of the State" (Article 46.1).

Similarly, Zambrano referenced the Health Code also promulgated in 1971 (Official Registry No. 158, dated February 8, 1971), which contained a number of "rules of mandatory application" for "any public or private matter or action" (62). In relation to "Environmental Sanitation," the code reads, "No person may dispose of solid, liquid, or gaseous wastes into the air, soil, or water, without prior treatment that makes them harmless to health" (Article 12, cited in Zambrano's judgment, 63) and "industrial waste may not be discharged, directly or indirectly, into creeks, rivers, lakes, irrigation ditches, or any other watercourse for domestic, agricultural, industrial, or recreational use unless previously treated using methods that make them harmless to health" (Article 25, cited in Zambrano's judgment, 63). That no specific laws existed to regulate specific hydrocarbon activities "is no hindrance," Zambrano noted, "to the prevailing obligation that such substances must be handled under sanitary conditions that eliminate such risk" (63). These norms are mandatory for all persons, natural and legal, as article 16 expressly clarifies: "All persons are required to protect the water springs or water basins that serve the water supply" (63). Likewise, the Health Code affirmed, "No one may discharge directly, or indirectly, harmful or undesirable substances in such a way that they could contaminate or affect the sanitary quality of the water" (Article 17, cited in Zambrano's judgment, 63).

In 1972, the Republic of Ecuador promulgated the Ley de Aguas (Water Law, Official Registry No. 69, dated May 30, 1972). Zambrano noted that article 22 states that "all contamination of the waters that may affect human health or the development of flora or fauna is prohibited" (Zambrano's judgment, 64). Following article 1, this applied to all waters—maritime water, surface water, groundwater, and atmospheric water in the national territory, in all their physical states and forms. Consequently, Zambrano concluded, the "right to use the waters granted to the operator in the Tenth clause of the 1964 concession contract, which established the Concessionaire's right to use the waters without depriving them of their potability and purity, was also subject to the provisions of this law" (64). And the Water Regulations (Official Registry No. 233, dated January 26, 1973) defined "contaminated water" as "all [water] that, flowing or not, shows deterioration of its physical, chemical or biological characteristics, due to the effect of any element . . . and that as a result is totally or partially limited for domestic industrial, agricultural, fishing, recreational, and other uses" (Article 89). Additionally, "harmful change," the law states, is "that which is produced by the effect of contaminants or any other action susceptible to causing or increasing the degree of deterioration of the water, changing its physical, chemical, or biological qualities" (Article 90, cited in Zambrano's judgment, 65).

In response to Chevron's defense that Texaco had broken "no laws," Zambrano noted that although "no laws" in effect at the time detailed "numerical provisions" (73) regulating oil operations, the "spirit of the legislation invoked in this ruling is clearly to prevent contamination, not to authorize it" (66).[11] "Neither said lack of 'parameters' nor state supervision provided exempt Texaco from its obligation to comply with the legislation in effect, which required the oil company to operate using mechanisms to avoid harm to the flora and fauna, and to refrain from removing from the water its qualities of potability and purity in order to comply with a clear, express mandate. This does not require any regulation or parameter whatsoever" (68–69). As Zambrano clarified, "lack of regulations or numerical standards does not render ineffective the other laws promulgated in that period"; the answer "cannot be to ignore the law, as Chevron's defense proposes, since the lack of regulations cannot be understood as implicit permission to defile the water, or engage in practices that have placed human health at risk" (70).[12]

Geoengineering Practices and Knowledges

Zambrano cited the above norms to establish the legal framework through which to assess whether Texpet, as operator of the consortium, violated the law in effect at the time of its operations. However, the possibility existed that Texaco was not aware that its operations could have a negative impact. Zambrano felt it important to understand and register the contemporaneous knowledge that Texaco had of any potential harm that its operations could cause. Predictably, experts portrayed opposed perspectives of best practices and standards used by the oil industry at the time. Consequently, Zambrano placed particular emphasis on one piece of evidence he considered impartial in defining the "technical principles" of the oil industry.

In 1962, the American Petroleum Institute (API), the preeminent industry association, published *The Primer of Oil and Gas Production*, which outlined the operations standard of a "good oil company."[13] The tenth chapter of the primer, "Special Problems," was coauthored by Karl C. Brink, then director of research for Texaco's Research, Environment, and Safety Department. It states, "The management and disposal of produced water requires extreme caution, not only due to the possible damage to agriculture, but also to the possibility of polluting lakes and rivers that provide water for human consumption as well as for irrigation" (cited in Zambrano's judgment, 82). Given that the primer was published prior to the onset of Texaco's operations in Ecuador and that a Texaco senior scientist collaborated in its publication, Zambrano wrote, "We conclude not only that Texaco had prior knowledge of the injury it could cause, since a decade earlier its own officers were writing books warning of this effect, but that this was the state of technical knowledge according to the American Petroleum Institute" (82). Not only was Texaco aware that its operations generated risk of harm, but the design of its operations in Ecuador violated the company's own directives for treating formation waters.

Another collection of documents similarly proved noteworthy for Zambrano. In March 1972, Texaco submitted an application to the US Patent Office for an "invention that belongs to the field of underground disposal of liquid waste." In June 1974, the Department of Commerce granted Texaco Inc. patent no. 3,817,859 for an invention called "Waste Water Treatment Method."[14] The text of the patent states, "Certain effluent streams from industry are waste that have no apparent use. These streams must be disposed of, but doing so in or close to the ground surface may cause considerable contamination problems. Furthermore, the treatment of those fluids in such a way that they can be legally and harmlessly discharged into streams of sources of water is, in most of the cases, excessively expensive. . . . [A] solution is to inject these fluids inside the underground formations whose geologic characteristics prevent the possibility of contact with the surface or underground fresh water formations."[15]

In June 1980, a district superintendent of Texaco Inc., D. W. Archer, sent a letter to René Bucaram, then general manager of Texpet, in response to concerns about oil pits and the pollution they were causing. The Ecuadorian government had requested that Texaco conduct a study to determine the feasibility and costs of eliminating the pits. Claiming that the risk of contamination is "minimal," the Texaco executive "recommends not covering, enclosing or lining the soils of the [waste] pits."[16] Economics loomed in his explanation as he "underestimate[ed] the problems or harm these [waste] pits would cause" (cited in Zambrano's judgment, 162). In Archer's words, "The current pits are necessary for efficient and economical operation of our drilling and workover programs and for our production operations. The alternative for using our current pits . . . [comes] at a prohibitive cost.[17] Although not quoted in Zambrano's ruling, the last sentence in Archer's letter read: "Therefore, we recommend not to fence, coat or fill the pits and to continue using siphons."[18] Internal Texaco memos from corporate personnel in Ecuador and the United States similarly discussed contamination problems and potential concerns. It would seem the corporation chose not to obviate potential harm, despite having the capacity to do so.

But was Texaco aware that effluents siphoned from its waste pits were causing harm? Zambrano noted that in addition to Texaco cowriting API guidelines, the court record contained individual testimonies and written correspondence suggesting that it did. To name one example, in March 1983, then governor of Napo, Ney Estupiñan Recalde, sent a letter to Bucaram that read, "Dear Manager of Texaco, the people are clamoring about the grave harm being caused in the area of Shushufindi as a result of the dumping of hydrocarbon wastes they are subjected to by workers of the CEPE-Texaco Consortium and that is polluting their waters, rivers, streams, and creeks. . . . Therefore, I most respectfully take this opportunity to request that you provide the appropriate means to prevent this harm from continuing, which, as will not escape your enlightened attention, in the end will result in incalculable repercussions to the ecosystem and, especially, the agricultural sector."[19] For Zambrano, this letter and other community complaints (both oral and written) made to Texpet constituted "warnings and request[s] made to the defendant company that are difficult for this Court to ignore" (80). They were clear indications "of a harmful situation in progress" and were a "timely request . . . to cease the activities that were causing harm," all the while predicting "incalculable repercussions" for local life-forms (80). That "Texaco Inc. had the knowledge and technical capacity to prevent such harm at a reasonable cost" meant that "the harm was not only foreseeable, but also avoidable" (80).

During the trial's early years, Chevron's legal defense regularly minimized the amount of effluents that its operations seeped into the environment. That changed, however, when the top legal representative of Texpet, Dr. Rodrigo Perez Pallares, in an open letter to the head of a respected Ecuadorian news magazine, *Revista Vistazo*, wrote that "15.834 billion gallons of [formation waters] were dumped in Ecuador between 1972 and 1990 during the entire period the consortium was operated by Texaco."[20] In and of itself, this information was not new, proprietary, or confidential. It formed part of the well-production log records that Petroecuador kept of Texaco's operations; and many thought a more accurate figure would be higher. Over a decade earlier—prior to the initial filing of the lawsuit against Texaco in 1993—environmental legal scholars publicly campaigned around Texaco having dumped billions of gallons of toxic wastes into the environment.[21] This research formed part of the evidence submitted by the plaintiffs' lawyers to the US federal court in the 1990s and the Ecuadorian court in the 2000s. Yet in the spring of 2007, the Perez Pallares letter and its disclosure (published by several national newspapers in full) gained particular traction. In the midst of the litigation, Texpet's chief Ecuadorian lawyer publicly admitted that Texaco dumped nearly sixteen billion gallons of crude-laced formation waters into the Ecuadorian headwaters.

Most alarming about this datum—which Zambrano qualified as a "true fact" (112)—was not simply the quantity released but also the very "hazardousness of the substance dumped, that is, the hazards that may arise from dumping formation water into surface waters used for human consumption" (113). Because Texaco knew formation waters were harmful to the environment and human health, Zambrano reasoned that their disposal into Amazonian waterways was detrimental to the forms of life that these waters supported. "Moreover, if we consider the amounts of formation waters dumped in relation to the hazardousness of the substance dumped . . . it is evident that people using these water sources were exposed to the contaminants that were discharged into [them]" (113). Furthermore, Zambrano noted that because "formation waters contain hydrocarbon[s]" like "BTEX (benzene, toluene, ethyl benzene and xylene); PAHs (polycyclic hydrocarbons) and TPHs (total petroleum hydrocarbons)," elements known to "pose hazards to human health, the harm and risk become apparent" (113).

Yet despite being "legally obligated to exercise extreme care" and take "all the necessary measures and precautions" to prevent undermining the resources of the concession area, Zambrano stated that Texaco chose not to use available "technological measures to avoid dumping formation waters" into the environment (164–65). Having chosen to carry out its operation as it did, Texaco knew "of the potential harm it would cause" and "that it also had the knowledge and technical capacity to prevent" that harm "at a reasonable cost," such that harm was not only "foreseeable, but also avoidable" (80). Zambrano declared that the events recorded "lead us to the conviction that the system implemented by Texpet for the treatment of its waste did not eliminate or manage risks in an adequate or sufficient manner, but rather in an economic [manner]. As it was designed, the pit system allowed for waste to be discharged into the environment, following a decanting process. That is, THE SYSTEM WAS designed TO DISCHARGE THE WASTE INTO THE ENVIRONMENT, to be economically expedient" (165–66, emphasis in original). Moreover, the company "did not adequately address risks of harm, but rather externalized them" (166). The use of pits to dispose of industrial wastes was an engineering decision "motivated on merely economic grounds" while simultaneously "underestimat[ing] the potential harm implied for the environment and third parties" (162). The effect of "managerial decisions based on costs" was that Amazonian peoples had "to pay the real costs of such decisions" (162).

The nearly sixteen billion gallons that Texaco dumped between 1972 and 1990 were a consequence of engineering: the wastes, the pits, the goosenecked pipes all "inevitably contaminat[ed] the natural water sources of the region on which the local population depended" (166). These operational "protocols," Zambrano maintained, dumped "a far from inconsequential amount of dangerous substances" into the ecosystem, "notwithstanding the fact that the law stipulates specific prohibitions" to pollution (166). "The dumping of formation waters directly into the ecosystem . . . constitutes without a doubt a definite harm, legally proven and publicly acknowledged by the legal representative of Texaco Petroleum Company. And its cause lies in the acts attributable to the defendant who . . . was solely responsible for the technical aspect of [the] consortium's operations" (166).[22] Zambrano concluded: "Given the legally required duty of Texpet to prevent such harm under the historic legislation in effect during the period it operated the consortium, in the opinion of this presidency, the acts of the defendant clearly constitute grossly negligent conduct" (175).

Assessing Injury

Yet how to assess the permutations and intensities of harm that Texaco's operations had on local biotic and human communities? Given the controversy over the chemistry of crude and the epidemiology of crude oil's effects, Zambrano was deliberative and slow in parsing these disputes. Schematically, he divided his discussion into examining evidence on the existence of chemical compounds in the environment and their effects, the epidemiological evidence of health effects, and local inhabitants' experiences of their own and their families' health.

CHEMICAL INJURY

In considering the chemistry of harm, Zambrano began by noting that the court reviewed over one hundred expert reports—experts explicitly hired by each party, experts appointed by the court but nominated by the parties, and experts appointed by the court and not nominated by the parties. But in reviewing these reports, Zambrano observed that the conclusions reached "contradict each other despite the fact that they refer to the same reality" (94). Consequently, dispensing with the "personal assessments and opinions of all the experts" (94, 119), the court did "not consider . . . the conclusions presented by the experts in their reports" (94) and instead considered only "the technical content of their reports" (94, 119). That is, independent of the experts' deduced opinion, Zambrano formed his own assessment, "in accordance with the rules of sound judgment" (94, 119). And he based his assessment on an "evaluation of the results of the laboratory analyses of the samples taken by the experts" (94). Taken together and assembled collectively, Zambrano maintained that the data in these reports were trustworthy and reliable and led to "the conclusion that there are distinct levels of contaminating elements that derive from the hydrocarbon industry" (95-96). Furthermore, the laboratory analyses demonstrated "the existence of environmental harm that originated from the petroleum exploitation carried out during the operation of the concession" (94).

However, Zambrano is quick to clarify (and repeatedly so) that in evaluating the data in experts' reports, the court did not use or consider the different "legal references" that various experts deployed to establish "toxicity threshold" levels (74, 96, 119). As noted earlier, Ecuador had not established numerical legal standards guiding the allowable limits of industrial discharge during the time that Texaco operated. And, consequently, the court did not use the parameters emphasized in each expert's report as the basis on which to determine a violation or liability. Rather, current Ecuadorian and international threshold levels served as "another reference parameter, among all those provided by the parties," that allowed Zambrano to determine "the possible existence and magnitude of any environmental harm" (74). These parameters placed in relief "the current condition of the environment in question" and attuned "our own measure of reality" (74).

Over the next thirty pages, Zambrano's ruling extracted elements from experts' reports to develop his conclusions. Zambrano noted that the court took into consideration thousands of samples: Chevron's 2,371 samples (50,939 results); the Ecuadorian plaintiffs' 466 samples (6,239 results); the court experts' (not nominated by either party) 178 samples (2,166 results); and further Chevron-nominated court experts' 109 samples (1,547 results) (99). What this data revealed, Zambrano observed, was "not given in black and white, but rather in a series of shadings" (99), such that his task was "not a simple mathematical process," nor was it merely "accounting" (99). Rather, his task was to carry out what he called a "statistical evaluation, such that the result depends on what the sample represents" and could accommodate "a certain degree of error" (99).

Zambrano stressed that he included only the "valid samples" in his assessment and specifically did not consider those samples provided by Cabrera—an expert whom Chevron fiercely opposed and one of the reasons for its US RICO litigation. So, upon excluding Cabrera's work, "truly of interest" for Zambrano in his assessment was "that all the elements of the sampled universe [be] known, and that all these elements [had] the real possibility of being included in the sample" (99). Despite differences in degrees, Zambrano noted that all the reports, whether presented by plaintiffs or the defense, "show the presence of different concentrations of hydrocarbons and/or elements used in drilling and maintaining oil wells" (96). Beginning with TPHs and the various controversies over what TPH levels indicate, Zambrano reasoned that "even if [TPH levels are] not a precise indicator of health risks, [they are] a good indicator of the environmental condition in general, in terms of hydrocarbon impacts" (101). While important indicators, TPH concentrations were "most appropriate," however, when "considered together with the other evidences" (101). That is, other factors needed to be considered when weighing the adverse health effects of hydrocarbons.

Significantly, along with gross TPH concentrations, the samples from the fifty-four site inspections plus those of the court-appointed experts (not including Cabrera) revealed the presence of various explicitly hazardous elements in the environment—these included polycyclic aromatic hydrocarbons (PAHs), heavy metals (chromium VI, cadmium, barium, zinc, and lead) present in drilling muds, and anticorrosive agents used in drilling and maintaining wells. Citing specific concentration levels and the state of knowledge from the World Health Organization (WHO), the International Agency for Research on Cancer (IARC), and the US Environmental Protection Agency (EPA), Zambrano noted, in turn, the damaging effects of these elements on biological life-forms and human health: chromium VI is a known carcinogenic agent for human beings (98, 110–11); cadmium can seriously irritate stomach and respiratory systems and is a probable human carcinogen (110); barium compounds can have harmful effects on health and could even cause cancer (111); lead, in addition to disrupting neurological function and reducing cognitive ability, can cause cancer (109); and mercury can damage the brain and kidneys and is a possible carcinogenic agent in humans (109).

With respect to PAHs, Zambrano sidestepped engaging with the complex science of crude—as did the majority of the court proceedings—and instead considered PAHs as an undifferentiated matter. PAHs were important to consider because formation waters typically contained them. Thus, despite assays not detecting significant concentrations of light PAHs, and tests never conducted to detect large PAHs, these hydrocarbons were important because of their long-assumed presence in the environment. As Zambrano noted, PAHs are "potentially carcinogenic" (108) or shown to have "carcinogenic effects" (101). This was particularly significant because, decanted into streams along with formation waters, PAHs, being "not very soluble in water[,] . . . can persist attached to suspended solids and migrate long distances, even without degrading" (101).

The multiple samples extracted from below the pits further supported conclusions that contaminants from hydrocarbon operations travel. A significant number of these samples contained high levels of TPHs, which, Zambrano reasoned, was "a good indication that the pits are a potential source of the hydrocarbon contamination of groundwater" (117). To further underscore this point, Zambrano quoted one of Chevron's experts who commented on soil cores extracted during a judicial inspection: "'As you can see, the soils here are laden with clay. The smell of petroleum emerges at these horizontal distances [from the pit], even when [boring] deeper, even when [boring] farther. . . . As the slope gradually descends, some samples are cleaner than others. But the further [the auger] enters the stronger the smell of petroleum. And you can smell it in any sample.' "[23] Waste-pit seepage was not only nearly ubiquitous in the samples, but also historic. Zambrano highlighted this by ascribing "full evidentiary weight" to a 1976 memo between Texpet personnel detailing that underground and water contamination had "caught the notice of the authorities." It read, "Seepages of crude from pits [associated with] well Lago Agrio 15 have contaminated a nearby stream at the site where a spring originates. These waters are contaminated along their course until they reach the Aguarico [River], and have caused damage to a farm belonging to a settler."[24]

In attempting to ascertain the extent of the effect of hydrocarbon activity, the court recognized that not all the soil in the former Texaco concession was contaminated. But, without doubt, "the quantity and consistency of the data gathered during the 54 judicial inspections" (106) constituted "a representative sample of the universe of sites operated by Texpet when it was in charge of the concession" (106). That is, given the "similarity of the results from the inspections"—regardless of whether a site had been part of Texaco's remediation plan between 1995 and 1998—the insights garnered from the judicial inspections could "be extrapolated" to all Texpet sites, given that the technology and practices did not vary across the concession. For the court, this strongly suggested that the "presence of hydrocarbons buried at the pits entails a risk to the environment and eventually to flora, fauna and human health, since groundwater could become contaminated, thus becoming a risk to the health of people who come into contact with these waters" (117). "Undoubtedly," Zambrano argued, results from the inspections "constitute a manifestation of the existence of environmental damage because they reference hazardous elements that were introduced in hazardous quantities into the ecosystem as a consequence of the hydrocarbon practices that Texpet as operator employed" (119).

It bears noting that Zambrano forcefully reiterated his position with respect to one of Chevron's arguments. He stated, "Beyond any irreverent argument by some lawyer who intends to deprive the law of its meaning . . . the lack of regulations or parameters regulating the dumping" of wastes into the ecosystem "does not in any way signify an implicit authorization to dump this hazardous substance into the environment" (99). Although industrial wastes were not specifically regulated, "elementary standards of justice, legality, decency and respect for human life" dictated that Texaco should have handled hazardous substances in accordance with the legal provisions in effect. Indeed, Texaco was mandated to foresee and prevent dangerous by-products of its operations from entering the ecosystem and affecting well-being. The presence of such substances "that could place people or the ecosystem at risk," in addition to being a violation of law, "would constitute evidence of legal harm, which, as such, brings with it the obligation to make reparations" (97).

HARMED COMMUNITIES

The plaintiffs' claim, Zambrano noted, that Texaco's contamination led to adverse health consequences, "including elevated rates of cancer, miscarriage, high infant mortality and genetic deformities" (125), is "the most complex and urgent of all the issues brought before this court" (125). To grapple with that complexity, Zambrano leaned on and drew from a variety of evidence. He began by invoking the international accords that underscore the prime importance of health—the Universal Declaration of Human Rights, the International Convention on the Elimination of All Forms of Racial Discrimination, the Convention on the Elimination of All Forms of Discrimination against Women, the Convention on the Rights of the Child, and the American Convention on Human Rights. Quoting the United Nations Committee of Economic, Social, and Cultural Rights, he observed that health is "'a fundamental human right that is indispensable to the exercise of all other human rights. Every human being has a right to enjoy the highest possible level of health to enable him to live in dignity'" (126). And Zambrano reasoned that since the right to health was "a necessary component of the right to life, . . . an assault on people's health is tantamount to an assault on their lives" (126).

For Zambrano, the profundity of this reasoning multiplied when one considered the plaintiffs' unique living circumstances. Texaco's oil operations occurred not only in a zone designated for hydrocarbon extraction, but also in a place where the very "lives of people and their cultural integrity is firmly associated with the health of the land" (125). The vast majority of inhabitants of the region depended heavily on natural water sources for their daily existence. Springs, streams, rivers, and wells were, and still are, for many their sources for consumption and bathing, cleaning, fishing, and playing. Consequently, that industrial wastes were found to reside in the concession's environment—in its soils and waters—and had the capacity to travel (both through time and space) suggested, according to Zambrano, that "severe consequences" (125) to the well-being of human and nonhuman communities were nearly inevitable.

Throughout the litigation, Chevron recognized that local inhabitants experienced dire challenges to their health. The corporation repeatedly insisted, however, bringing report after report assessing the socioeconomic status of the Amazon, that poverty was the cause of the residents' compromised health. Poor sanitary conditions and inadequate nutrition were responsible for the ailments that local peoples experienced; illness was not a consequence of Texaco's oil operations. Assessing these reports, Zambrano reasoned differently. Since "other provinces with similar poverty indexes" are not confronted with the same health concerns, "it is not poverty that directly causes mortality, but rather a common denominator" (130). Poverty is not the source of ill health; rather, contamination exacerbates the predicaments of those who live in poverty—who, dwelling in regions with no water infrastructure, rely acutely on forest water systems and have scant access to health services. Provoking reflection, Zambrano drew attention to the conclusions of one expert report: "The analyses carried out on fish tissue determined the presence of total [petroleum] hydrocarbons in fish in values far above the maximums allowed in water" (130). Given that products deriving from petroleum "'are highly harmful to health'" (quoting from the US Agency for Toxic Substances and Disease Registry), hydrocarbons' persistence in aquatic ecosystems "'could evolve into a nutrition problem . . . [affecting] the health of those who consume [fish].'"[25]

Furthermore, Zambrano discussed at length two studies that signal how oil operations could negatively shape animal and human health. The first, "Yana Curí," was prepared by the vicarage of Aguarico's department of health in collaboration with the London School of Medicine and Tropical Hygiene (131).[26] Led by Miguel San Sebastián, the "Yana Curí" research was explicitly oriented toward understanding oil operations' potential "adverse effects on health" (131). Zambrano highlighted different dimensions in order to build his case. To begin, the review of literature established a basis for caution. Animal studies "demonstrate that 'exposure to crude oil can cause lesions in different organs, cancer, reproductive defects, and even death,' both in domestic animals and in wild animals" (132). Epidemiological studies on oil workers in the United States and China suggested a relationship between hydrocarbon exposure and acute myelogenous leukemia (132) and, in general, "an elevated risk of serious and irreversible effects on [workers'] health" (133). Comparing the health of individuals living within five kilometers of an oil installation with those living farther away, the authors of "Yana Curí" noted how "difficult [it was] to establish a relation between contamination by oil and its impact" (133), given the lack of information on past levels of contamination and exposure and the diversity of possible effects over time. Notwithstanding, the report claimed that "women living near oil wells and stations present worse overall health conditions than those women living far from these oil wells and stations" (133). Zambrano concurred with the logic of "Yana Curí."

The second study that Zambrano cited was the study discussed in chapter 2.[27] The study had analyzed data from the National Tumor Registry in Quito of cancer cases from individuals residing in Amazonian cantons in which oil operations had not taken place and those residing in cantons in which oil exploitation had occurred over the past twenty years. Without claiming any causality, the study found a statistically significant increased incidence of cancer in cantons where oil operations were long-standing. In establishing a connection between exposure to hydrocarbon activity and worsened states of health, this study, the court noted, produced "statistical data of highest importance" and can be used "as a measure for evaluating the medical probability of an illness being caused by a given agent" (135). Although "the data on its own may not suffice to establish a causal relationship" (135) between the risk of contracting cancer and living among long-standing oil operations, they did indicate "an association," which the court would analyze in conjunction with other evidence.

The question of association versus causality loomed large, given that Chevron engaged multiple experts to undermine San Sebastián's research as methodologically weak and unable to "establish causal relationships" (per Michael Kelsh's testimony during the trial on February 27, 2010; cited in Zambrano's judgment, 135). The court was fully aware, Zambrano noted, that biases influenced the statistical data in San Sebastián's work and that this study had limitations. He already indicated that, given their geographic and socioeconomic marginalization, the majority of ill people living in proximity to Texaco's former oil operations probably never went to Quito to be diagnosed and treated in the only hospital providing histopathological services. And the cancer data recorded in the National Tumor Registry in Quito were "in all probability underestimated" (San Sebastián, cited in Zambrano's judgment, 134). Zambrano suggested that a letter submitted during one of the more emotionally fraught judicial inspections at Sacha Sur suggests how to proceed. He was referring to the letter signed by the fifty scientists across the world supporting San Sebastián's work, indicating that in epidemiology no study is perfect. Present at the Sacha Sur inspections, one author of the international letter noted that the response to critique is not to ditch the study but to conclude that it compels "a level of suggestive suspicion" that encourages one "to assume the principle of precaution" (Jaime Breihl, cited in Zambrano's judgment, 136).

Zambrano reminded that the lawsuit, the legal claim, against Chevron was for reparations to environmental harm and its effects. The fact that "no particular injuries or harm have been proved is irrelevant," given that the complaint does not ask for indemnification for a specific injury or a specific health problem of a given individual. Rather, the plaintiffs' request with respect to health is "to contract, at the defendant's expense, specialized persons or institutions to design and carry out a plan for the health improvement and monitoring of the inhabitants affected by contamination" (138). At issue, Zambrano opined, was "the existence of harm to public health and whether this harm is directly related to the reported environmental impacts for which reparation is required" (138–39).

At this point in his ruling, Zambrano shifted the tenor of his argument. Conventionally, he noted, a judgment "normally require[s] unfailing proof," as indicated through chemical samples and epidemiological studies "of the existence of harm" (145). However, "the evolution of the law," with its emergent logics, "has allowed for the use of other probative means" (145). Amplifying "probative means," Zambrano stressed the importance of paying attention to what local inhabitants said about their well-being. And he gave extensive space in his ruling to quotes that recited local peoples' testimonies. Zambrano noted that those who testified "present[ed] a very poor picture of their own health," each recounting parallel stories that related their experience after "drinking water contaminated by hydrocarbons" (139) or coming into contact with "the products deposited in the pits and other oil facilities" (143). Testimonies reiterated how dependent local peoples and their animals were on natural sources of water, using waters (out of necessity) that were contaminated, unbeknownst to them, by Texaco's industrial wastes as it seeped through hydraulic networks (142–43). And, in Zambrano's mind, there was no doubt that "different forms of exposure" (146) to toxic-laced waters were the "cause of their health problems" (143).

Zambrano elaborated by quoting another epidemiological study (which he had discounted earlier in the ruling): "The direct contamination of rivers—an indispensable source of water for the majority of the families—is one of the worst problems" (147). Ingestion and absorption of contaminants occurred not only when humans and animals drank river waters, or when people used them for cooking, bathing, and washing clothes, but also when they ate aquatic life that was part of a polluted trophic chain. The report continued: "Exposure to and consumption of water from these rivers produce[d] skin diseases, intestinal and vaginal infections, and in many cases, cancer in women, principally to the uterus, ovaries and breasts, and, in general, the throat, stomach, kidney, skin, and brain" (147).[28] Furthermore, broader ecological and economic forces harmed people; when "wild, domestic, and farm animals" came into contact with hydrocarbon contamination, their death or weakened condition undermined their "productive capacity and the quality of food for people" (152). Zambrano determined that, given "the danger posed by the substances dumped and all the possible mediums of exposure," the contamination found in Texaco's former concession "puts the health and life of people and the ecosystem at risk" (147).

Importantly, Zambrano underscored that a testimony did not represent an isolated claim. Together, the statements were a profound collective conviction that people's health, their being, their families, their animals, their livelihood had been adversely affected by Texaco's former operations. In fact, words pronounced under oath during the judicial inspections gave rise to what Zambrano termed a "procedural truth"—"all the statements received are identical, without a single statement indicating the contrary" (139). The "overwhelming coincidences among all the declarations" strongly support "the thesis that what we have here is continuous harm caused by contamination, and not by random factors" (152).

Further reflecting on the testimonies given by rural residents, Zambrano argued that although their statements "are not decisive and irrefutable evidence that there is a health problem among these citizens" due to oil operations, "nevertheless, their value also cannot be dismissed," especially given "the impressive coincidence among the facts described in these statements, without one single statement or declaration to the contrary" (144). The corroboration was significant; Chevron also called rural peoples to testify. Zambrano continued:

> The experts who have participated in this case have submitted reports to this court in their fields of expertise. However, these experts have not experienced what it means to live in this environment, and they have no deeper historical knowledge than that which they have found in specific documents. Not one of the experts who participated in this case knows the historical reality better than those who have lived here. Consequently, these statements [made by local peoples] will be considered with the value they deserve, together with the other evidence submitted by the parties, and in accordance with the rules of sound judgment. (144)

And while the court recognized that

> the citizens who gave testimony during the judicial inspections are not doctors or health care professionals—as was left clear by Chevron's Counsel of Record, Adolfo Callejas, who, when examining Mr. Carlos Cruz Calderón at [the] Yuca 2B judicial inspection, asked him: "How do you know that these are toxic [wastes]? What studies have you made to determine that they are toxic?" . . . this court is inclined to think that the coincidences in the testimonies corroborate what has been said and lead us to think that the suffering narrated in these statements is real. (144)

Basis of Obligation

The question remained as to the source of legal obligation. Zambrano began with the Ecuadorian Civil Code, which states, "An act that has injured or harmed another is a source of obligations."[29] Of first note, unlawful acts extend beyond "the personal acts or omissions of the responsible party who intentionally or culpably causes harm . . . (as dictated by articles 2241, 2242, 2242, 2244 and 2245 of the Civil Code noted above)" (75). They similarly encompass "harm caused . . . from the things that are their property or to which they help themselves (articles 2250, 2251, 2253, 2254 and 2255)" (75). Consequently, Zambrano inferred, "direct fault on the part of the responsible party is not required in every case"; rather, fault could be "assumed through the acts of third parties or harm caused by things the responsible party makes use of" (74).

Expanding this point with greater precision, Zambrano cited at length "the most studied judgments in Ecuadorian law," rendered by the First Civil and Commercial Chamber of the Supreme Court of Justice, on October 29, 2002.[30] Zambrano relied heavily on this ruling; its "careful and brilliant analysis" served as the authority through which to understand liability and the basis of obligation. In fact, pages of the Sucumbíos ruling are verbatim iterations (often quoted but sometimes not) of the supreme court's "masterful explanation" of what Zambrano called a "new type of liability" (82). In the sphere of extracontractual responsibility, "what [was] important" for this new liability was "not the manner of conduct but rather the consequences of the conduct" (82) itself.

Crucial sections of the Supreme Court decision echoed insights from the likes of Ulrich Beck (1992), Michel Callon and colleagues (2009), and Sheila Jasanoff (1995, 2012). Zambrano quoted: "The current world and that of the near future, with its extraordinary and steady accumulation of risks, calls for a more vigorous defense of human values, as a result of a science that is both all-providing and all-threatening at the same time. The multiplicity of actual contingencies of dangers and risks that currently seem uncertain because they are not yet realized . . . led to a slow evolution of elements and knowledge that has helped the most advanced legal systems enter into a risk-distribution mechanism whereby the risk victim would not be left unprotected" (82). This predicament, the ruling continues, had given rise "to risk theory, according to which whoever uses and takes advantage of any benefit-yielding medium generates social risks, and therefore must assume liability for the injury thereby caused" (83). This is the "risk of advantage, with its origin in the Roman maxim *ubi emolumentum, ibi onus* (where benefit is found, responsibility follows)" (83); the counterpart to benefiting from activity is compensating for any injury caused.[31]

Continuing, the Supreme Court noted that "the risk of a thing is a legitimate danger and socially accepted as the counterpart to the social or economic benefits that are entailed by the operation, use, or utilization of the hazardous elements" (83). In particular, Zambrano followed the Supreme Court's logic with respect to article 2356 of the Civil Code, which contemplates extracontractual civil liability for risky or hazardous activities. Reversing conventional logics of liability, the Supreme Court's interpretation of this article ruled that, when it came to hazardous activities, "negligence is presumed," a move that, in turn, "relieves the victim of the need to provide evidence of negligence [and] lack of care or skill"; rather, the burden of proof rested on the defendant "to prove that the event occurred as a result of force majeure, an act of God, the intervention of an extraneous element or the exclusive fault of the victim" (cited in Zambrano, 75). Zambrano clarified that Article 2356 of the Civil Code—changed subsequently to Article 2229—was referenced in the lawsuit as the grounds for claiming harm caused by oil operations. That is, when extracontractual civil liability is at issue, a "purely objective liability" was what presides; proof of fault or intentional misconduct was not required: "it is enough for the injury to be a direct consequence of the event that gave rise to it" (83).

The Supreme Court clarified that the "theory of objective liability" was not "widely accepted in the laws of most countries and in the case law of foreign courts" (83). In most jurisdictions, "fault of the liable party must be proved" (83). But, the court asserted, since "the burden of proof of fault is almost impossible or very difficult for the victim to meet in most cases," a number of jurisdictions determined it "necessary to reverse the burden of proof" (83), such that "whoever uses and takes advantage of a risky thing is the one who must prove that the injurious act occurred as a result of" (83) actions other than their own. In other words, fault is presumed for whoever uses or takes advantage of processes whose risk causes harm.

A growing number of jurisdictions, including France, Argentina, and Colombia, embraced this logic, particularly in case law, the Supreme Court of Justice noted. And Zambrano adopted this legal logic, too, given that Texaco's operations to explore for, exploit, process, and transport hydrocarbons constituted high-risk and high-hazard activities. Citing one of Colombia's foremost twentieth-century legal scholars and Supreme Court justices, Arturo Valencia, Zambrano continues:

> Modern times, especially the twentieth century, created a new and ripe source of harm: those caused by hazardous activities or operations, which originate in the use of all types of vehicles, machinery, and new energy. . . . For these kinds of harm, fault-based criteria have proven insufficient in obtaining compensation, since the cause of most accidents is unknown. This is why it is often said that modern man "uses forces whose nature and power he does not himself know." . . . Consequently, it is necessary to establish a new category of liability for these kinds of harms [that reasons that] the owner of the operations or industry must respond directly for harm whose cause stems from that industry or operation, such that he can only be released from liability if he can show that the harm was not caused by his operation, but rather by an outside factor (force of nature, third-party liability, or by the victim himself). The owner of the operation cannot be allowed to be released from liability by proving the simple absence of fault.[32]

The Supreme Court noted that Colombia's Article 2356 is what Article 1384 is for France—a provision used to resolve legal problems arising from technologies in our modern era "that involve inevitable and uncertain dangers" (86). Without having to *prove* malice or fault, the liability of a person is presumed for the acts of things over which he has custody. Fault is imputed and may only be refuted by proving exemption from liability—that the harm is not a result of the handlings or operation of things. The absence of guilt through proof does not exonerate the custodian of things (or owner of the operation) from liability. If anything, "the regime favors the victim of the harm, who need only prove the harm and its accompanying causal nexus in order for his claim of harm to succeed" (86).[33]

Chevron had claimed that "the hydrocarbon activities that allegedly caused the harm . . . were legitimate activities authorized by law" (86). However, Zambrano reasoned, Texaco's operations were "lawful activity," which simply implied that "the risk of the thing is a lawful danger and socially accepted as counterpart to the social and economic benefits" (86). This did not mean, he noted, that the party conducting legitimate activity was exempt from liability. On the contrary, law had come to establish the "presumption of guilt of the person that uses and takes advantage of the risky thing by which the harm was caused" (86). Considering that "the production, industry, transport, and operation of hydrocarbon substances undoubtedly constitute high-risk or hazardous activities" (86), Zambrano deemed it "imperative" to deploy "this new type of liability" (86).

As the Supreme Court recognized, establishing a "causal nexus"—the degree to which one action has caused another—was anything but straightforward: "In most cases, the facts do not appear as pure and simple, but rather, on the contrary, they are mixed or combined with other occurrences, or even conditioned by different events, or favored or limited by other concurrent,

FIGURE 53 Donald Moncayo (*left*) and William Lucitante (*right*) amid a portion of the case file in the offices of the Frente de Defensa de la Amazonía. Photo by author.

underlying, or pre-existing actions" (87). Such is the predicament of the Texaco case; events, facts, and actions were never presented alone but always combined, conditioned, favored, and limited by a series of other circumstances. Complexities abounded. As such, Zambrano echoed Ecuador's Supreme Court and asserted that in establishing a causal nexus he would abide by the "theory of sufficient causation"—meaning that the court placed trust in the judge's capacity to use "discretionary powers" in ascertaining "when a harmful action is likely to generate liability," with the "duty to provide adequate grounds for his decision" (87).

Juris Prudence

In his March 2011 clarification order, Zambrano further commented on his practice of welcoming and valuing expert and vernacular insights.[34] As is its duty, he noted, the court considered "the testimony of all the noteworthy individuals who participated" in the legal process at the parties' request.[35] However, "doubts still remain[ed]" as to the reasoning behind many of the Chevron experts' statements. In every case, he continued, Chevron brought "a well-paid, recognized foreign expert" to Ecuador to testify and give their expertise. But, in "most cases, these experts did not bother to become familiar with the region" over which they expounded. By and large, their statements were "limited to analyzing documentary information (provided by Chevron) to reach their conclusions, which are in all cases similar."[36] Zambrano cited examples from the epidemiologists whom Chevron contracted; their reports and testimonies indicated an "absence of harm and health risks" associated with Texaco's operations and underscored "the limitations" of work by San Sebastián and colleagues. Although these opinions were "considered valuable," Zambrano wrote, "it is not possible that those words would convince me to ignore the dozens of testimonies given by residents, [testimonies] that are all consistent and narrate a similar story." Zambrano continued, "It is precisely the humility of these individuals who, even though they are not experts or doctors or PhDs, do share a single history and a single condition as victims, which cannot be feigned or fabricated."[37] The very "authenticity" of local peoples placed, in his opinion, "the testimony of the foreign experts in question, who, buttressed by their academic degrees and the study of the documents submitted by Chevron, refute—from afar—the suffering of local peoples."[38]

Uncertainties surrounding environmental contamination and its imbrication with the well-being of living forms have generated intense social, technical, and legal controversies around the world (Fortun 2001; Murphy 2006; Petryna 2002; Wylie 2018). Whether from toxic chemicals, nuclear wastes, or electromagnetic fields, the effects of contamination—especially the effects of long-term, low-dose exposure to substances suspected to be harmful— often slide into a realm of scientific ambiguity. That tendency invariably intensifies when those whose interest it is to ensure uncertainty (pace "doubt is our product") enter the arena. This condition collides with the normative fashion of resolving environmental-health problems. Here, controversies around the environment and health proceed in a linear, sequential fashion: "scientists tell the truth and establish certainties," while judges, regulators, and politicians draw "the obvious conclusions" and transpose the analyses presented to them into concrete, programmatic decisions (Callon, Lascoumes, and Barthe 2009: 204). Within this logic, "science" is what "enables uncertainty to be removed" (204). Yet amid conditions of disputed partial-knowns, restricting action to a techno-rational analytic is singularly debilitating (Fortun 2001; Latour 2014).

Zambrano's ruling exceeded this rationality and interrupted its conventional decision-making logics. Given the evidence garnered during the judicial inspections, there was no shortage of expertise. Expert commentary flooded the case file with hundreds of scientific reports. Nonetheless, Zambrano maintained in his March 2011 clarification order that these scientific conclusions were far from definitive. For instance, "statistical bias" marred all the epidemiological studies submitted by the parties.[39] How could it be otherwise? These studies "collect[ed] information from official sources"— health facilities—which, based in Quito and having no presence in the Amazon, only had "limited access to information" of the region, which, in turn, "undermine[d] the value of the data" upon which the studies hinged.[40] Similarly, Zambrano stated in his judgment that the contradictory conclusions reached by experts underscored "the irrefutable lack of scientific certainty" (Zambrano's judgment, 89) about the chemistry of crude and its effects on the environment.

But, as Zambrano clarified on more than one occasion, the function of the expert—his or her reports and opinion in the case file—is "to assist" the judge "in reaching a resolution that appropriately takes into consideration unfamiliar sciences or arts" (37). While an expert's report may support the judge in rendering a decision, the judge is "not bound to accept a report against his own conviction" (38), nor is he "obliged to abide by" (37) its conclusions. That is, "what an expert says or fails to say does not tie the hands of the judge" (37). Rather, following Ecuador's Civil Code, the judge "is authorized to appraise all evidence" and is responsible for "assessing the contents of a report in accordance with the rules of sound judgment, giving it the value it warrants" (37). And, he reminded, the Code also gives "a preponderance of value" to the "immediacy of the judge in the proceedings" (38)—meaning the judge's tangible engagement in the judicial procedures and inspections. This "immediacy" instills confidence in the "discretion" of the judge to "apply his reasoning and sound judgment" (39) to assess expert reports in relation to the universe of evidence brought to the fore during the legal proceedings.

In rendering his decision, Zambrano noted that significant evidence before the court—the risk associated with industrial wastes, the risk associated with hydrocarbon elements, a number of the epidemiological studies, and, importantly, personal testimonies—coincided in determining the presence of a human-health risk. That "coincidence" determined "satisfactorily that there are legal/scientific bases for reasonably linking the health claims made by inhabitants of the region *with* the oil contamination that derives from the Texpet's activities," despite no one factor or class of evidence being "attributed with either direct causation or exclusive responsibility" (170–71). The plaintiffs' legal claim, Zambrano reminded, did not ask "for reparation of harm to the health of specific individuals" (170). It sought, instead, "the elimination or removal of the contaminating elements that still threaten the environment and the health of the inhabitants."[41] Thus, the court was concerned with contamination "without having to determine precisely which element caused harm" (88–89), and it was concerned about health without the need for "a particular injury or harm [to] have been proved" (138). At issue was the "presence of a public-health problem" (170) as a consequence of environmental damage attributed to Texaco's operations.

Given the universe of evidence presented to the court, Zambrano argued that it was "sufficiently proven that an impact on public health exists and that this impact has a reasonable medical probability of being the result of people living in the concession area having been exposed to substances discharged by Texpet into the ecosystem" (170). Because formation waters have "known potential for harm" (which the industry long knew) and because the predicted harm is "fully in accordance with the ailments found among the inhabitants of the zone" (170), there was, Zambrano argued, "the culpable creation of the unjustified risk of a hazardous situation" because harm was "foreseeable" (88). The judge maintained that "the mere existence of harm would be sufficient to attribute a causal nexus between the harm and the hazard that had been created" (88–89). The hundreds of scientific reports and testimonies forming the "record" demonstrated "that thousands of human beings have been effectively exposed to risk [posed by] soil and water contamination, the presence of which in the environment constitutes a substantial factor that was principally caused by Texpet's activities" (170). Zambrano wrote, "In this Court's opinion, considering the dependence of the area's inhabitants on the natural water sources and the discharges made by Texpet . . . it is more probable [than not] that exposure to the effluents discharged by Texpet into the environment produced an adverse impact on [people's] health" (170).

Upending conventional hierarchies of expert/layperson, science/politics, fact/value, Zambrano momentarily stabilized indeterminacy by flattening the sphere of evidence and inviting forms of knowing, otherwise dismissed, into the debate. This de-authorized the omnipotence of expertise and gave legitimacy to local people as valid interlocutors with consequential effects. Intimate knowledge registered cumulative intensities and empowered a proximate and iterative understanding of harm. It emerged from an episte-mological order significantly different from that which generated certainty out of a precise, detached, and reductive science. In the 2011 ruling, injury manifested through substantial presence, not only objective research; from palpable experience, not only theoretical probabilities; from specific tangi-bles, not only abstract correlations. And in the process, campesino and Indig-enous voices could not be muted as inarticulate distortions, nor could they be ignored as anachronistic aberrations. Zambrano's ruling interrupted the prevailing relations of knowledge and power and enacted a different compo-sition for effecting truths.

His judgment wrestled with the extent to which corporate behavior vi-olated the law at the time, the extent to which that violation posed deep uncertainty to the soils and waters and the human and other-than-human life-forms dependent on them, and the extent to which harm could be said to exist. In line with Ecuador's recent environmental case law, Zambrano's de-cision marked an understanding that scientific knowledge of crude toxicity was inherently unsettled. This emergent Ecuadorian environmental reason-ing folded together knowledge, imprecision, and ignorance—one in which indeterminacy, uncertainty, and probability proffered an expanded connec-tive space for those who shared a stake in the epistemological rules for what counts as harm (cf. Petryna 2002). Together, the ruling and the seven-year trial allowed for and compelled iterative procedures that relationally materi-alized the presence, proximities, and intensities of potential harm to public health.

Delocalizing Stabilities

The logic of Zambrano's decision resonated with concerns central to an evolving precautionary principle, though he invoked the phrase only once. As Ulrich Beck (1992) warned decades ago, contrary to modernist fantasies of ever-grander certitude and control, science and technology have proffered both extensive capacity and mounting risk. In fact, ignorance may be weightier and more consequential than claims to know. Public controversies increase the visibility of these uncertainties and the many machinations around them. Zambrano repeatedly clarified that Ecuadorian case law considered "the oil industry to be a high-risk activity" that raised the probability of harm to Amazonian ecologies—despite the chemical indeterminacy of oil operations and their wastes, despite limited understanding of the relationship between crude and health, despite the virtual nonexistence of exposure data, despite the complexity of the lived circumstance in the region. In the context of hydrocarbon activity that gambled at the expense of human and nonhuman collective well-being, Zambrano's decision took risk seriously— despite an emerging symptomology lacking a precise etiology.

Zambrano grounded his final rendering in precaution—a framing that obliges considered action in the face of scientific uncertainty. From its early development in post–World War II Germany, the notion of "precaution" has been linked to an "ethics of responsibility" (Hans Jonas, cited in Callon, Lascoumes, and Barthe 2009: 198). Confronted with technoscientific uncertainties, Hans Jonas wrote, "While waiting for the certainties resulting from [scientific] projections to become available—especially in view of the irreversibility of some of the processes unleashed—prudence is the better part of valor and is in any case an imperative of responsibility" (199). Fifty years later, the European Commission enacted legislation to make the principle law, asserting, at the turn of the twenty-first century, "The precautionary principle applies where scientific evidence is insufficient, inconclusive or uncertain and preliminary scientific evaluation indicates that there are reasonable grounds for concern that the potentially dangerous effects" that a substance or activity may have "on the environment, human, animal or plant health may be inconsistent with the high level of protection chosen by the EU."[42]

As detailed in the 2002 Ecuadorian Supreme Court ruling that Zambrano cited, European jurisprudence seeped across hemispheres into a number of Latin American legal systems. However, it bears remembering that North-South environmental organizations instantiated the precautionary principle into the international legal sphere in 1992 with the Rio Declaration on Environment and Development. Moreover, hints of this principle were codified into Ecuador's tort law in 1861. More will be said about this in chapter 7, but here let me note that Article 2260 (presently Article 2236) of the Ecuadorian Civil Code gave an individual (or a group of individuals) the right to lodge a complaint against someone whose "imprudence or negligence" produces the "threat" of a "contingent harm"—meaning the risk of a future harm.[43]

One of the precautionary principle's foundational ethics insists on dislodging and inverting the onus of proof around plausible dangers. In conventional tort law and proceedings, the victim of harm is responsible for proving that the harm exists and that it exists as a consequence of the fault of a specific person, activity, or substance. The onus of proving fault and injury rests on those who are the recipients of harm. Under the precautionary principle, the reverse holds; it is the responsibility of a proponent of an activity to establish that the proposed activity has not (or is very unlikely to have) resulted in significant harm. In tort proceedings, the onus is placed on the executor of an activity to show that harm resulted from some other cause. Challenging the conventional onus of proof also challenges that normative decision-making logic: which asserts that when fault is precisely determined, when truths have been scientifically secured, then action can ensue. The precautionary principle insists, by contrast, that the obligation to mitigate suspected dangers need proceed through a wide-ranging exploration of controversies and beyond the legitimacies conventionally warranted to address them. Precaution demands a broad investigation into what is and can be known as well as who can legitimately partake in that determination.

Zambrano's ruling transposed this ethic into the judgment of an environmental-contamination suit. The court affirmed that rural plaintiffs did not have to prove fault—they needed only to demonstrate that harm had occurred, that it would most likely occur in the future, and that harm was reasonably believed to be the effect of Texaco's practice of decanting hydrocarbon wastes into the environment. It was incumbent on Chevron to prove that Texaco's hydrocarbon operations were safe and/or that any harm that might have ensued from them was an effect of force majeure. Under Zambrano, the precautionary principle disrupted the science-leads-to-action equation and embraced expanded regimes through which harms were recognizable. In this sense, his ruling deepened the challenge long posed by Amazonian, non-Indigenous and Indigenous, popular organizing around health concerns—the need to interrogate the unquestioned authority of science and the assumed composition of the polity.

Clearly, Zambrano sought to instantiate "the difference between legal causation and scientific causation" (Zambrano's judgment, 89), a concern echoed in Bruno Latour's discussion of the distinction between law and science. Latour writes, "When Roman lawyers intoned the celebrated adage *res judicata pro veritate habetur*, they were declaring that what had been decided should be taken *as* truth, which means, precisely, that it should in no way be confused *with* the truth" (2010: 238; Jasanoff 1995, 2007). For Zambrano, legal causation emerged through a distinctive combining of evidence before the court. That compositional method was definitively misunderstood by the US Court of Appeals for the Second Circuit when it stated that Zambrano "reached his decision without 'assess[ing]' discrete 'facts,' without following 'concrete, express legal rule[s].'"[44] It emerged, I suggest, from the frictive tensions inherent within the unknowns of science and the prehensive residue of a sensorium-in-the-plural. Precaution afforded a moral ethic of "consequential procedures" (Callon, Lascoumes, and Barthe 2009: 191) whereby recognized limits to knowledge and recognized risk ruptured the normative distribution of who embodied a legitimate interlocutor (producer and holder of facts) and disrupted the normative distribution of the excesses or overflows of late capitalism.

Bringing forth evidence from statutory principles, chemical assays, epidemiological studies, geoengineering practices, geographic surveys, visceral experience, and, profoundly, narratives of physical and existential trauma, Zambrano's ruling orchestrated this dispersed and heterogeneous record into a novel relational composition. It selectively enfolded nascent affinities among diffuse knowings—of analytical results, of statistical values, of scientific pronouncements, of fervent pleas, of experiential overload—that gathered in clouds of concurrent impressions. Separated from familiar signifying relations, these diffuse knowings formed constitutive elements in a delocalized stability that enacted the transubstantiating effect of the benzene ring. Dislocated stabilities triggered an emergent force that was other-than that of its constituent parts and through which its constituents (as prior isolates) no longer held as absolutes. Zambrano's ruling did not articulate the binary form, the Lavoisierian equation, of relating and equating entities and their qualities. Rather, his juris prudence—his taking of considered judicial measures—enacted an asymmetrical relationality that brought into existence qualities and capacities that had yet to exist.

A precautionary ethos destabilized the constitution of knowledges and the demos, dislocating prior comfort zones of authority, legitimacy, and reason. Controversies about uncertainty, Callon and colleagues suggest, "are powerful apparatuses for exploring and learning about possible worlds" (2009: 28). They invite investigation of and challenges to the presumed certainties about "our knowledge of the world" and the "composition of the collective" (119). Displacing the traction of expertise, Zambrano's ruling forged the productive conditions whereby uncertainty, indeterminacy, and nonclosure obliged an expanded mode of discerning, registering, and modestly affecting injury through combinatory power. The judicial inspections spurred, and Zambrano's ruling momentarily retrieved and honored, a particular kind of valenced assembly—conjuring in potentia other modes of abiding—where the effect of combining was as much to extend alternative configurations of the sensible as it was to challenge the material conditions of the present (cf. Callon, Lascoumes, and Barthe 2009; Latour 2004). It is precisely through this transformed conjugation of causality, agency, and emergent truths that Zambrano's ruling rattled entrenched distributions of power and knowledge. Along with a monetary liability, this is what fundamentally poses a threat to Chevron.

In stipulating the "measure of redress for harm," Zambrano affirmed in his judgment that in accordance with "the precautionary principle, all agents potentially harmful to health and the environment should be removed" (Zambrano's judgment, 124). In Texaco's former oil concession, this would entail remediating and cleaning up surface waterways and their sediments; reclaiming lost habitat by recovering native flora, fauna, and aquatic ecosystems; establishing potable water systems; creating a health system designed to address contamination health problems; implementing Indigenous reaffirmation programs among ethnic groups who lost their lands; and removing and remediating all contaminated soils and materials in and around waste pits. As the life of the lawsuit extended to its countermode in the US federal court, this last stipulation would prove highly controversial. Zambrano's judgment calculated that 7,392,000 cubic meters in 880 pits were to be remediated for a cost of $5,396,160,000—an area and amount that was illegible and inconceivable to the US court and, as such, easily absorbed and reconstituted into Chevron's corruption narrative.

NEVER

In 2013 and 2014 (and perhaps for longer), the opening web page for the law firm Gibson, Dunn & Crutcher (Chevron's principal external counsel) appeared as follows:

FIGURE 54 Image of the home page of the website of the Gibson, Dunn & Crutcher law firm (gibsondunn.com), circa 2014.

Bonding Veredictum

Of Corporate Capacity and Technique

6

ON FEBRUARY 1, 2011—two weeks prior to Judge Nicolás Zambrano's $9 billion ruling against the Chevron Corporation—Chevron filed a counterclaim in US federal court that quickly became umbrellaed under the Racketeer Influenced and Corrupt Organizations Act (RICO).[1] Lodged against those who had sued the corporation in Ecuador—forty-eight named Ecuadorian plaintiffs and their lawyers, legal advisors, and experts, who all constituted the collective called the LAP—the countersuit alleged that the 2011 Ecuadorian ruling was obtained through unethical, corrupt, and illegal means.[2] Three years later, on March 4, 2014, Judge Lewis Kaplan, a federal court judge for the US District Court for the Southern District of New York in whose court the case had landed, ruled on the high-profile RICO case. Kaplan determined that the 2011 Ecuadorian ruling had indeed been procured through bribery, fraud, and extortion, and as such it was illegitimate. He issued an injunction foreclosing the enforcement of the $9 billion ruling (initially worldwide and ultimately) in the United States and imposed a constructive trust for Chevron's benefit on any property that those who sued the corporation might receive from enforcement of the ruling elsewhere in the world. On August 8, 2016, the US Court of Appeals for the Second Circuit upheld the lower court's 2014 ruling. And on June 24, 2017, the US Supreme Court declined to review the case.

How was it that a US district court judge presiding over a seven-week bench trial—who speaks no Spanish, has never been to Ecuador, and is not familiar with Ecuadorian law or legal procedures—declared invalid a judicial decision reached in Ecuador after seven years of litigation? A judicial decision thrice upheld by that country's higher courts. And how was it that the US Court of Appeals for the Second Circuit upheld the US district court ruling? The doctrine of "comity of nations" holds that sovereign nations mutually respect each other's executive, legislative, and judicial acts.[3] Furthermore, when in 2001 the same New York district court remanded the lawsuit initially filed against Texaco in 1993 to be tried in Ecuador, it did so under three conditions—one being that the ruling of the Ecuadorian court would be enforceable in the United States and that the corporate entity would satisfy any Ecuadorian final judgment, subject only to invoking a limited defense set forth in New York's Foreign Money Judgment Recognition Act.[4]

A morass engulfs the connective tensions among the 2011 Ecuadorian judgment, the 2014 US district court judgment, and the 2016 US court of appeals judgment. The task of distilling this morass is monumental. Indeed, with a case file rivaling that in Lago Agrio, the RICO litigation deserves a treatise of its own. Chevron's rendition of how the Lago Agrio ruling came about is well recited and now embossed in US law. This rendition says that the plaintiffs' legal team conjured up and succeeded in implementing a plot to bribe Judge Nicholás Zambrano into allowing them to ghostwrite his 2011 legal decision in their favor. The twists and turns are elaborate, and Chevron's story, which Judge Kaplan effectively renders in his near-five-hundred-page ruling, is persuasive. Persuasive, that is, if that is all one knows.

As I will argue in this chapter, there is a deep irony to Kaplan's ruling—a ruling that professes not only *jurisdictum*—the declaring of the law—but also, being juryless, *veredictum*—"the declaration of the truth of the matter in issue" (Black 1968: 1732).[5] Being both judge and jury, Kaplan dictated strict parameters around what would and would not be litigated in his courtroom. The case before him was a RICO case alleging bribery and ghostwriting. At issue was whether a foreign ruling finding Chevron liable for a sizable sum was, or was not, ill-gotten. It was not a claim concerning environmental contamination. Consequently, Kaplan barred from the trial proceedings any legal argumentation about Texaco's contamination in its former concession and promised to hold in contempt any party that sought to do so. The irony is that Chevron's strongest evidence of fraud hinged on contamination—and in fact, hinged on inaccurate, troublesome, and misdirected interpretations of that contamination.

As I will argue here, because the corporation's evidence of bribery and ghostwriting was circumstantial (and often bordered on fantastical), Chevron's corruption allegations recurrently referenced contamination—and indeed misreferenced its complex dimensionality—in order to generate "facts." Yes, there was the testimony of an alleged bribery scheme. This testimony was recounted by Lago Agrio ex-judge Alberto Guerra, whom Chevron had bought to the tune of over $2 million and whose performance on the witness stand reflected his hundreds of hours of being coached by corporate lawyers.[6] There were fuzzy photocopies of receipts containing blurry, hard-to-read signatures for two bank deposits of $1,000 each. There were parts of LAP documents appearing in Zambrano's ruling that purportedly were never submitted to the court. And there were decontextualized and creatively spliced video clips and emails of LAP counsel engaging in supposedly nefarious acts. These pieces of evidence, however, made for a very weak-linked racketeering ring hardly proving fraud. And Chevron had no smoking gun. The corporation had surfaced no version (full or partial) of the purportedly ghostwritten 2011 ruling, despite the fact that Kaplan had rescinded the LAP's attorney-client confidentiality privileges and had given Chevron access to the "universe" of over a decade's worth of internal legal strategizing correspondence, documents, musings, diaries, text messages, and more among LAP lawyers.[7]

In other words, Chevron had no "hard" evidence of fraud. And because of this, the corporation consolidated the appearance of a racketeering scheme by deploying select expert analyses. Generated by a scientific expert, one of Chevron's own expert counsel, and two digital forensic experts, this expertise consolidated a comprehensive truth of conspiracy, fraud, and corruption by dislodging the referent of its analysis (the condition of contamination) from its prior meaning and contexts. Significantly, the stability that Chevron sealed through this delocalized evidence hinged on misattributions, partial understandings, and misplaced graspings of the science of contamination and the intricacies of contamination within the Ecuadorian litigation.

With tragic poignancy, legal technique (or what Annelise Riles calls "the technical aesthetics of law" [2005: 976]) deployed and abused the strictures inherent in what a RICO filing entailed. That is, even though Chevron's filing of its case under the RICO Act directed Kaplan to excise from his court arguments about whether vast expanses of crude-laced pestilence haunts the soils, waters, and bodies of rainforest dwellers, a good part of Chevron's RICO arguments, I suggest, concocted a worlding that distorted and derided that condition.

This chapter expands on three moments prominent in the RICO trial in which Chevron generated facts of conspiracy, fraud, and corruption by variously enrolling misleading conclusions about processes deemed key to ascertaining crude contamination in the Ecuadorian litigation. Each moment served to consolidate Chevron's narrative of manipulation and ghostwriting. Like the electrons that detach from their originary atomic orbitals in order to forge a new molecular orbital, these moments created the coherent bonds of a LAP racketeering ring and thus consolidated Chevron's fraud worlding. That is, Chevron staged these three moments in Kaplan's courtroom to conjure evidence, where there was none, that would entrench the conviction that the corporation was the victim of racketeering. Were the US court to have had a more nuanced understanding of the chemistry of crude, or the intricacies of the Ecuador litigation and its judgment, or the multiple ways the chemistry of hydrocarbons had contravened prior contractual remediation, then it would most probably not have been persuaded by the opinions of Chevron experts and lawyers. That is, the trial moments that I discuss here warranted more discerning scrutiny.

The first moment addressed below revolves around David Russell, whom the LAP counsel hired at the onset of the Ecuador litigation as an environmental consultant. His relationship with the LAP lawyers long soured, Russell was called as a witness by Chevron to testify that scientific *conspiracy* was core to the LAP's litigating method. The second moment revolves around Nicolás Zambrano, who had traveled to New York to defend his 2011 judicial opinion. Chevron lawyer's staging of Zambrano's testimony upon cross-examination sought to demonstrate *fraud*—that Zambrano was not the author of the 2011 Ecuadorian ruling, given his ignorance of the most basic facts central to a judgment he claimed to have penned. The third moment concerns the provenance of the data that the author of the 2011 ruling purportedly used to calculate the largest portion ($5 billion) of the liability imposed on Chevron. The effect of testimony by two Chevron experts (James Ebert and Spencer Lynch) was to prove that *corruption* was at the heart of the largest monied category in the $9 billion liability. Specifically, Chevron's experts claimed (despite Zambrano's declarations to the contrary) that the Lago Agrio ruling relied on a report that the LAP lawyers had secretly orchestrated in order to substantiate a huge dollar penalty.

At each juncture, Kaplan found, in turn, conspiracy, fraud, and corruption—precisely the effect that Chevron lawyers sought to achieve. However, as I will show, these conclusions were not inevitable. A deeper understanding of the chemistry of crude, a better grasp of the logics of legal assessment in Ecuador, and a richer appreciation for the presence and distribution of contamination would have allowed Kaplan to reach different conclusions.

Analytically, Zambrano's 2011 ruling instantiated a delocalized stability by dislodging "evidence" from normative logics and scientific legitimacies in order to actualize the urgency of a deep relational context. Kaplan rendered his ruling by differently enacting the delocalized stability of an aromatic compound. Kaplan's ruling constituted evidence of fraud where it did not exist by subsuming elements within the confines of a racketeering ring and fraud worlding without realizing how in being unfastened from their complex rainforest relationalities these elements had been distorted to form something they were not. That is, Kaplan found fraud—even the truth of fraud—by synthesizing evidence within the constraints imposed both by him and upon him by the strictures of the RICO Act. As will soon become apparent, testimony and documents referencing the Lago Agrio litigation, once detached from their constitutive import and enmeshed in the realm of a RICO infringement, assumed (through the exquisite lawyering of corporate attorneys) a seemingly inevitable conspiratorial significance—one of con artists, fraudsters, swindlers. Thus was the power of a delocalized stability. With different court strictures and a sufficiently financed defense team, other arguments could have persuasively dismantled key portions of Chevron's corruption claim.

Conspiratorial Alchemy

David Russell was an environmental engineer. In 2003, the legal team then heading up the litigation against Chevron contracted Russell to formulate an estimate of what remediating Texaco's former concession would cost.[8] As the founder of an environmental remediation firm, Russell had years of experience assessing and overseeing the remediation of hydrocarbon-contaminated sites. Russell traveled to Ecuador in October 2003 as the litigation against Chevron was just beginning in the Lago Agrio court. Within a couple of weeks, he had a projected cleanup cost. The estimate, when released to the public, sent shock waves—$6 billion would enable the complete environmental remediation of Texaco's oil operations.

From the moment Russell's remediation estimate went public, Chevron declared the multi-billion-dollar price tag absurd. And it claimed that LAP lawyers were using this pseudo-scientific assessment to threaten Chevron and extort a settlement. Ten years later (in collaboration with a David Russell who had become disgruntled with his former LAP clients), the $6 billion cleanup estimate proved crucial in setting the stage for Chevron's corruption narrative. The worlding intrigue that Chevron elaborated (in alliance with Russell) claimed that LAP lawyers exaggerated the presence of contamination by creating evidence where it was not. The goal was to depict LAP lawyers as deeply unscrupulous. And so began the opening chapter of Chevron's smear campaign against the entire LAP legal and scientific team, focusing specifically on Steven Donziger. Significantly, Chevron's claim of how the LAP exaggerated the extent of Amazonian contamination hinged on misrepresentations of the chemistry of crude.

Poised on the RICO witness stand in mid-October 2013, Russell related his version of events that transpired ten years prior. The broad outlines of Russell's story were detailed in his "witness statement," crafted weeks earlier under the guidance of Chevron's lawyers. In October 2003, Russell spent a week-plus in and around Lago Agrio surveying alleged contaminated sites and then a few days holed up in the Gran Lago Hotel calculating an estimated cost of a comprehensive environmental cleanup. However, as the LAP's "chief environmental scientist," Russell soon realized that his $6 billion estimate was "wildly inaccurate and had no scientific data to back it up."[9] The lawyers for the LAP meanwhile had paraded the $6 billion tab through the media as if it were "an accurate, scientifically supported cost estimate" in the hopes that making "exaggerated claims publicly" would "pressure Chevron into a settlement."[10] Singling out Donziger, Russell argued that the LAP lawyers "did not really care about the scientific or technical evidence" and in fact "didn't care about evaluating actual conditions in Ecuador and determining a scientifically-based cost estimate."[11]

On the witness stand, Russell's story came damningly alive. Teasing from his memory, Russell reckoned that he saw forty-five sites—some of which he physically inspected (i.e., assessed on foot to gain a sense of their size, concentration, engineering, and hydrology) and some he eyed out a car window at speeds of forty to fifty miles an hour. Calculating a number of assumed parameters, Russell devised his estimate.

During cross-examination, a lawyer for the LAP defense asked Russell why he called his $6 billion remediation projection "very rough." To which Russell responded, "There are a large number of unknowns. While I attempted to define unit costs very carefully—and by unit costs I mean, for example, the cost to produce a unit of water or something similar—the quantities were . . ." Russell's voice trailed off and he paused. With dramatic effect, he continued: "I used the word 'SWAG'—which I won't expand upon but to say it's a scientifically based wild guess. . . . [Given] the amount of unknowns and the lack of information that I had with regard to not only levels of contamination but [also] the extent of those levels of contamination, it [was] a best estimate." Kaplan interjected.

COURT: "SWAG" is an acronym, right, in common use?

RUSSELL: Yes, sir.

Q: And what do the letters S-W-A-G stand for?

A: Scientific Wild Ass Guess.[12]

The courtroom buzzed with murmured chuckling.

Continuing his story, Russell apparently became disillusioned in late 2004 with the cost estimate he had produced a year earlier. In an email to LAP counsel, he stated, "We cannot provide anyone with a realistic cost estimate" (as if, at that point, that was news) because "we don't know the extent of the soil contamination or the magnitude or the extent of the groundwater contamination."[13] On the stand, Russell expanded: "I believe, to the best of my recollection, that I indicated that we could not provide any sort of accuracy for a cost estimate other than the SWAG that I had provided in 2003."[14] As of late 2004, the judicial inspections, he recalled, were not producing "analytical results consistent with high levels of contamination which may have been in need of remediation."[15] In fact, the "gathered data [seemed] to indicate that the costs appeared to be substantially lower."[16]

Minutes later, Chevron's lawyer asked Russell once more why he was "disillusioned" with his initial estimate and was claiming that the cleanup price tag was bogus. Russell responded: "The amount of data that I saw had indicated that it did not square up to . . . it was not representative of the assumptions that I made back a year earlier by that point—[with] the levels of contamination we might expect . . . would be in need of remediation. So, the contamination was not there."[17]

Delving deeper, Chevron's lawyer extracted an exhibit—an email Russell sent dated November 4, 2004. The email referenced a meeting in New York between him and senior LAP lawyers (Cristóbal Bonifaz, Alberto Wray, and Steven Donziger) in which they discussed results from the chemical analyses performed on samples extracted from the first judicial inspections in Ecuador. Addressing Russell on the witness stand, Chevron's lawyer asked him to explain a phrase in the email: "the analysis for BTEX and GRO would be counterproductive to the case."[18] Russell responded: "the data that we were finding from some of the analysis . . . the fact that we're finding BTEX (which is benzene, toluene, ethylbenzene, and xylene) and GRO (which is gasoline range organics) are much more indicative of contamination from Petroecuador rather than from Texaco because these compounds are volatile and degrade quickly in hot, wet, warm environments such as in the jungle."[19]

Kaplan followed up: "Now, first of all, remind me, please, what the acronyms mean."[20] Spelling out the molecular compounds, Russell explained that they "are classified as aromatic hydrocarbons [and] have some immediate health effects on exposure. For example, benzene is a carcinogen."[21] At which point, Russell oddly contradicted himself: "at the time, [the data] did not show any of these compounds at the types of levels which would be indicative of any sort of health effects."[22] He fumbled over his words. "I'm sorry," he continued, "the presence—we *were* finding these compounds. There we go. I'm sorry. I was confused. We found BTEX and GRO, and that was indicative of recent contamination rather than contamination which would have been ten or perhaps twenty years old from Texaco."[23] And did a specific directive emerge from the meeting? Russell: "We stopped analyzing for those compounds. We started instead substituting a less reliable measure which was total petroleum hydrocarbons [TPH]. . . . The analytical problem with TPH is that TPH methods currently in use can show up naturally occurring [TPH] compounds as an indication of petroleum, so give you a false positive."[24]

In Kaplan's analysis, Russell professed truths. As such, Russell's testimony, with its purported scientific backing, served as a foundational starting point for Kaplan's 2014 ruling delegitimizing the $9 billion liability that the Ecuadorian court had imposed on Chevron. In Kaplan's 485-page judgment, he felt that duplicity, conspiracy, and extortion pervaded the actions of the LAP counsel. And Russell provided the fodder for how this scientific duplicity and deceit began. Kaplan's juris-diction states that Russell's $6 billion "SWAG estimate" was a "drive-by" "guesstimate" and used as "a key weapon . . . to exert pressure on Chevron and convince the company—and the world—that the damages in the Orienté [*sic*] were substantial and the threat of an enormous judgment against it was real."[25] Furthermore, the LAP "avidly used Russell's $6 billion figure in the media to generate leverage," all the while knowing "that it could not withstand serious analysis."[26] Clear extortion. Invoking Russell's courtroom testimony, Kaplan spun a tale of the LAP lawyers becoming aware that their scientific analyses were incriminating the wrong entity—Petroecuador instead of Chevron. The directive? The LAP technical team "'stopped analyzing for those compounds [BTEX and GRO, and] started instead substituting a less reliable measure which was total petroleum hydrocarbons,' or TPH."[27]

Misunderstanding the chemistry of crude, Kaplan improvised: "The methods the team used to test for TPH, however, were unable to distinguish between TPH attributable to recent activity and activity that occurred a considerable period earlier."[28] "Moreover, they were subject to a further problem, namely that 'TPH methods currently in use can show up naturally occurring compounds as an indication of petroleum, so give you a false positive.'"[29] Kaplan's implication was that the LAP lawyers conspired such that their technical team engaged in a racketeering scheme of dubious and deceitful scientific analysis.

A profusion of false and confused half-truths clouds Kaplan's legal assessment. Where to begin? Perhaps first by pointing out that with respect to the 2011 Ecuadorian ruling, Russell was insignificant. As was his 2003 $6 billion clean-up estimate; that calculation was never submitted to the Ecuadorian court as a piece of legitimate science. Furthermore, contrary to Kaplan's understanding, Russell was not *the* "chief environmental scientist" for the LAP. Rather, Russell served that role for fewer than six months between July and December 2004. At that point, the judicial inspections were just getting underway and, consequently, Russell's access to laboratory analyses of soil and water samples from the inspections as a whole was limited, making his scientific assessment (even if it were correct) remarkably partial. One might assume, given all the attention Kaplan paid to Russell in his ruling (citing him by name ninety-six times), that Russell and his $6 billion estimate were crucial to the LAP and consequential to the 2011 Ecuadorian ruling. Neither is the case. The $6 billion cleanup estimate was of no consequence to the Ecuadorian litigation or judgment—and no one in Ecuador was under the illusion that it was. The $6 billion figure was one environmental engineer's estimate, quickly generated in building a legal and public relations strategy. All scientific analyses that were submitted to the Lago Agrio court needed to emerge from the judicial inspections, which only began in August 2004. The LAP lawyers knew, just as well as Russell, that those inspections needed to be completed first in order to garner an accurate understanding of the effects of Texaco's operations.

Furthermore, given the authority Kaplan invested in Russell's words, one would assume his scientific assessment—as articulated above—was accurate. It is off. Let me explain.

By late 2004 and early 2005, the time period in which Russell was refuting the validity of his $6 billion estimate, he had (at the most) reviewed the chemical analyses of four judicial inspections (Sacha-6, Sacha-21, Sacha-94, and Shushufindi-48)—four out of a then-anticipated 112 total inspections. But Russell's contention that the LAP were encountering high levels of BTEX and GRO does not square with the historical specificity of those four sites— each of which generated (as was always the case) different and distinct data. One well, Sacha-94, which had five waste pits associated with it, was drilled and operated exclusively by Texaco. Texaco drilled Sacha-94 in spring 1981, brought the well into production in autumn 1982, and closed the well in February 1985. Texaco drilled Sacha-21 in autumn 1971, started production in summer 1972, and Petroecuador closed the well in December 1995—nearly ten years prior to the judicial inspections. Consequently, it would have been impossible for samples from all four sites to have evinced BTEX and GRO as Russell suggested.[30]

It would seem discussions among the LAP legal and scientific team on testing for petroleum hydrocarbons in former Texaco oil operations were more complex than Chevron's and Russell's worlding of connivance and conspiracy was letting on. Recalling the basic chemistry of crude oil helps crystallize some matter out of the murk that Russell's testimony generated. For context prefigures what types of analysis may more accurately capture the complexity of contamination.

Crude oil, one might recall, is made up of several hundred different hydrocarbon compounds. Among these hundreds of hydrocarbon molecules, only a couple of dozen have been sufficiently studied to determine their toxicological capacities, the majority being aromatics composed of twenty-four or fewer carbon atoms. BTEX are the lightest of these toxic aromatics, each composed of one benzene ring, while the rest are within the class of PAHs (polycyclic aromatic hydrocarbons), meaning they are composed of more than one benzene ring.

Over time and across space, crude oil changes, sometimes dramatically, both chemically and physically such that the crude that persists in the environment years after initial extraction is not the crude that existed initially. What persists is by and large hydrocarbon compounds with heavy molecular weight. Analyses that test for BTEX, GRO, or even DRO (diesel range organics)—hydrocarbon compounds comprising six to twenty-four carbon atoms (C_6 to C_{24})—cannot by and large account for these larger molecules. As such, testing for these three groupings—precisely the hydrocarbon ranges for which Chevron tested—could not account for the range of hydrocarbons in pits that were decades old. As discussed in chapter 1, those heavier, more complex, and previously thought inert aromatic hydrocarbons are much more problematic than assumed. While the regulatory and industry science sought to tighten and refine the assessment of harmful hydrocarbons by testing for particular fractions, that very effort dismissed, even undermined, the recognition that heavy hydrocarbon compounds are more biologically available than previously anticipated and are far more detrimental to cellular function than heretofore scientifically understood. Given the sites and context in which contamination exists in the Ecuadorian Amazon, Kaplan's contention that testing for TPH was "less reliable" is simply misinformed.[31]

Because the soil and water samples extracted during the inspections came from waste pits associated with oil wells drilled between 1967 and the mid-1980s, the crude oil—along with the formation waters, solvents, and drilling muds—still residing within those pits was old. As, too, were the hydrocarbons and affiliate drilling wastes that had seeped into the surrounding soils and rivulets. Given that twenty- to forty-year-old crude was a primary source of alleged suffering among the LAP, it made little sense to test for the presence of BTEX or GRO. Russell knew that. As he noted, these hydrocarbons are volatile; when exposed to oxygen, volatile organics dissipate and break down. There was nothing nefarious about the LAP lawyers choosing to focus on TPH. To be sure, the raw data from the laboratory tests that the LAP scientific experts ordered did indeed enumerate the levels of BTEX, the sixteen PAHs, and thirteen fractions found in all soil and water samples they extracted. Binder after binder in the Lago Agrio case file attest to the reams of chemical panels generated from laboratory assays that included data on light aromatic hydrocarbons. As would be anticipated from decades-old sites, these levels were by and large negligible. The LAP technical reports focused on what they deemed the crucial cause for alarm—TPH and metals. TPH *was* the regulatory index for hydrocarbon contamination in Ecuador. Russell's suggestion that background ("natural") levels of TPH would render a "false positive" is specious; given the crude-tainted soils from which samples were extracted, that would never be the case. And he knew that too.

Kaplan's contention that a conspiratorial intent informed the LAP lawyers' decision to test solely for TPH, instead of BTEX and GRO, is, frankly, problematic. Even if that is what they did—which it was not—one could state with more confidence that Chevron's testing method (determining TPH levels by adding GRO-TPH and DRO-TPH) underestimated the degree of hydrocarbon contamination in Texaco's former oil concession. In a footnote, Kaplan noted that one witness provided a perspective contrary to Russell's. This was Steven Donziger, toward whom Kaplan was less than sympathetic. Kaplan's footnote reads as follows:

Donziger testified that "the conclusion of the conversation [in Manhattan] was that if we were looking for a sample analysis that would more precisely evidence the scope of Texaco's contamination, testing for total TPH was the more appropriate test to use. . . . Accordingly, we adopted a focus on sampling for TPH rather than BTEX or GRO, although we kept a balanced portfolio of chemical analyses." [DX 1750 (Donziger Direct) ¶114.]

The Court does not credit this testimony. It is contrary to Russell's testimony on this technical point, a point on which his testimony was not challenged. Donziger, for reasons discussed below, is not a credible witness.[32]

Delegitimizing Juris-Diction

Randy Mastro has a skewering reputation. Chevron needed a mastermind to launch its New York counteroffensive legal strategy. Gibson, Dunn & Crutcher was the right place for Chevron to be. And Mastro, a seasoned senior partner, was the right person for the job. Before joining Gibson Dunn in 1998, Mastro worked for New York City mayor Rudolph Giuliani, where, as deputy mayor, he spearheaded a number of "initiatives to remove organized crime."[33] And prior to that, he served as assistant US attorney in the US attorney's office for the Southern District of New York, where he prosecuted a number of mafia cases, allegedly dodging death threats.[34]

On a frigid Tuesday morning in November 2013, Mastro cross-examined Zambrano in the US courtroom. Zambrano had arrived in New York the Friday before, whereupon that evening a lawyer representing the Republic of Ecuador briefed him on US legal procedures and his imminent two-day (Saturday and Sunday) deposition by Gibson Dunn lawyers. Zambrano had never been to the United States. He did not speak or understand English. He was unfamiliar with the practices and procedures of the common law tradition. But the harsh cold melted like butter on his skin in comparison to the icy welcome Mastro soon delivered. The story, of course, was already written—Nicolás Zambrano was "the Ecuadorian judge who claimed to have authored the fraudulent Ecuadorian judgment . . . but who, instead, allowed the judgment to be ghostwritten in exchange for a bribe."[35]

In the courtroom, Mastro could be theatrical, intimidating, and downright wicked. As one peer joked (according to Mastro's online corporate biography), "You do not want to meet Randy down a dark alley. But you REALLY don't want to meet him in a lighted courtroom, . . . going against him must be like wrestling an alligator."[36] In Chevron's RICO claim, Mastro's goal was to undermine Zambrano's credibility. His chosen saboteur's tactic was to conduct a pop quiz about the 188-page, single-spaced ruling Zambrano purportedly wrote nearly three years earlier.

Mastro laid the foundation: "Did you work long hours to prepare the judgment, sir? . . . Did you pour your heart and soul into working on that judgment, nights, weekends? . . . And nobody helped you do the research you needed to do to write and author the judgment, correct, sir? . . . Thank you, sir. So you know that judgment because you authored every word, correct?"[37] Then came the real questions.

QUESTION I

MASTRO: So, sir, please tell us what the judgment says on page 107 is "the most powerful carcinogenic agent considered in this decision." Please tell us what that was. Sir, I don't want you to look at the judgment. . . .

[Two lawyers for the LAP team raised objections. Kaplan overruled letting them speak.]

MASTRO: Sir, do not look at the judgment. Please tell us . . .

COURT: The record will reflect that the witness has been leafing through the document.

MASTRO: Please tell us what the judgment says is "the most powerful carcinogenic agent considered in this decision."

[Once more a lawyer for the LAP objected.]

COURT: I've already ruled. . . . Sir, the fact of the matter is I'm right here and when the question was first put, the witness reached in front of him and picked up the exhibits before him which include the judgment.

Now, I'll stand amended to this extent. I'm not sure which piece of paper he was leafing through because there's a big pile of them, but one can draw inferences.

MASTRO: Please, sir, answer that question.

[. . .]

ZAMBRANO: First of all, I grabbed the judgment but it's in English. And I tried to look for it here but I wasn't even able to find the page.

COURT: OK.

ZAMBRANO: So, therefore, it's not that I reviewed even that page. And regarding what I'm being asked, I don't recall exactly, but if you give me the names, perhaps I could remember. [Query by interpreter.] The hexavalente is one of the chemicals that if it is exceeded in its limits, it becomes cancer causing, carcinogenic.

INTERPRETER: And just for the record, the witness used the word E-X-A-V-A-L-A-N-T-E.

[Kaplan intervenes to inform Zambrano on US courtroom etiquette: he is not to speak beyond the questions that are being asked.]

MASTRO: So, Mr. Zambrano, just to be crystal clear, can you tell me what substance the judgment says is, quote, "the most powerful carcinogenic agent considered in this decision," yes or no?

ZAMBRANO: I don't recall.[38]

QUESTION 2

MASTRO: And can you tell me, Mr. Zambrano, what report the judgment says at page 134 is the "statistical data of highest importance to delivering this ruling," yes or no? Yes or no? He's leafing through the document, Your Honor.

COURT: Mr. Zambrano, stop! Put the document down and answer the question. The record will reflect that the witness was going through the document. Would you hand me what you were going through please, Mr. Zambrano? The witness has just handed me Plaintiff's Exhibit 400, English.

MASTRO: He also has 399.

COURT: I understand. That's what he handed me. That's what I saw him looking at. Now, I think it might be better for this line of questioning to take the documents away from the witness stand and end this controversy.

[LAP legal representatives once again object.]

MASTRO: Mr. Zambrano, I repeat my question. Can you tell us, sir, what report the judgment says is the "statistical data of highest importance to delivering this ruling?" Yes or no?

ZAMBRANO: Yes.

MASTRO: What is that, sir?

ZAMBRANO: The report by expert Barros.

[. . .]

MASTRO: Sir . . . I'm going to hand you Plaintiff's Exhibit 399, which is the Lago Agrio Chevron judgment in Spanish. . . . Sir, isn't it a fact that the report, statistical data of highest importance to delivering this ruling, is the San Sebastián "Cancer in the Amazon in Ecuador"? Do you see that, sir, page 134, study entitled "Cáncer en la Amazonia Ecuatoriana, Cancer in Ecuadorian Amazonia," that's what the judgment says is the report of the statistical data of highest importance to delivering this ruling, correct, sir?

ZAMBRANO: Yes.[39]

QUESTION 3

MASTRO: . . . Sir, can you tell us what theory of causation the judgment says its author agrees with on page 88 of the judgment? Do not look at the judgment, sir. Do not look at the judgment, please. Can you tell us without looking at the judgment what theory of causation the judgment says its author agrees with on page 88 of the judgment? Yes or no, sir.

ZAMBRANO: I don't recall.[40]

QUESTION 4

MASTRO: Isn't it a fact, sir, that when I deposed you this past weekend, you couldn't even tell me what TPH stands for, correct, sir?

[A LAP legal representative objects, and the court sustains the objection, stating that Mastro should provide the relevant page of the deposition.]

[. . .]

MASTRO: Thank you, Your Honor. Zambrano deposition, page 30, lines 13 through 21:

Q: Tell us what TPH stands for, sir.

A: I'm trying to. Well, it pertains to hydrocarbons, but I don't recall exactly.

Q: It is a simple question, sir: What does TPH stand for? Do you know or not?

A: Yes.

Q: Do you want to tell us what TPH stands for?

A: No.[41]

[Toward the end of the day, Mastro interjected one more pop-quiz question.]

QUESTION 5

MASTRO: Sir, do you know what the English word "workover" means?
[...]

ZAMBRANO: Could you please write it down for me so that I can see what it is?

MASTRO: I'd be happy to. Your Honor, I will approach the witness.

COURT: Make your handwriting better than mine.

MASTRO: So that everyone can see, I have written the word "workover." [Having printed the word in large block letters, Mastro fanned the sheet of paper toward the gallery like a magician teasing his audience with one last trick.]

MASTRO: Sir, can you tell us what that English word "workover" means?

ZAMBRANO: I don't know what it means in English.

MASTRO: Sir, I'd like to show you page 21 of the judgment on the left—

ZAMBRANO: But.

MASTRO:—and page 5 of the draft "fusion memo" [a purported proprietary document from the plaintiffs] on the right. Did you know before comparing these two documents today that the highlighted text in these documents is the same?

ZAMBRANO: Could you please repeat the question?

MASTRO: Did you know before comparing these two documents today that the highlighted text in these documents is the same, yes or no, sir?

ZAMBRANO: No.

MASTRO: Sir, directing your attention to the highlighted word "workover." Do you know how it is, sir, that an English word that you couldn't even identify in court today ended up appearing in the draft "fusion memo" and then in the judgment?
[...]

ZAMBRANO: No.[42]

The afternoon press squealed over Zambrano's ineptitude and seeming confusion over basic information: (1) "benzene" was the "most carcinogenic agent"; (2) a study by San Sebastián contained the "statistical data of highest importance"; (3) "the theory of sufficient causation" was *the* "theory of causation"; (4) TPH stands for "total petroleum hydrocarbon"; and (5) the term "workover" refers to the work of maintaining an oil well. Mastro proved his case. In his ruling, Kaplan declared that "the Court finds that Zambrano did not write the Judgment."[43]

My concern here is not Mastro's trial tactics; while masterful, these tactics are not unique to the RICO litigation. Yes, he engaged in arguably devious lawyering, performing what a number of legal anthropologists (Jacquemet 2009) observe as practiced cross-examination techniques. Yes, Mastro excelled in reducing answers to yes/no, controlling the significance of a witness's speech, casting doubt on a witness's competence and candor, and producing inconsistencies in a witness's responses by posing convoluted questions that made translation imprecise.

What concerns me deeply, however, is how Mastro's performance rendered a distorted understanding of the Ecuadorian litigation and its constitutive dimensions. Below, I decompose each question to show how Kaplan's conclusion could have been otherwise. Specifically, Mastro's tactics are profoundly disturbing on two accounts. First, Mastro posed his deceptively clear-cut questions as if they were key to the Ecuadorian litigation, and he implied that Zambrano's ineptness on the stand unquestionably indicated that he did not know the ruling. Yet, as I show below, the questions that Mastro posed were neither central to the Ecuadorian proceedings nor significant in the 2011 ruling. Second, Mastro paraded his "pop quiz" to demonstrate Zambrano's incompetence and to demonstrate that Mastro knew the "correct" answers. Yet this was largely not the case. Although Mastro's answers correctly mirrored specific words printed on specific pages in the text of the 2011 judgment, they were far from correct in the context of the Ecuadorian litigation. Rather, Zambrano's responses, which to Mastro and Kaplan denoted ignorance, far more clearly resonated with the substance of the Ecuadorian proceedings and ruling. So a little analysis.

QUESTION I (DECOMPOSED)

". . . Please tell us what the judgment states on page 107 is 'the most powerful carcinogenic agent considered in this decision.'"

Zambrano's response was, "Hexavalente." Granted, this is not a complete response, but Zambrano was correct. In fact, only someone who understood the ruling would have said this. Benzene—which, by the way, has a hexagonal planar structure—is indeed universally understood to be the most carcinogenic compound in crude oil, and as noted earlier, the oil industry has long known of the danger that benzene posed, despite working for decades to foil its regulation. But in considering the litigation in Ecuador as a whole, hexavalent chromium proved to be the most ubiquitously carcinogenic agent found during the judicial inspections. Chromium VI (as it is also known) is not a derivative of crude oil but rather is found in the drilling muds and solvents used in the drilling for and extracting of oil. Its presence in soil samples extracted during the inspections was deemed highly toxic. By contrast, only rarely was benzene present in samples.

More disturbing, however, is the implication of Mastro's question. It suggests the judgment determined that one agent was distinctively more dangerous than others in reaching its decision. That was not the case. Nor was it implied in the larger written context from which Mastro extracted his questioning phrase. This particular sentence comes after a long discussion assessing the consensual scientific understanding of distinct hydrocarbon compounds. The sentence in Zambrano's original judgment reads as follows: "Having said this and in view of the danger [posed] by certain contaminants, we begin a grid of reference by [looking at] sample results that contain levels of benzene, noting that benzene is soluble in water, and that although it can be found naturally in the environment, is the most powerful carcinogenic agent of those evaluated in this judgment."[44]

At this point in his ruling, Zambrano was setting up a framework through which to assess a multiplicity of compounds evinced in the chemical analyses of soil and water samples extracted during the inspections. He was not making a declaration about contaminated conditions or claiming that benzene was a component of those conditions. Fourteen sample results (all from samples extracted by Chevron experts) did register detectable levels of benzene. In the context to the 64,000 chemical-sampling results examined, the presence of benzene did not figure in assessing contamination in former Texaco industrial sites, and it was not a factor in Zambrano's judicial decision. Mastro's staging wanted Kaplan and his courtroom audience to believe that Zambrano did not even know the name of the most carcinogenic agent that informed his purported ruling. If such an agent existed, it was hexavalent chromium.

QUESTION 2 (DECOMPOSED)

". . . What report the judgment says at page 134 is the 'statistical data of highest importance to delivering this ruling?'"

Zambrano's response was, "The report by expert Barros." According to Mastro, that answer was wrong. The correct answer, in Mastro's mind, was a study by San Sebastián that analyzed the relationship between cancer and residential proximity to oil operations (discussed in chapter 2). However, were one to reflect on the 2011 ruling as a whole—instead of extracting a sentence and demanding its completion—Zambrano's reply is correct. The expert report by Geraldo Barros (an expert nominated by Chevron) was of ultimate importance in determining the judgment. As I discuss in further detail below, Barros's work played a key role in establishing the largest chunk of monetary liabilities—US$5,396,160—within the ruling. Barros's reports were essential in deriving the total cubic volume of contaminated soil and calculating its remediation cost. If the aim was to undermine Zambrano's credibility and prove him wrong, then Mastro is correct; Zambrano did not fill in the blank in the sentence on page 134 correctly. If, however, the aim is to appreciate Zambrano's broad understanding of the Ecuadorian judgment as a whole, then Zambrano's response is both accurate and vastly more insightful.

Zambrano does make note of the work of San Sebastián and colleagues, but he does so in the space of six pages, citing him six times. By contrast, Zambrano carefully references and uses the work of Barros in reaching his ruling, citing Barros twelve times over a spread of ninety pages (on pages 91, 99, 122, 125, 131, 180, 181, 182, and 184). Therefore, in multiple senses, Barros's work was of the highest statistical importance in determining the 2011 Ecuadorian ruling.

QUESTION 3 (DECOMPOSED)

"Sir, can you tell us what theory of causation the judgment says its author agrees with on page 88 of the judgment?"

Zambrano's response was, "I don't recall." The correct answer, according to Mastro, was "the theory of sufficient causation." Page 88 of the 2011 Ecuadorian ruling reads as follows (according to Chevron's translation, and I add my own emphasis here): "4. Theory of sufficient causation. This theory, *with which we agree*, is the one toward which the majority of writers on legal doctrine and the case law of foreign courts are inclined."[45]

Effective reading, however, always entails more than pronouncing words. Scratching only the surface demonstrates that Chevron's attorney was confused. Contrary to Mastro's contention, the author of the 2011 ruling is not voicing his own opinion on page 88, taking a stand, or arguing a point. Rather, he is quoting the voices of others. The "we" who "agree" does not refer to Zambrano but instead to the judges of Ecuador's highest court, Supreme Court of Justice, First Chamber of Civil and Commercial Claims. As mentioned in chapter 5, the Supreme Court's 2002 ruling in *Delfina Torres Vda. de Concha v. Petroecuador* is one of the most-studied and most-cited recent Ecuadorian legal decisions. And the Lago Agrio judgment recites substantial portions of the ruling word for word—sometimes citing "properly" and sometimes not.

The longer passage from which Mastro's question is extracted reads as follows:

> Theory of sufficient causation. This theory, with which we agree, is the one toward which the majority of writers on legal doctrine and the case law of foreign courts are inclined. It consists in leaving the analysis of the matter of when the harmful action is likely to generate liability on the part of the perpetrator of the harm in the hands of the judge, which means that any general rule can be ignored and trust is placed in the discretionary powers of the judge. Finally, according to this theory, before anything else, a criterion must be set to establish liability in an objective analysis related to the external character that links it to the causal nexus.[46]

Although the quote in Zambrano's original judgment is not appropriately closed at "nexus," quotation marks do signal the beginning of this specific discussion of four theories of causation.

The larger section—over ten pages of the 2011 judgment—from which Mastro extracted this pop-quiz question served to establish a *method* (as discussed in the previous chapter) through which Zambrano might arrive at his ruling. Zambrano was not outlining *his* own opinion. Instead, in this section of his judgment, Zambrano thought it wise to acknowledge "what legal doctrine says about criteria for judging liability" and to examine "different theories" before "reviewing the evidence."[47] Contrary to what Mastro's pop quiz implied, "the theory of sufficient causation" was not *the* theory upon which the Ecuadorian opinion rested. As chapter 5 demonstrated, what most intrigued Zambrano—the theories he ultimately put most to use—were European-influenced emergent theories in Latin America concerned with "the unjustified risk of a hazardous situation" and the "continuation of the harmful behavior."[48]

QUESTION 4 (DECOMPOSED)

"Tell us what TPH stands for, sir."

Zambrano's response was "hydrocarbons, but I don't recall exactly." In Ecuador, the word "hidrocarburos" is regularly used to refer to fossil fuels, primarily petroleum. Every schoolchild in Ecuador learns that hidrocarburos are important natural resources, whose extraction is facilitated by the Secretaría de Hidrocarburos, which oversees approximately 35 percent of the country's exports. There is no question that Zambrano understood that TPH referred to the presence of crude oil. The English acronym TPH does appear throughout Zambrano's ruling in talking about chemical readings (i.e., "900,000 mg/kg TPH"). Thus while Mastro sought to underscore Zambrano's ignorance, Zambrano's response draws attention to his awareness of what was at stake in the acronym, and his lack of dexterity with English.

QUESTION 5 (DECOMPOSED)

"Sir, do you know what the English word 'workover' means?"

Zambrano's response was that he did not know. The term "workover" appeared four times in the English translation of Zambrano's ruling. It appears only twice in the Spanish original. Each time it appears in the original, it is in quotation marks with a translation offered after its first appearance: "'workover' (*soporte y mantenimiento*)."[49] Two points are noteworthy. First, in Zambrano's ruling, it is not the term that is important, it is the context that is critical. On the two consecutive pages on which "workover" appears (pages 20 and 21 of his judgment), Zambrano traced correspondence in the 1970s and early 1980s between US-based executives of Texaco. These internal Texaco memos had been obtained by LAP lawyers in the mid1990s when New York district court judge Broderick permitted the Ecuadorian plaintiffs discovery of some Texaco documents. These in-house company memos and correspondence formed part of the LAP's larger legal argument for piercing the corporate veil, an essential part of being able to hold a parent corporation legally responsible for the actions of its subsidiary. Zambrano used these documents in his judgment to demonstrate that lower-tier executives repeatedly solicited permission and monies from higher-tiered executives to engage in distinct oil activities in the Ecuador concession.

Mastro argued that the term "workover" and a section of text in Zambrano's ruling also appeared in an internal LAP memo called the "fusion memo." Because, according to Mastro, the "fusion memo" was (allegedly) never submitted to the Lago Agrio court, he maintained that the only way for it to get into Zambrano's ruling was if the LAP legal team put it there. *Fusión* in Spanish means "merger," and the "fusion memo" was the LAP's legal response to Chevron's argument that the Ecuadorian plaintiffs had sued the wrong entity—that is, that they should have sued Texpet or, barring that, Texaco, which retained its juridical personality even after the ChevronTexaco merger.

The contents of the fusion memo were no secret. The LAP legal team began articulating its arguments for lifting the corporate veil—the filaments that separate a corporation from its shareholders and, consequently, a subsidiary from its parent company—beginning on the first day of the trial in October 2003. At the Aguarico-2 judicial inspection on June 12, 2008, the LAP lawyers enumerated the argument laid out in the merger document and submitted the "fusion memo," along with a number of accompanying exhibits, to the Lago Agrio court.[50] Although the exhibits attached to the memo were properly filed, it appears that the memo itself was misfiled and according to Chevron, it did not exist in the 230,000-page case file. For better or worse, the court record was rife with errors in relation to the inspection on June 12—where unnumbered and misnumbered pages interrupted the narrative, and sections from tens of thousands of pages earlier in the court record reappeared in the legal file regarding that June day.[51]

That the memo was misfiled or missing in the official court record speaks as much to clerical error by a court that was overwhelmed with the immensity of the litigation against Chevron as it does to the LAP's suspicions that subterfuge was afoot.[52] It was not uncommon for parties to submit documents after hours under the judge's or secretary's door, whereupon, though registered, they became lost in paper stacks until later. Given that the memo was a key component of the LAP's response to Chevron's interminable arguments about the sanctity of the corporate form, there is no reason to believe the memo was not submitted to the court and every reason to believe it was. None of these considerations, however, entered into Kaplan's deliberations.

In his 2014 judicial decision exonerating Chevron and condemning the LAP, Kaplan wrote,

> Chevron contends that Zambrano did not write the Judgment, that the LAPs prepared it, and that the LAPs bribed Zambrano to decide the case in their favor and to sign the judgment they had prepared. The evidence concerning those contentions and its analysis are [sic] extensive. The Court here summarizes its findings before proceeding to the detailed discussion of how it reached them.

The first major point is that the Court finds that Zambrano did not write the Judgment, at least in any material part. . . .

In Part IX.A, the Court examines Zambrano's trial testimony and finds that it was not credible. Zambrano neither could recall nor explain key aspects of the 188-page opinion despite his claim that he alone wrote it. He was a new judge with very little civil experience, so much so that he admittedly had another former judge ghostwrite orders for him in civil cases. He was unfamiliar with—and on occasion bewildered by—certain of the most important concepts and evidence with which the opinion dealt. His testimony was internally inconsistent and at odds with other evidence in the record. He was an evasive witness. Finally, Zambrano had economic and other motives to testify as he did. His livelihood, what remains of his reputation after having been removed from the bench, and perhaps even his personal safety hinged on his protecting the legitimacy of the $18 billion Judgment [the amount obtained when Chevron did not publicly apologize] by claiming authorship. . . .

Zambrano testified at trial. He claimed that he "was the one who exclusively drafted" the Lago Agrio Judgment, that "no one . . . helped [him] to write the judgment," and that he did all the research for the Judgment. He flatly denied that he considered anything that was not in the official court record.

The Court rejects Zambrano's claim of authorship, let alone sole authorship, as unpersuasive for a host of reasons.

1. Zambrano Was Unfamiliar with Key Aspects of the Judgment He Signed

Even at the most general level—that is, without considering the inconsistencies between Zambrano's deposition (taken days before his trial testimony) and his trial testimony, the internal inconsistencies in his trial testimony, and the inconsistencies between his testimony and other evidence—Zambrano was a remarkably unpersuasive witness. As an initial matter, Zambrano was unable to answer basic questions about the Judgment that he ostensibly wrote and that he came to New York to defend.

The Judgment states that "benzene . . . is the most powerful carcinogenic agent considered in this decision." But when Zambrano was asked "what substance the judgment says is, quote, the most powerful carcinogenic agent considered," he could not recall. Instead, he said that "[t]he hexavalente is one of the chemicals that if it is exceeded in its limits, it becomes cancer causing, carcinogenic."

Zambrano was asked also which report the Judgment stated is "statistical data of highest importance to delivering this ruling." He responded "[t]he report by the expert Barros." But the Judgment stated that the "Relative Risk established in" the study entitled Cáncer en la Amazonia Ecuadoriana "is statistical data of highest importance to delivering this ruling." . . .

Zambrano was unable also to recall the theory of causation on which the Judgment relied. And, although the English word "workover" appears twice in the Judgment, Zambrano testified that he does not speak English, did not know what "workover" means, and could not explain why the word was in the Judgment.

TPH—which stands for total petroleum hydrocarbons—appears over 35 times in the Judgment. Indeed, the Judgment awards plaintiffs over $5 billion for TPH cleanup. But when Zambrano was asked at his deposition what TPH stands for, he testified that "it pertains to hydrocarbons, but I don't recall exactly."

Zambrano's inability to recall every detail of a 188-page decision of course would not itself prove that he had not written it. But the aspects of the Judgment he was unable to recall were not insignificant details—they included the identification of a substance for the presence of which the Judgment awarded $5 billion, the identity of a substance that the Judgment described as the most powerful carcinogenic agent it considered, and the source of the most important statistical data. It is extremely unlikely that a judge who claims to have spent many months reviewing the record and to have written this lengthy and detailed decision would not recall such important aspects—especially when, as will be seen, that Judgment was hailed by the president of Ecuador as the most important decision in the country's history.[53]

As chapter 5 showed, Judge Nicholás Zambrano's 2011 judgment is an intricately argued, less-than-transparent, single-spaced, 188-page document. If anything stands out as unique in the ruling, it was that science neither had a determinative capacity nor was the final arbiter of truth. There was no ultimate carcinogen discovered. There was no single epidemiological study that determinatively proved the health effects of contamination. There was no unifying theory of causation capable of capturing the vast complexity of the lawsuit. TPH, although indicative of the ubiquitous presence of hydrocarbon wastes in Texaco's former concession, never was the indicative measure of toxicity. As will be evident below, the $5 billion to which Kaplan referred was based on cubic meters, not TPH per se. Rather, Zambrano's ruling had to wade through the opposing claims about the toxicity of crude oil, the conflicting epidemiological metrics for calculating health effects, and the plurality of criteria for judging liability and causality—all in line with Ecuadorian statutes and legal philosophy.

A year after Kaplan wrote his decision, in late April and early May 2015, forensic studies examined and cross-examined during weeks of hearings before the tribunal of the Permanent Court of Arbitration determined that the "Final Judgment" document in the Ecuadorian litigation was created on Zambrano's computer in his locked office, beginning on October 11, 2010; that the text size of the same document grew incrementally over the next four months; that the document was saved over four hundred times; and that it was uploaded to the court's internet system on February 14, 2011, and then distributed to the relevant parties.[54] This is the only version of the judgment found in any form, in any place, before it was made public.

Counting Contortions

On August 8, 2016, the US Court of Appeals for the Second District upheld Judge Kaplan's decision, solidifying the district court opinion ever more in perpetuity. An important discussion within the appellate court's ruling focused on Kaplan's "finding" that the 2011 Ecuadorian ruling relied extensively on one particular report—the Cabrera report. This purported reliance was crucial because it determined two-thirds of the total Ecuadorian judgment award—that is, $5.4 billion of the total $8.6 billion in damages. And the evidence that proved this finding, that the $9 billion ruling relied heavily on the Cabrera report, was a pit count, or a specific number of contaminated waste pits (whether open, covered, or remediated) that existed in Texaco's former concession.

The Cabrera report was a key component of Chevron's RICO case—and indeed was a fulcrum upon which the corporation had built its legal strategy. In particular, the company's displeasure with the Cabrera report spurred it to file a number of §1782 proceedings in the hope of directly and indirectly undermining it.[55] Briefly, US federal statute 28 USC §1782 allows a litigant in a legal proceeding *outside* the United States to apply to a US court to obtain evidence for use in foreign proceedings. Chevron invoked this statute with the aim of garnering evidence in the United States that the judicial process in Ecuador was tainted. The scale of this discovery campaign was breathtaking. Between 2009 and 2010, what had become Chevron's army of lawyers submitted over twenty-five requests to obtain discovery materials from thirty different parties in fifteen federal courts and jurisdictions across the country. The aim was to amass evidence that the corporation hoped would derail the Ecuadorian case and reveal evidence of fraud. The Cabrera report was central to that effort.

Richard Cabrera was an engineer appointed by the Ecuadorian court to gather a team of experts to conduct a "global expert assessment," ultimately known as the Cabrera report. Cabrera's charge was to write a report that synthesized all the data garnered during the judicial inspections (and additional inspections that his team completed) and, were damages to be assessed, to provide the monetary cost for remediating contamination in the concession. As was their right under Ecuadorian law, the lawyers for the LAP requested that the court order this global expert assessment. And as was the norm in the Ecuador litigation, the party that requested an expert assessment was the party that paid, had some say over who would take on that job, and could meet with the expert. After over a year of investigating, Cabrera presented his extensive report—over eight hundred pages with appendixes—in spring 2008.

Soon after the release of the film documentary *Crude*, however, Chevron cried foul. The corporation claimed that the LAP had collaborated and colluded with the Cabrera team and, in fact, had hired an environmental firm, Stratus Consulting, to conduct the scientific analysis and to draft significant portions of the Cabrera report. Using the §1782 proceedings, Chevron obtained discovery documents and subpoenaed depositions that detailed just that: scientists working with Stratus had contributed to penning large portions of the report that appeared under Cabrera's name.

The deciding Ecuadorian judge, Nicolás Zambrano, knew about Chevron's allegations of collusion between the LAP counsel and the Cabrera report. As noted, Chevron was already deeply engrossed in its §1782 US proceedings when Zambrano presided over the case, and the corporation had filed numerous complaints of collusion in Zambrano's court. Attentive to these concerns, Zambrano stated in his judgment that in "addressing" Chevron's "motion that Cabrera's report not be considered," yet without suspending proceedings in order to investigate the corporation's accusations, "the court accepts [Chevron's] petition that said report not be taken into account to issue this verdict."[56] In response to Chevron's postruling submission, Zambrano reiterated in his clarification order of March 4, 2011, that "the court decided to refrain entirely from relying on expert Cabrera's report when rendering judgment. If the defendant feels that it has been harmed because the court refused to void the entire case against it in response to the alleged fraud in expert Cabrera's expert assessment, which is allegedly demonstrated by those videos, the court reminds the defendant that its motion was granted, and that the report had NO bearing on the decision."[57]

Chevron's conspiracy narrative, however, was committed to demonstrating that this was not the case. Chevron's lawyers argued, in fact, that the Cabrera report was crucially instrumental in determining Zambrano's judgment and its damages amount. Chevron's theory hinged on one number: "880." In Zambrano's judgment, 880 was the number of waste pits that existed within Texaco's former concession area. According to Chevron, Zambrano obtained this number from, and only from, the Cabrera report. The corporation concretized this determination through the opinions of two experts: James Ebert (an expert in spatial and photographic evidence) and Spencer Lynch (an expert in digital forensics). The experts' declarations referenced and depended on one another. Their task was to debunk Zambrano's assertion that he did not consider Cabrera's report when determining the dollar amount put forth in the judgment.

Zambrano had concluded in 2011 that, in assessing "the presence of hazardous elements resulting from the operations of Texpet," the concession area contained "880 pits (proven through aerial photographs certified by the Geographic Military Institute which appear throughout the record, analyzed together with the official documents of Petroecuador submitted by the parties and especially by the expert Gerardo Barros, and aggravated by the fact that the defendant has not submitted the historical archives that record the number of pits, the criteria for their construction, use or abandonment)."[58] That is, Zambrano detailed the sources that he had referred to in order to determine that 880 waste pits existed in Texaco's former oil concession.

Chevron sought to undermine that contention by framing a conundrum: the Ecuadorian judgment and the Cabrera report both determined there to be 880 pits in Texaco's former concession. How was that possible? Ebert was to opine on the chances of that coincidence and offer his expert knowledge on "the ability to accurately count pits from the aerial photographs" in the Lago Agrio case file. Ebert affirmed: "In my expert opinion, it is impossible that the author of the Ecuadorian judgment and the author of the Cabrera report could independently review the hundreds of aerial photographs in the record and reach the exact same conclusion that there are 880 pits requiring remediation. Thus, I conclude to a reasonable degree of scientific certainty that the 880 pit-count in the Ecuadorian judgment is based on the information in the Cabrera report rather than an independent analysis of the aerial photographs in the record."[59] And thus, without considering what the ruling and the Cabrera report actually said, and without evaluating other evidence in the 230,000-page Lago Agrio case file, Zambrano is determined to have relied heavily on the disgraced Cabrera report, despite having twice clarified that was not the case.

Chevron's framing was specious. The Cabrera report never recites, let alone even mentions, the number "880." Instead, the Cabrera report maintained that 916 pits had been located. That is hardly the same number. To accommodate that discrepancy, Chevron devised a theory for how it was that Zambrano deduced 880 from 916. Enter Lynch, the digital-forensics expert. After analyzing the 2011 judgment and the Cabrera report, with special attention to its Annex-H section, Lynch determined that the author of the 2011 judgment came to the number 880 through selective subtraction. That is, if one were to take away *some* of the pits that had prior connection with the remedial action plan (RAP)—as a reminder, the RAP was the Texaco remediation plan sanctioned under the 1995/1998 couplet contract discussed in chapter 3—then indeed one could reach the number 880. Those pits included ones that, according to RAP indications, were pits constructed, modified, used by, or otherwise the responsibility of Petroecuador (a total of seventeen), pits with "no impact detected" (a total of eighteen), and one area that was determined not to be a pit (a total of one). After subtracting 36 from 916 one arrives at the magic number, 880. To complicate matters further, the table that Lynch submitted to US district court depicted the accounting of RAP tallied by Stratus Consulting calculated a total of 917 pits, not 916 pits as Cabrera had.

As Lynch explained in his direct testimony for Kaplan's court, "The Ecuadorian Judgment did not include 'no impact' figures or similar entries or those related to 'Petroecuador' and 'Petroproducción.' Therefore, I sorted the 'COMENTARIO DEL RAP' column and removed all references to these entries as shown in Figure 34. The result was 880 records—the same number that appeared in the Ecuadorian Judgment."[60] In reply to a Gibson Dunn lawyer, Lynch testified, "Relying on Mr. Ebert's or Dr. Ebert's opinion that it [the pit count of 880] was not as the judgment describes based on the aerial photographs, the only source that I have seen is an original version of Anexo H1, an Excel version, and then Anexo H1 itself. And my opinion is that it is more likely than not, given the analysis that I performed and the data that I had available to me, that [the pit count] was derived from Anexo H1 or the original Excel version."[61]

TABLE 5 **Data Counts from Stratus Compilation**

Comentario del RAP	Full Count	Revised Count
Cerrada previamente [Previously closed]	21	21
Construida después [Built after] del 6/30/90 por Petroecuador	3	
Construida después [Built after] del 6/30/98 por Petroecuador	2	
El propietario no permitió el paso [Land owner prohibited access]	3	3
Impact below action levels	1	1
Modificada después [Modified after] de 6/30/90 por Petroecuador	6	
No detectó impactos [No impact detected]	18	
No determinada como piscina [Not determined to be a pit]	1	
Petroecuador construyó sobre la piscina [Petroecuador built on top of the pit]	1	
Petroproducción usó la piscina [Petroproducción used the pit]	1	
Petroproducción solo descargar basura [Petroproducción used as garbage pit]	1	
Piscina cerrada [Closed pit]	1	1
Pit was graded and revegetated	1	1
Plantación de maiz [Used for corn plantation]	1	1
Remediación complete [Remediation complete]	156	156
Responsabilidad de Petroecuador [Responsibility of Petroecuador]	1	
Revegetada [Revegetated]	1	1
Soil TPH below action levels	1	1
Usada como piscina para peces por la comunidad [Used for aquaculture by the community]	2	2
Usada por la comunidad local [Used by the local community]	15	15

(continued on next page)

Table 5 continued

Comentario del RAP	Full Count	Revised Count
Usada por Petroecuador [Used by Petroecuador]	1	
Used as a municipal landfill	2	2
Utilizada por Petroproducción como piscina de quema [Used as a burn pit by Petroproducción]	1	
(blank)*	676	675
Grand Total	917	880

* The Charapa-4 pit is not part of the former concession area.

Source: Spencer Lynch's witness statement, October 7, 2013, p. 43, Chevron Corp. v. Donziger, *ECF 11 Civ. 0691 (LAK), DI 1584-1 (S.D.N.Y.).*

But why would Zambrano, or the alleged LAP ghostwriters, consider any data from the RAP to be accurate, let alone reliable enough to determine a comprehensive pit count?[62] Throughout the 2011 judgment, Zambrano specifically notes that the chemical analyses from the judicial inspections undermined the validity of Texaco's mid-1990s remediation and that the RAP was not valid. That is, the judicial inspections between 2004 and 2008, and further expert reports in 2008 and 2009, repeatedly indicated that the levels of hydrocarbons in soils and waters stated in the RAP were less than reliable, if not suspect. Indeed, as noted in chapter 3, representatives of Texaco/Chevron (along with several government officials) were under investigation for contractual fraud, for ratifying that a certain level of remediation occurred when it had not. Given that, it seemed extremely unlikely that Zambrano would ever use data from the RAP in order to calculate a precise count of waste pits.

Moreover, Lynch's expert analysis exposed how little he knew about what he was dealing with. On the one hand, if the RAP indications were seen as the authoritative source for accurately determining the number of pits in the former Texaco concession, why would one stop at the two categories that Lynch noted? In addition to the eighteen "no detectable impact" pits and the seventeen "Petroecuador" pits, why not also subtract the fifteen "used by the community" pits? They similarly appeared to have been negligibly impacted since these pits were being used. More telling, Lynch did not seem to realize that the "RAP commentary" was designed to address only one-third of the wells drilled by Texaco. Consequently, if the 2011 judgment sought to reach an absolute number of pits using Lynch's methodology, then one would apply parallel subtraction practices to the pits associated with the other two-thirds of the wells in the concessions. If one were to assume eighteen "no detected impact" pits were associated with each one hundred wells, the total pit count for the entire concession would be 844—and that is without taking into account any pits purportedly associated with "Petroecuador" or "Petroproducción."

For an expert, Lynch was impressively unfamiliar with how the RAP was conceived, let alone the ways that the litigation in Ecuador profoundly challenged its integrity as a serious remediation effort. There is not one place in the ruling where the RAP is invoked as a source of reliable data. Using RAP data to calculate a method for reaching the magic "880" number made no sense. This was fantastic selective subtraction. It was also Chevron's witting crafting and deployment of expert ignorance to build and reify a corruption narrative—one showing that the author of the 2011 judgment, despite assertion to the contrary, used the Cabrera report that Chevron had already demonstrated was tainted by the alleged swindling hands of the plaintiffs.

Eliding all these irregularities, Kaplan wrote the following in his "Appendices to Opinion":

> In sum, Lynch and Ebert collectively testified that (1) the 916 pit count in the Cabrera Report, once adjusted in a very common sense way to eliminate the 36 "pits" that either were those of PetroEcuador [*sic*] or required no remediation, was 880, (2) the pit count in the Judgment was 880, and (3) neither the pit count in the Judgment nor that in the Cabrera report could have been determined accurately from the aerial photographs upon which each purported to rely. They further concluded that, as a practical matter, it is impossible that these two documents could have reached the net count of 880 pits independently on the basis of examination of the aerial photographs, which was the sole stated foundation of each. . . .
>
> [D]efendants have not identified any possible source in the Lago Agrio record for the Judgment's 880 pit count, other than the Cabrera Report, save for the claim that Zambrano reached that figure independently by counting what appeared to him to be waste pits on low resolution aerial photographs. The Court finds that hypothesis to be incredible given both the quality of the photographs and Zambrano's lack of credibility. . . .
>
> If there were a source in the record other than the Cabrera Report that supported the pit count figure—which was the basis for the largest com-

ponent of damages—the LAPs would have cited it. But they did not. And that logically suggests that there was nothing in the Lago Agrio record to support the pit count except the Cabrera Report, adjusted to eliminate the PetroEcuador [*sic*] and the "no impact" pits.

The Court finds that the 880 pit-count in the Judgment came directly out of the Cabrera Report, adjusted only for the PetroEcuador [*sic*] and "no impact" pits. It further finds that the circumstances discussed by Ebert and Lynch, whom the Court credits, make it impossible that the pit count in the Judgment came from anything but the Cabrera Report.[63]

Kaplan's conclusions baffle anyone familiar with the Ecuadorian litigation. First, the 2011 Ecuadorian judgment clearly stated that aerial photography was *not* the only source for determining the number of waste pits in Texaco's former concession. Second, the exact number of Texaco waste pits present in the northern Amazon is not known—except by Chevron. But a broad consensus agrees that two to five waste pits are associated with each of the more than three hundred wells that Texaco drilled. Far from extravagant or fantastical, this number positions the 880 count on the conservative side. Were the lawyers for the LAP defense allowed to introduce evidence from the Lago Agrio litigation into Kaplan's court, they would have been able to cite multiple sources that verified this. One Chevron expert, John Connor, identified 148 pits over the course of conducting inspections (both official and unofficial) at 45 Texaco well sites. This means that Chevron's own expert determined there to be on average 3.29 pits per well site. Given that Texaco drilled, at a minimum, 330 oil-exploration wells, this leaves an estimated 1,085 pits in the concession.[64]

The material reality behind the 880 number further disturbs. For, in finding a logic by which the "same number" is obtained, Kaplan ignored what that number meant. Given that the US court found that a Cabrera-derived 880 pit count served as the basis for finding Chevron liable for over $5 billion, one would assume that the 2011 Ecuadorian judgment and the Cabrera report shared an understanding of what a "pit" actually represented, of what it consisted. This was not the case.

Remarkably, the focus on making a pit count of 916 equal 880 had elided the gaping disparity between what a pit count for Zambrano and what a pit count for Cabrera actually signified. Importantly, the 2011 judgment does not base Chevron's $5 billion-plus remediation liability on a pit count per se. It was based on a calculation of cubic meters of contaminated soil and remediation levels. My point: the 880 pit count in the 2011 judgment was not in

TABLE 6 Surface Area of Concession Soils Requiring Remediation (>1,000 ppm TPH)

Soil Surface Area	Wells	Stations	Total
Total surface area of pits	691,000 m²	77,500 m²	769,000 m²
Surface area of pits requiring remediation	553,000 m² (80% of pit soils)	77,500 m² (100% of pit soils)	631,000 m²
Surface area of soils outside of pits that requires remediation (50% of pit soils)			316,000 m²
Total surface area of soils requiring remediation			947,000 m²

Source: Cabrera Vega 2008: "Annex N: Soil Remediation Costs," 1.

itself significant. Rather, it was significant because it served as the basis from which to calculate a three-dimensional space of contamination. And the way Zambrano calculated that three-dimensional space was radically different—nearly by two times—from that calculated by Cabrera.

The Cabrera report used multiple indicators to calculate area (this method is detailed in its appendix N). Of the 916 pits in appendix H, 89 of those pits were located at production stations and 828 were associated with oil wells. Computing data on TPH levels (recorded in the judicial inspections) and metric area (as determined from the judicial inspections and aerial photography), the Cabrera report considered 100 percent of the station pits (that is, all 89 waste pits accounting for 77,500 meters squared [m²]) needed remediation, and "80 percent of the pits at wells" (or 662 waste pits accounting for 553,000 m²) needed remediation. To that was added an additional 50 percent (316,000 m²) of the total area (631,000 m²) to account for leaching, spills, and/or accidents. Because the Cabrera report determined that soils were contaminated at a depth of four meters, the cubic volume to be remediated was 3,788,000 m³ (total area: 947,000 m² × 4 meters).

By contrast, the 2011 judgment determined that there were 880 pits in the concession and calculated each pit as having an average pit-area size of 2,400 m² (60 m × 40 m). An additional area of 1,100 m² was added to account for spill, seepage, leaks, and/or accidents. To calculate volume, the ruling estimated that remediation should be at a depth of 2.4 meters for a total of 8,400 cubic meters (m³) per pit. When that figure is multiplied by 880 pits, the total volume of soils needing remediation equals 7,392,000 m³.

To calculate the cost of remediation, the judgment first referred to the calculations of Gerardo Barros, the court-appointed expert nominated (and paid) by Chevron. This was the expert who, Zambrano noted in response to Mastro's pop quiz, had "statistical data of highest importance" in determining the 2011 judgment. Barros had provided a range of per-cubic-meter remediation costs. Zambrano took the mean of the range that Barros detailed—$365 per cubic meter—and multiplied it by the volume, rendering a total of $2,698,080,000. This was the cost for remediating contaminated soils to a level of 1,000 parts per million (ppm) TPH. However, since the LAP legal claim "requested the removal of all elements that could affect their health and lives" and "that the level of remediation should aim to leave things in the state in which they were prior to the operations of the consortium," doing so "would cause the per-cubic-meter cost estimated from the information proposed by expert Barros to increase."[65] Drawing from another court expert nominated by the LAP (Douglas Allen), Zambrano observed that the cost of remediation "practically double[d]" depending on the level of cleanup (i.e., remediating to a 100-ppm TPH level was nearly twice the cost of remediating to a 1,000-ppm TPH level). Consequently, the judgment states that "on the basis of this specification, the quantity that this presidency estimates to be necessary for the remediation of soils shall not exceed FIVE BILLION THREE HUNDRED AND NINETY-SIX MILLION, ONE HUNDRED AND SIXTY THOUSAND DOLLARS (USD$5,396,160,000.00) and shall tend to recover the natural conditions of the soil impacted by Texpet's activities."[66]

The $5 billion-plus price tag, representing nearly two-thirds of Chevron's total liability, was not "derived" from a simple-minded pit count. Rather, volume calculations (for cubic meters) and remediation levels were the basis for the calculation. It is for this reason that Zambrano responded, "The report by expert Barros," when Mastro had asked him "what report" contained the "statistical data of highest importance to delivering this ruling."

Delocalized Stability

In its August 2016 decision, the US Court of Appeals for the Second Circuit upheld and further reified Kaplan's "findings." It determined that the 880 pit count came from the Cabrera report; it determined that Zambrano did not write the 2011 ruling; and in "light of Zambrano's 'astonishing[] unfamiliar[ity] with important aspects of [the judgment's] contents,' along with the 'evasive[ness] and internal[] inconsisten[cies]' in his trial testimony and the differences between his trial testimony and 'his deposition just days before,' the district court found that "Zambrano did not write the Judgment issued under his name."[67]

The rulings of the US federal courts were delocalized stabilities. Here, court "findings" or assertions of fact came to collectively constitute the truth that Zambrano did not write the 2011 precedent-setting judgment against Chevron. These findings, however, could be affirmed only by latching onto specific elements of the Ecuadorian litigation, detaching them from their embeddedness in extensive relations, and fusing them into the worlding of a racketeering ring. I have sought in this chapter to demonstrate how that was the case and recompose those severed connections.

Chevron's RICO case contained many more elements than what I have unfolded above. Other dimensions derive from the "universe" of images and documents garnered during the §1782 US proceedings against the filmmakers of *Crude* and the LAP's US-based advisory lawyer, Steven Donziger. Chevron's lawyers dexterously exploited slips and missteps by an at-times anxious, always underfunded, small LAP legal team as it struggled to keep its head above water whilst confronting Chevron's legal leviathan. And, unrelentingly, corporate lawyers have gone after Donziger (the only US-based lawyer to have continued to worked on this legal saga since 1993) only to secure a misdemeanor conviction derivative of Kaplan's erroneous findings of fraud (Sawyer and Ofrias forthcoming). Despite access to the "universe" of internal documents disclosing the LAP lawyers' legal strategizing, Chevron counsel never, not once, unearthed a draft of the 2011 judgment among the LAP files, let alone a hint that ghostwriting was ever considered.

This chapter has delved into three moments in the RICO trial that, according to Chevron, demonstrated conspiracy, fraud, and corruption: Russell's testimony, whereby the LAP lawyers' alleged manipulation of scientific data attested to conspiracy; Zambrano's testimony, whereby his supposed lack of knowledge about the 2011 judgment attested to fraud; and Ebert's and Lynch's testimonies, whereby their statements that the 2011 judgment purportedly relied on the tainted Cabrera report attested to corruption. When explored with care, however, each moment exposes how Chevron conjured a reality about a process that never took place in the way or with the implications that the company claimed. Chevron's corruption worlding and Kaplan's 2014 judicial decision were delocalized stabilities—where elements torn from prior context and events in Ecuador were ensconced in, and thus served to fortify, a new unshakable relational configuration of extortion and a racke-teering ring. The circumstances of the moments explored in this chapter (of Russell, of Zambrano, of 880 pits) were clearly not the same in their Ecua-dorian context versus the US litigation context. And it was a distortion for Chevron to claim, and the US court to concur, that they were the same. In the US courts, Chevron's litigating devised a method whereby the detached elements (Russell, Zambrano, 880 pits) became absorbed in, and constitu-tive of, an increasingly entrenched and seemingly inevitable configuration of fraud that they claimed permeated the Ecuadorian legal process. That the US courts misattributed the Ecuadorian judiciary as corrupt reflected its own parochial and imperious manner.

Chevron's legal team forged and propelled this delocalized stability in great part due to its mastery of "the technical aesthetics of law" (Riles 2005: 976), that is, its skilled maneuvering of and within the US legal system. In the presence of a judge intrigued by, if not sympathetic to, the corporate line, the corporate legal team, together with a particular court chamber and legisla-tive act, rendered legal proceedings that maligned, denigrated, and traduced a historic foreign environmental lawsuit with the ruse of conspiracy. The al-legation that the $9 billion Ecuadorian ruling was procured through fraud, bribery, and extortion enabled Chevron to shape a RICO trial that "proved" fraud by sidelining the ubiquity of corporate contamination. That is, Chev-ron simultaneously proved the illegitimacy of the $9 billion ruling and ren-dered its own actions in Ecuador beyond the law. With Kaplan holding in contempt of court any introduction of evidence concerning contamination in Ecuador, the legal team for the Ecuadorians was unable to mount an hon-est defense against Chevron's allegation. Russell had scant basis to insinuate that the LAP manipulated data. Zambrano's pop-quiz answers testified more to his connection (than disconnection) with the 2011 judgment. Finally, the 880 pit count was not (indeed, could not be) derived from a tainted expert report; Kaplan was mistaken in claiming that it was "impossible that the pit count in the [2011] Judgment came from anything but the Cabrera Report."[68]

The legal proceedings pursued under the RICO action were highly circum-scribed and overdetermined—by what sorts of evidence the court made ad-missible, by the hubris and want of humility of US law, and by the capacity of monied litigating to directly and dramatically affect courtroom theater. Ka-plan's 2014 decision proclaimed the second largest oil company in the United States to be the "victim" of extortion and fraud committed by marginalized forest peoples and their greedy lawyers in their zeal to secure the 2011 Ecua-dorian judgment. Deploying highly problematic "findings of fact," the US

District Court for the Southern District of New York and the US Court of Appeals for the Second Circuit determined that an adverse foreign ruling against Chevron had been procured through corrupt means and that the 2011 Ecuadorian judgment was prohibited from being enforced in the United States. Delocalized stabilities transmogrified an environmental contamination lawsuit into a civil RICO wrong.

TETHERED

In May 2013, John Keker, the counsel defending the LAP lawyers in the New York RICO case, submitted a memorandum to Judge Kaplan informing him that it was impossible for him to continue to defend Donziger and colleagues under the circumstances. Keker's court submission offers a glimpse into Chevron's litigating tactics.

UNITED STATES DISTRICT COURT

SOUTHERN DISTRICT OF NEW YORK

CHEVRON CORPORATION, Plaintiff, v. STEVEN DONZIGER, et al., Defendants.	Case No. 11-CV-0691 (LAK)

MEMORANDUM OF LAW IN SUPPORT OF KEKER & VAN NEST LLP'S MOTION TO WITHDRAW AS COUNSEL FOR DEFENDANTS STEVEN DONZIGER, THE LAW OFFICES OF STEVEN R. DONZIGER AND DONZIGER & ASSOCIATES, PLLC

. . .

I. INTRODUCTION

It is with regret that undersigned counsel is forced to make this motion to withdraw. This is an extraordinary case, which has degenerated into a Dickensian farce. Through scorched-earth litigation, executed by its army of hundreds of lawyers, Chevron is using its limitless resources to crush defendants and win this case through might rather than merit. There is no sign that Chevron wants a trial on the merits. Instead, it will continue its endless drumbeat of motions—for summary judgment,[1] for attachment,[2] to reinstate long-dismissed claims,[3] for penetration of the attorney client privilege,[4] for contempt and case-ending sanctions,[5] to compel discovery already denied or deemed moot,[6] etc., etc.—to have the case resolved in its favor without a trial. Encouraged by this Court's implacable hostility to Donziger, Chevron will file any motion, however meritless, in the hope that the Court will use it to hurt Donziger. Donziger does not have the resources to defend against Chevron's motion strategy, and his counsel should not be made to work for free to resist it.

In the fourteen months since the stay was lifted in February 2012, this case has been litigated at a feverish pace, which has increased at an exponential rate. The docket sheet shows

---

1 See, e.g., Dkt. Nos. 396, 483, 583, 744.

2 See, e.g., Dkt. Nos. 353, 404.

3 See Dkt. No. 782.

4 See, e.g., Dkt. Nos. 475, 562, 656, 850, 1031.

5 See Dkt. No. 893

6 See, e.g., Dkt. Nos. 1018, 1074.

81 entries for the first quarter of 2012; 43 entries for the second quarter of 2012; 81 entries

for the third quarter of 2012; 114 entries for the fourth quarter of 2012; and 261 entries for

the first quarter of 2013. April has yielded another 130 docket entries, plus a three-day ev-

identiary hearing. Letters, emails, discovery responses, meet and confer calls, and other

non-docketed materials have all followed the same trajectory of relentless increase. Chev-

ron served over 210 document requests to Donziger, many with subparts, as well as dozens

of pages of interrogatories, and 1,228 requests for admissions, many with multiple subparts.

Donziger's counsel has spent thousands of hours in recent months dealing with Chevron's

seemingly limitless discovery demands. Now, with the addition of two special masters and

their associate and assistants, there is the additional burden of responding to their various

letters, orders and emails. Keker Decl. ¶¶ 5 & 6.[7]

Defense counsel has sought the Court's intervention to control and manage what has be-

come unmanageable. See Keker Decl., Ex. A (March 1, 2013 letter to Judge Kaplan). The Court

did not respond to this letter and indeed, in the weeks since March 1, has made matters worse

by consistently and cumulatively increasing the litigation burden on defendants: allowing

dozens of fact depositions to occur from Park Avenue to Peru;[8] ordering over objection the ap-

pointment of two very expensive special masters with burdensome procedural requirements;[9]

ordering a three-day evidentiary hearing in New York on a Chevron motion seeking sanctions

because an Ecuadorian went to court in Ecuador to clarify his Ecuadorian attorneys' responsi-

bilities under Ecuadorian law;[10] threatening defense counsel that they were to "proceed

---

7 "Keker Decl." refers to the Declaration of John W. Keker, filed concurrently herewith.

8 Dkt. Nos, 882, 910, 941.

9 Dkt. No. 942.

10 Dkt. No. 997.

at their own risk" and "at their peril" when they wrote to the court and special masters that Donziger could not pay his lawyers, much less the exorbitant fees of the special masters;[11] forcing Donziger's counsel to spend hundreds of thousands of dollars on attorney time responding to a motion for summary judgment, which counsel begged to be put off until after discovery closed,[12] only to rule on the motion by announcing that it was denied without prejudice until after discovery closed, and could be reinstated later;[13] and ordering not one seven-hour day, but three days of depositions (on top of 16 previous days of deposition) of Donziger.[14]

This Court's hostility towards Donziger, already the subject of a motion to recuse and currently the subject of a pending mandamus proceeding, has in recent weeks become even more pronounced. When Chevron complains about defense counsel's possible role in an Ecuadorian lawyer seeking an order from an Ecuadorian court, the Court responded by ordering a three-day evidentiary hearing in New York, but when defense counsel complained about Chevron making blatantly false statements, the Court responded by accusing defense counsel of "bickering" and "venting of [counsel's] spleen."[15] During the recent evidentiary hearing, and in its order demanding Donziger's presence at a deposition for three days, the Court has made plain that its mind is made up, and its hostility toward Donziger is implacable.

Chevron's litigation tactics, which this Court has endorsed and encouraged throughout these proceedings, notwithstanding the dictates of Federal Rule of Civil Procedure 1, have made the costs of this litigation unsustainable to Donziger. Simply put, Donziger cannot

---

11 Dkt. No. 999, 1055.
12 Dkt. No. 780.
13 Dkt. No. 1063.
14 Dkt. No. 1060.
15 See Dkt. No. 1055.

afford to pay what is required to litigate effectively against a hostile wealthy corporation in a hostile court. As set forth in the accompanying Declaration of John W. Keker, Donziger has since September 2012 fallen into significant payment arrears such that Keker & Van Nest is now owed more than $1.4 million in unpaid fees and costs, including for work presently being conducted. Keker Decl. ¶ 9. More significantly, to even stay alive in this case, without appearing at depositions or other frills, through discovery and trial will cost another six to ten million dollars in attorney time and costs—an amount about equal to what we estimate Chevron is paying its lawyers each month. See id. ¶ 10. There is no reasonable prospect of payment of the current receivable, nor of payment of the future fees and costs anticipated to be incurred through trial. See id. ¶ 11.

Keker & Van Nest therefore seeks to withdraw as counsel. Mr. Donziger will represent himself and his law firms for the remainder of the pretrial phase of this case. If he is able to hire (or re-hire) outside counsel for trial, he will do so. But for now, his counsel is unwilling to continue on a pro bono basis under the current conditions, and should not be made a slave to this impossible situation.

[...]

IV. CONCLUSION

For all the foregoing reasons, Keker & Van Nest respectfully requests that the Court enter an order permitting Keker & Van Nest to withdraw from its representation of Donziger in this action, and allowing him [to] represent himself and his law firms in this action.

Respectfully submitted,

KEKER & VAN NEST LLP

Dated: May 6, 2013

By: /s/ John W. Keker
_____
JOHN W. KEKER (*pro hac vice*)
ELLIOT R. PETERS
JAN NIELSEN LITTLE (*pro hac vice*)
MATTHEW M. WERDEGAR (*pro hac vice*)
633 Battery Street
San Francisco, CA 94111-1809

Attorneys for STEVEN DONZIGER, THE LAW OFFICES OF STEVEN R. DONZIGER AND DONZIGER & ASSOCIATES, PLLC

DERISION

Below is a short description that I wrote when I accompanied Donald Moncayo to a waste pit that Texaco had purportedly remediated in the mid-1990s. My first encounter with Texaco's waste pits was in June 1988. This account comes from my most recent visit in August 2019.

How could a body not react?

The first pit lay fewer than twenty meters west of the Aguarico-2 wellhead. At an opening down a slippery forest path, Donald Moncayo Jimenez dipped his shovel into surface soil. Unearthed emerged an amalgam of dense black crude and biotic matter. Donald had long been a local icon shuffling the curious—from college students, to US representatives, to Hollywood celebrities, to national and international film crews—on "toxi tours" of Texaco's former concession. Having worked for Texaco by maintaining its wells for nearly a decade, Donald knew how the company had worked its operations. Soon after the litigation against Chevron began in Lago Agrio, Donald began working with the Union de los Afectados por Texaco.

Aguarico-2 was one of the well sites that Texaco remediated according to the 1995/1998 couplet contract. The well was drilled in 1972 and solely operated by Texaco until capped in 1982. Aguarico-2 had three waste pits. According to the certificate of completion, all three pits were remediated successfully and were included in the acta final—that document that released Texaco for being liable for any claims made by the state in the future.

FIGURE 55
Donald Moncayo's
hand after
skimming the
surface waters
of a covered and
seeping waste
pit at the oil well
Aguarico-21. Photo
by author.

The stench of crude oil caused muscles to flinch, pores to tighten, the eyes to wince. But when Donald dug only slightly deeper, the waft of crude oil consorting with drilling mud forced the body to recoil.

"A theatre of derision. A mockery of a cleanup"—that's what Donald said we were witnessing. "It's inhumane," he continued. "Regardless of how many times I tour these sites, I'm outraged by these acts." His voice paused and picked up midsentence: "an intolerable actuality replete with racism."

Metamorphic Reprise

Valence in the Mixt

FOR MANY, IT WAS A STORY FORETOLD. That the Republic of Ecuador's defense in the international arbitration was damned before it began. Yet the outcome seared regardless. On August 30, 2018, a tribunal of the Permanent Court of Arbitration (PCA) in The Hague submitted its judgment in an arbitration claim that Chevron had filed against the Republic of Ecuador.[1] Composed of three arbiters, the tribunal concluded that the 2011 Lago Agrio judgment had represented a "denial of justice" and that "no part" of the ruling "should be recognised or enforced by any State with knowledge of the Respondent's said denial of justice."[2] The said "denial of justice" resulted from the Republic of Ecuador having violated articles of the US-Ecuador bilateral investment treaty (BIT), signed on August 27, 1993, and entered into force on May 11, 1997. Broadly speaking, the tribunal ruled "denial of justice" because the republic "violated its obligations" by not upholding the release it had granted the corporation under the 1995/1998 couplet agreement regarding remediation. Were the republic to have honored its couplet-contract obligation then it would never have issued and upheld a ruling against Chevron in 2011 that was "corruptly" obtained.

How does a legal body in The Hague designed largely to address commercial disputes come to render as illegitimate and unenforceable the ruling of a sovereign judiciary in a civil contamination lawsuit? A ruling that was pronounced in 2011 and upheld on appeal by three higher appellate courts—the Lago Agrio Appeal Division (2012), the National Court of Justice (2013), and the Constitutional Court (2018). More precisely, how does an intergovernmental arbitral forum (the PCA) determine that a sovereign state's judicial system unlawfully held a corporation accountable for environmental contamination after years of litigation, oceans and continents away? And how does corporate behavior deemed ecologically reckless by a constitutionally independent judiciary morph into that foreign sovereign having breached a BIT?

Bringing this book to a close, this chapter reflects on how Chevron's claim before the PCA transformed a controversy over whether a multinational corporation contaminated the environment into a controversy over whether a nation-state violated an international trade agreement. Troubled by this metamorphosis, I attend here to the valence—the relational tensions—that rendered the composition of the PCA tribunal's arbitral truth. Thinking with valence, chemical bonding, and the mixt brings attention to how elements, when brought together in unique orbital relations, have the capacity to produce extraordinary entities within which the former elements no longer abide. They exist in potentia, not in actuality. In chemical reactions, elements rarely escape this passage. The suggestion extended here is that, given matter's wily capacities, such can be the case as well in social worldings. Chevron's arguments and the PCA tribunal's judgment are alarming manifestations of this. Consumed within unfamiliar relational configurations, questions of contamination, of crude's chemistry, of ailing life-forms, and of Ecuadorian law transmogrified into a new stable form: the breaching of a commercial compact. Despite its new substantive mattering, that novel entity—violation of obligation—was, however, never without that which it was not, that which it no longer manifested and yet on which it depended. Such is the conundrum of the mixt and Chevron's metamorphic reprise.

Chevron filed its dispute claim before the PCA in 2009.[3] This was the year in which the corporation was forging its corruption narrative and filed its first §1782 action (the little-used US federal statute [28 USC §1782] discussed in chapter 6). In the claim before the PCA, Chevron argued that Ecuador had breached the 1997 BIT on two counts. First, Chevron claimed that the Republic of Ecuador violated the investment treaty when it both failed to execute the terms of the 1995 settlement and the 1998 acta final agreement (analyzed in chapter 3) and failed to dismiss the Lago Agrio lawsuit from its judiciary *tout court*. Second, Chevron claimed that Ecuador violated the BIT—in not ensuring what the treaty called "fair and equitable treatment"—by subjecting the corporation to a legal process and judiciary in the Amazon that allegedly forsook due process, was mired in corruption, and was rife with collusion against the corporation.[4] By infringing on specific clauses of the BIT, Chevron maintained, the Republic of Ecuador violated the corporation's "contractual, legal, and Treaty rights by failing to protect [Chevron] from, and affirmatively seeking to subject them to, the claims and liabilities from which Ecuador previously released" the company.[5] Furthermore, the claim read, the "enforcement of a Lago Agrio judgment imposing any liability on Chevron for environmental impact or remediation effectively will eviscerate [its] contract and Treaty rights."[6]

The tribunal's 2018 response to Chevron's claim is a long, methodical, five-hundred-plus-page ruling based on a decade of arbitral litigation that entailed numerous legal submissions by the parties, multiple scientific studies, months of closed-door hearings of legal arguments and expert testimony, and a site visit (by the three arbitrators, legal counsel, and scientific experts) to four sites in Ecuador allegedly contaminated by Texaco. As with all rulings, the tribunal's parallels the claim before it. Yet in contrast to Kaplan's page-turner, the tribunal reached its conclusions through the sober and systematic analysis of separate frames (i.e., principal issues, principal legal texts, the Ecuadorian judgment and appeals, questions of jurisdiction, the merits of Chevron's claim, forms of relief, and operative measures). As with the Racketeer Influenced and Corrupt Organizations Act (RICO) case, this international arbitration warrants a treatise of its own. However, in this chapter, I focus on one core dimension of the tribunal proceedings and ruling—the "violation of obligation"—and the forces that enabled it.

Reactants

Turning a contamination dispute into an investment dispute required considerable labor. First, a couple of steps were needed to trigger that potential transformation, which entailed enfolding the work of two precise reactants. Chevron's claim before the tribunal hinged on the couplet contract (composed of the 1995 settlement and 1998 acta final) and that agreement's supposed unquestioned integrity. As chapter 3 signals, however, the 1995/1998 couplet contract had long been besmirched. Years of judicial inspections during the Lago Agrio litigation demonstrated that Texaco's remediation program (remedial action plan, or RAP) had been carried out ineffectively. So Chevron needed to convert the tainted couplet contract into a respectable and weighty agreement.

Key to this transformation was the corporation's allegation that the Lago Agrio plaintiffs (LAP) were fundamentally deceptive and corrupt. Enter here, again, David Russell and the infamous 880 pit count. Reiterating their arguments in the RICO case, Chevron's lawyers convinced the PCA arbiters that duplicity was at the heart of the LAP litigating strategies. Indeed, the tribunal ruled that the LAP's scientific and chemical claims fundamentally reflected connivance and corruption. Such a characterization turned connivance and corruption into reactants that transmuted the RAP and 1995/1998 couplet agreement into a newly cleansed compositional form. Let me explain. In chemistry, a reactant refers to those entities brought into a system to form a chemical reaction that in the process are consumed—meaning they no longer have agentive capacity once integrally distributed to form a new composition. Chevron's corruption worlding via Russell and the 880 pit count functioned in the same fashion. They recomposed the 1995/1998 couplet contract within a new orbital configuration of integrity and transparency.

Russell did not appear in person before the tribunal. Yet he figured as a pivotal character in the tribunal's 2018 decision. Russell's RICO testimony, which Chevron invoked in its legal briefs and arguments during tribunal hearings, was consequential. The tribunal determined (echoing Kaplan's 2014 findings, discussed in chapter 6), that Russell, as the LAP's "chief scientist," clearly proved that LAP lawyers were devious and unscrupulous. Rehearsing and extending Kaplan's confused understanding of crude hydrocarbons, the tribunal determined that the LAP conjured contamination where it was not. Once again, the specter of chemistry in all its misattribution served to condemn the LAP legal and technical team as manipulating science and as exaggerating contamination. And thus in the PCA proceedings, the LAP legal team members (who were not party to the proceedings) were beleaguered by a plagued of connivance from the get-go.

The most prominent "evidence" of the LAP's connivance was, of course, the infamous 880 pit count upon which the 2011 Lago Agrio judgment supposedly relied in determining the largest portion ($5.396 billion) of Chevron's financial liability. Chevron's theory, you may recall, alleged that the pit count of 880 came directly from the tainted Cabrera report. During the Lago Agrio litigation, the Cabrera report swirled in a storm of controversy as Chevron, through its first §1782 actions, determined that the LAP scientific team penned good portions of what was paraded as an independent court-appointed expert's analysis. Chevron argued that the 880 number not only reflected the LAP's penchant to inflate the extent of contamination, but also underscored the corruption at the heart of the LAP legal team and their collusion with the Ecuadorian court.

But the intent of invoking the 880 pit count extended deeper when brought to the tribunal. There, claiming that a count and contamination of 880 pits was bogus had the effect of resurrecting as relevant and legitimate the 1995/1998 couplet contract. As detailed in chapter 3, the 1995/1998 couplet agreement between Texaco and the Ministry of Energy and Mines negotiated the terms by which Texaco committed to cleaning up 113 designated oil production sites in its former concession. The parties signed a contract that established the parameters of this agreement in 1995 and, upon conducting surveys, the RAP then detailed the unique circumstances of the waste pits at each site—with a collection of pits designated as not needing remediation. As chapter 6 explained, Kaplan found that the 2011 Ecuadorian judgment had relied on the Cabrera report, even though the Lago Agrio opinion clearly stated that was not the case. And Kaplan's reason for determining that was the 2011 ruling's 880 pit count. Kaplan resolved that in ghostwriting the 2011 Ecuadorian judgment, the LAP reached that number, and could *only* have reached that number, by taking the 916 pits detailed in the Cabrera report and selectively subtracting specific pits that, according to RAP indications, did not need to be remediated.

In the context of the international arbitration, this had two effects. First, it affirmed that an entire world of corruption and exaggeration stirred beneath a simple number—880—with Chevron repeatedly asserting that extensive contamination was a "factual absurdity." Second, reenacting the purported calculus behind the 880 pit count reanimated with an unquestioned vitality the Texaco-ministry 1995/1998 couplet contract. In a brilliant legal move, Chevron positioned Texaco's 1990s RAP as playing a crucial role in determining the most substantial portion of the 2011 judgment's liability. That is, corporate lawyers made it appear as though the RAP posed a genuine obstacle within the Lago Agrio litigation and was a genuine hurdle; neither was the case. The Republic of Ecuador's legal team, of course, fiercely refuted this depiction. The Lago Agrio lawsuit was a civil tort claim against a corporation, and the mid-1990s corporate-state remediation agreement had no part in it—other than when Chevron lawyers invoked the agreement or when contamination levels unearthed during judicial inspections recurrently questioned the agreement's integrity. These efforts notwithstanding, the tribunal found that the conditions of the RAP agreement figured significantly in determining the 2011 Ecuadorian judgment, accounting for over $5 billion in cleanup costs. Significantly, the couplet contract itself surfaced unscathed. Not only did the RAP agreement appear uncompromised, but it emerged as a quintessential example of the near-sacrosanct power of the contract form.

Violating Obligation

As explored in chapter 3, the whole point of the 1995/1998 couplet agreement between Texaco and the Ministry of Energy and Mines was to negotiate a swap: Texaco committed to remediating designated contaminated areas in exchange for the Republic of Ecuador releasing Texaco (in all its corporate incarnations) from any future contamination claims by the ministry and Petroecuador. The 1995 agreement laid out the parameters for the cleanup; the 1998 acta final detailed the release. Chevron's argument before the PCA—also previously presented to the Lago Agrio court and the New York district court (although later withdrawn) when in 2004 the corporation was fighting against having its claim before the American Arbitration Association (AAA) dismissed—argued that the 1995/1998 agreement had already settled the corporation's liability for the effects of its oil operations in Ecuador and, as such, the corporation should not have been subject to the Lago Agrio lawsuit.

CHEVRON'S CLAIM

In Chevron's telling, the 1995/1998 couplet contract was a sanctified covenant that, when properly interpreted, would deliver the corporation of its legal afflictions. An accurate exegesis, the oil conglomerate claimed, would show the agreement's capacity to right judicial wrongs: namely, the Ecuadorian judgment against the corporation. Pace Chevron, the couplet contract certified that the corporation had already satisfied its obligation to address any negative impact that its operations had in the concession area. And in so certifying, the agreement exempted the corporation from any future claim associated with the consortium's operations. Having complied with the terms of the contract—or fulfilled its side of the bargain swap—Chevron was absolved by the state of all future liability. Consequently, Chevron argued, the entire Lago Agrio litigation and ruling were null and void, in addition to being fraudulent. The environmental effects of Texaco's operations had already been settled. As such, the Republic had violated *its* obligation under the 1995/1998 couplet contract by not holding up its side of the deal (i.e., by allowing the Lago Agrio litigation to proceed and be thrice upheld) and thus had breached its BIT.

To many observers, the logic of Chevron's argument was far from self-evident. The 1995/1998 agreement was between Texaco (with Texpet) and the Ministry of Energy and Mines (with Petroecuador); that is, the contract was between corporate and state entities. And it detailed specific remediation actions to take place in approximately one-third of the waste pits, strewn across Texaco's former concession. By contrast, the Lago Agrio litigation was between Indigenous and non-Indigenous forest dwellers and the Chevron Corporation: that is, the conflict there was between private citizens and a corporation. And the litigation sought the remediation of specific contaminated soils and water systems as well as the implementation of health-monitoring networks and facilities. Not only were the parties involved in each encounter not the same, but the sought-after and reached resolutions were not the same. They were temporally, judicially, and substantively incomparable settlement events.

However, the corporation's legal strategy before the PCA tribunal was to flip that perspective and make the 1995/1998 couplet contract and the 2011 judgment equivalent, even the same. Chevron's argument proceeded as follows. First, the 1995/1998 couplet contract and the Lago Agrio lawsuit revolved around the same material fact: Texpet's operations had a negative ecological effect that compromised the environment. Second, the contract and the lawsuit both invoked and sought to vindicate the same legal right: the right to a clean environment and health to be secured through remediation. Third, the contract and the lawsuit involved the same parties: on one side were the Republic of Ecuador (1995/1998) and the LAP (2003–2011), both representing the "community," and on the other were Texaco and Chevron, both constituting the same corporate entity.

Yet were these the same? The final element to effect formal isomorphism—the catalyst that triggered an isomeric relation—was an abstract, categorical, and flattened notion of legal rights being either "individual" or "diffuse." According to Chevron, the legal rights that both the 1995/1998 couplet contract and the Lago Agrio lawsuit sought to protect (the right to an uncontaminated environment and the right to health) were "diffuse rights" as opposed to "individual rights." Following this bifurcation, diffuse rights are public, collective, and indivisible, belonging to a grouping or class of individuals connected by circumstance (i.e., the right to live in a clean environment). By contrast, "individual rights" are discrete and divisible, belonging to a juridical person (i.e., the right to compensation for damage to property or person).[7]

Chevron argued that when the 1995 settlement agreement and 1998 acta final were signed, the Ecuadorian state was the only entity that could seek restitution for an infringement of diffuse rights.[8] The corporation acknowl-edged that a cluster of legal provisions (Articles 2214, 2215, 2229, and 2236 of the Civil Code) provided private individuals with the right to sue parties for personal injury past and future. Similarly, it noted that the 1998 Constitution (Articles 23[6] and 86) gave an individual the right to sue the state for not complying with the broad obligation to protect the environment. But, the corporation contended, at the time of the Texaco-ministry 1995/1998 couplet contract, there was no legal mechanism whereby individuals could gather collectively to file a lawsuit based on an alleged violation of a so-called dif-fuse right.

Furthermore, Chevron underscored that the 1995 agreement (in its Arti-cle 5.2) expressly references Article 19.2 of the Ecuadorian Constitution. Al-though the contract document cites only the numeral, not the text, of this constitutional article, Chevron specifically emphasized in its claim before the tribunal what Article 19.2 states: that all people have "the right to live in an environment free of contamination" and that it is "the State's duty to ensure that this right not be violated and to safeguard the preservation of nature."[9] Because, Chevron argued, the right to a clean environment is a "diffuse right" and because the Ecuadorian state was the only entity at the time entitled to protect and vindicate that right, then "the Government was necessarily rep-resenting the diffuse rights of its citizens in settling with [Texaco]."[10] That is, according to Chevron, in fulfilling its exclusive right and responsibility to protect a diffuse environmental right, the Republic of Ecuador acted as *parens patriae* in its capacity as the representative of the community and public interest when it negotiated and signed its agreements with Texpet.[11]

In the words of one Chevron legal expert, "A ruling resolving conflicts concerning diffuse rights has *erga omnes* effects, that is, it produces effects 'on everyone' or 'towards everyone' and does not merely affect those who partic-ipated actively in the proceedings."[12] Apparently, were this not to be the case, diffuse-rights litigation would have no closure; individuals or groups could claim they were not involved in prior proceedings, that prior rulings did not apply to them, and then endless litigation could ensue. Concurring, another Chevron legal expert declared, "The very notion of representative litigation requires such an approach. If representatives were not able to bind the en-tire community—including any of its members with similar standing to vin-dicate its diffuse rights—they would not really be representing it in a full sense. . . . Allowing reiterated vindication of the same diffuse rights would . . . encroach precisely upon the fundamental aims of efficiency and fairness that the doctrine of *res judicata* seeks to advance."[13] Res judicata, it might be remembered, dictates that what was previously resolved in a final judgment, either through settlement or judicial decision, shall not be litigated again.

And so, with the enumeration of a constitutional article (literally inscribing only "Art. 19.2" alongside a list of Ecuadorian laws), the 1995/1998 couplet contract absorbed and made inconsequential all other claims of rights violations that were not precisely or uniquely harming person and property, the so-called individual rights. Chevron's argument held this logic: because the LAP—like the Ecuadorian state in the 1990s—acted on behalf (as representative) of the affected community, the real party of interest was the "community." That is, the couplet contract and litigation shared the same subjective identity. In fact, the plaintiffs in the Lago Agrio lawsuit were a constitutive part of the "community" that the Republic of Ecuador had already represented during the 1990s remediation negotiations. And because the 2003 lawsuit—like the 1995/1998 couplet contract—sought to restitute alleged petroleum-induced environmental harm, the cause, reason, and right were the same. That is, the couplet contract and litigation shared the same objective identity. As such, the corporation claimed, the doctrines of parens patriae and res judicata deemed the lawsuit against Chevron null and void. The 1995 settlement and 1998 acta final had already settled the *community's* diffuse right to a clean environment and had released Texaco and its successors from further liability for any environmental harm caused by its oil operations.

THE REPUBLIC'S RESPONSE

Dichotomizing rights into two purportedly absolute categories made little sense to the Republic of Ecuador. The fact that the tribunal even entertained configuring Ecuador's legal landscape through this lens was highly troubling to the republic. Ecuador's National Court of Justice (a court of cassation) and Constitutional Court underscored as much in their rulings during Chevron's seven years of appealing the 2011 Lago Agrio judgment in Ecuador. As the republic's lawyers repeatedly indicated, contriving rights in this way aligned with Chevron's claim, but it did not comport with Ecuadorian law.

During the weeks of closed-door hearings before the tribunal in April 2015, the republic's legal team explained that the Lago Agrio complaint and the 2011 judgment relied on three provisions of Ecuadorian tort law outlined in the nation's Civil Code. These three provisions fall into two domains. Articles 2214 and 2229 resonate with conventional common law understandings of tort law: any person who has suffered harm caused by another has a right to file a legal action against the alleged perpetrator. If proven, said individual is liable for the discrete individualized harm that was perpetrated. But Ecuadorian law also honors another form of tort, or wrongdoing, which is embodied in Article 2236 (formerly numbered Article 2260) of the Civil Code. This tort form contemplates legal claims for preventing "the occurrence of a prospective harm" by compelling "the tortfeasor to remove that which creates the risk of prospective harm."[14] Article 2236 reads as follows: "As a general rule, a popular action [*acción popular*] is granted in all cases of contingent harm that threatens indeterminate persons because of someone's imprudence or negligence. But if the harm threatens only determinate persons, only one of them may file the action."[15]

This latter tort provision had no place in Chevron's binary world of tort law where, on the one hand, only the state could vindicate diffuse rights and, on the other, only an individual person could vindicate individual rights. That is, the popular action of Article 2236 had no place within the "partition of the sensible" (Rancière 1999) that constituted Chevron's dichotomous logic. Yet Article 2236—a long-standing tort provision in the Civil Code—figured prominently not only in providing the legal basis for the 2003 Lago Agrio compliant but also in providing a crucial basis upon which the 2011 ruling based its judgment. Chevron's characterization of Ecuador's tort law was incomplete.[16]

One lawyer for the republic, Tomás Leonard, explained the uniqueness of Ecuadorian tort law as follows. Say that contamination is present in an area and has "already caused the final injury to, say, a thousand people; the affected residents have the right to seek reparation from the tortfeasor under article 2214."[17] However, if that contamination also poses "the specter of harm in the future, harm to [individuals'] lives, harm to their health, their property, any one of those people affected by the threat of harm may assert a popular action under Article 2236."[18] These actions are related in important ways. A legal action under Article 2236 functions to prevent "harm from occurring to exactly the same rights" as those subsumed under Article 2214—the only difference is that the popular action is "prospective in nature."[19] That is, both forms of action are structured to vindicate plaintiffs' "individual rights"—meaning they both seek "to protect [plaintiffs'] health, their family, their livestock, their property."[20] Said otherwise, Article 2236 provides a cause of action to compel "the tortfeasor to remove the contamination that is creating a threat of contingent harm." And harm "to what" precisely? To plaintiffs' "individual rights, to their persons."[21]

Consequently, and crucially, Leonard affirmed, the individuals who file a claim under Article 2236 are "not seeking to vindicate any esoteric kind of right to the environment or a right of the environment to be clean from contamination."[22] Their claim is "to protect themselves."[23] Furthermore, as another republic lawyer, Nicole Silver, emphasized, "individual" in the context of Ecuadorian law refers to "individual rights, not harms." She continued: "That the judgment at times refers to public health, does not mean that the rights at issue in the Lago Agrio Case are diffuse ones or, indeed, that they are the same as the diffuse rights that Claimants [Chevron] allege were settled under the 1995 settlement agreement."[24] That the Lago Agrio judgment framed its damages award in terms of public health and remediation "does not morph the individual rights" at issues in the Ecuadorian litigation "into the diffuse rights" that Chevron allegedly "settled under the 1995 settlement agreement."[25] The damages award determined by the Lago Agrio court was commensurate with the risk looming over the plaintiffs and threatening their individual health, family, and property. That is, the rights at issue "stem from the subject harmed, not the Award meant to cure it"; rights emanate from the future harm imperiling individuals, not the specific environmental and health actions said to address that contingent harm.[26] And this was the unique condition of an Article 2236 action: it empowered individuals or groups of individuals to file a claim seeking to mitigate the risk of future harm to their individual person and attachments. But the sought-after mitigation—the legal remedy, if you will—by necessity would affect more than those individuals filing the complaint.

Codified in 1861, the Ecuadorian Civil Code is a systematized body of statutes and rules that govern the relations of persons, goods, property, obligations, and contracts. The Civil Code defines the norms of private law, determines the substantive rights of citizens and legal persons, and establishes the general rules of tort law in Ecuador. As was vociferously made clear since the day the Lago Agrio litigation began in 2003, articles in the Code outline the civil rights of the individual, and no entity, not even the state, has the authority "to dispose of citizens' rights" as defined by the Civil Code.[27] Furthermore, a number of Civil Code articles unequivocally "preclude the possibility that anyone, including the Government, may dispose by settlement of rights pertaining to third parties."[28] Consequently, the lawyers for the republic affirmed that a legal settlement cannot and could not have an "ergo omnes effect."[29]

Given that, the Republic of Ecuador argued that no party to the 1995/1998 couplet contract—neither the government, the ministry, nor Petroecuador—had the authority to "settled claims that any person in Ecuador. . . has a right to pursue under Article 2236."[30] Furthermore, "any attempt to do so would have rendered the agreement null and void by virtue of [the above-mentioned Civil Code] provisions."[31] No matter how vociferously Chevron's lawyers insisted that the Lago Agrio claim and ruling revolved around diffuse concerns, the Republic's lawyers insisted that the "right of action under Article 2236 is not within the scope of the 1995 settlement agreement."[32] And no matter how many times Chevron said it, the republic—following the decision of the Ecuadorian higher courts—"flatly rejected" Chevron's assertion that the Lago Agrio litigation concerned "so-called diffuse rights claims."[33] Rather, the republic insisted that the claims made in the Lago Agrio complaint and addressed in the 2011 judgment were not claims "asserted by citizens to protect some general public interest in the environment."[34] The LAP had asserted "traditional tort law" claims "to remove an imminent threat to their health and safety."[35] These tort claims were anything but the "diffuse rights" released under the 1995/1998 couplet contract.

THE TRIBUNAL'S JUDGMENT

In its 2018 decision, the tribunal sidestepped many of the legal predicates that the Republic of Ecuador outlined. Instead, the tribunal seemed entranced with Chevron's logic that hinged on a universe of rights divided in two: individual and diffuse. It would seem that in the tribunal's estimation this bifurcated rights universe precluded the need to consider any details of Ecuadorian tort law. For Chevron's bifurcated rights universe performed an isomeric function by collapsing the 1995/1998 couplet contract and the 2011 Lago Agrio judgment into the same, thus obviating the need to attend to the intricacies and singularities of Ecuador's Civil Code. In chemistry, isomers are compounds with an identical chemical formula but different atomic configurations and properties. What lends isomerism is not the substantive arrangements and capacities of composition, but rather an idealized formalistic code that is only deduced through reduction (i.e., benzene, whose chemical formula is C_6H_6, has a possible 217 isomers [Nagendrappa 2001], compounds sharing the same number and kind of elements despite enacting vastly different spatial arrangements, properties, and capacities). Coding two legal settlements (the 1995/1998 agreement and the 2011 judgment) through a bifurcated rights universe gave them a formalistic equivalence. This not only enabled the identity of the former to be the identity of the latter, but it also determined the identity of the latter to be redundant, inoperative, and null. The rights formula—"diffuse"—had already been achieved and settled.

Clearly the substantive elements, processes, and capacities composing the late twentieth-century corporate-ministry contract and those composing the early twenty-first-century subaltern-corporate lawsuit were hardly the same. Rather, formalistic identity resulted from the common invocation of a specific right—that articulated in Article 19.2 of the Ecuadorian Constitution: the right to live in an environment free of contamination. That both the 1995/1998 couplet contract and the Lago Agrio complaint and ruling mentioned this right only in passing, in fact, embedded in a sequence of other rights and concerns—with the 1995/1998 corporate-ministry contract never even detailing the article's contents—made no difference. In the tribunal's logic, simply naming Article 19.2 aroused its primacy above other summoned rights.

The tribunal's August 2018 ruling referred to earlier rulings to trace out its reasoning. In September 2013, the tribunal rendered its first ruling (the "First Partial Award on Track I") in the international arbitration. Track I of the arbitration focused on the purpose and effect of the 1995/1998 couplet contract. In its 2013 ruling, the tribunal expended concerted time on discussing Article 19.2 and the nature of its right. To begin, the tribunal determined that the right to live in an environment free of contamination was a "diffuse right." And since the constitutional article says that the state is to safeguard this right, the tribunal concluded that in 1995 "the Respondent [Republic of Ecuador], and only the Respondent, had the legal capacity to make and settle a diffuse claim under Article 19-2."[36] If the republic "could not make and then settle a diffuse claim under Article 19-2, no-one else could."[37]

Furthermore, the tribunal reasoned that in settling a "diffuse claim under Article 19-2" the state settled the claim "forever": And the "right to make an environmental claim based upon the diffuse right under Article 19-2 against the Releasees remained settled 'forever.'"[38] That is to say, the tribunal maintained that "no such diffuse claim could be made in the future against any Releasee."[39] This was the case, according to the tribunal, because the diffuse right granted under article 19-2 was "'indivisible': it was either settled in full or not at all."[40] And the tribunal "reject[ed] entirely" the idea that the diffuse right in Article 19-2 could "exist in separate parts, to be exercised by multiple claimants at different times with successive diffuse claims": such a scenario, the arbiters reasoned, would make "any effective final settlement or adjudication of such claims illusory."[41] Consequently, the tribunal determined that "the scope of the releases" outlined in the 1995/1998 couplet contract did "not extend to any environmental claim made by an individual for personal harm in respect of that individual's rights separate and different from the

Respondent."[42] However, the couplet contract did "have legal effect under Ecuadorian law precluding any 'diffuse' claim against" Chevron or its affiliates "under Article 19-2 of the Constitution made by the Respondent and also made by any individual not claiming personal harm (actual or threatened)."[43]

A couple of years later, in March 2015, the arbitral panel released a second ruling on Track I. In part, the 2015 ruling was meant to clarify its earlier ruling in 2013.[44] It explained that, in establishing that the scope of the release of the couplet contract did not include "any environmental claims made by an individual in respect of personal harm (or damage to personal property) violating that individual's rights, separate and different" from the state's, the tribunal was identifying "a category of claims that will be here referred to as 'individual' claims. Under Ecuadorian law, an individual claim belongs to that individual with the remedy personal to that individual; and it is not a diffuse claim."[45] Similarly, the tribunal explained that in stating that the release of the couplet contract "precluded any diffuse claims . . . under Article 19-2 . . . made by the Respondent and also made by any individual not claiming personal harm or damage to personal property (actual or threatened)," the tribunal was identifying "a category of claims that will be here referred to as 'diffuse' claims. Under Ecuadorian law, a diffuse claim may belong to a community of indeterminate people with the remedy indivisible; and it is not an individual claim."[46] Tellingly, the next paragraph notes: "The tribunal emphasises that the terms 'individual' claims and 'diffuse' claims are used in this decision to denote categories of claims that the tribunal has identified as relevant to its legal analysis of the Parties' respective cases in this decision. These English linguistic (but not legal) terms, as here used, are not otherwise intended by themselves to bear any definitive technical meaning under Ecuadorian, international law or any other law."[47]

In its August 2018 final decision, the tribunal reiterated its rights distinction: "between an individual claim for personal harm by a Lago Agrio Plaintiff (not being a diffuse claim) and a diffuse (or collective) claim. The former is not affected by the 1995 settlement agreement; but the 1995 settlement agreement precludes the latter, expressly so in regard to Article 19.2 of the 1978 Constitution."[48] Unabashedly, the tribunal concluded that four Ecuadorian courts all had misinterpreted their own national sovereign law. In turn, the Provincial Court of Justice of Sucumbíos, the Sucumbíos Appeal Division, the National Court of Justice, and the Constitutional Court each "deprive[d]" the "1995 settlement agreement . . . of any practical meaning, making it a one-sided and open-ended commitment undertaken unilaterally by . . . Texaco" and ultimately Chevron.[49]

In a brazen move eliding the "orthodox approach" in international law that extends "comity or due respect" to a nation's judicial branch, an international panel of three arbiters contradicted the rulings of Ecuador's two higher courts. Despite acknowledging that "the considered judgment of any municipal court applying its own municipal laws" is best practice for understanding "the content and application of that law," the tribunal members appeared to believe that they knew Ecuadorian law better than Ecuador's most esteemed justices, attorney general, and legal scholars.[50] Steeped in the realm of commercial law, the tribunal decision stated that "the tribunal regrets its inability to follow the Constitutional Court's interpretation and application of the 1995 settlement agreement in regard to Chevron's liability under the Lago Agrio Judgment."[51] The tribunal pronounced that it was "driven to conclude . . . that the decision of the Constitutional Court is inconsistent with the effect that the treaty [BIT] requires to be given to the 1995 settlement agreement."[52]

Overriding the need to account for Ecuador's Civil Code and tort law, the tribunal funneled its ruling through a conjured universe of bifurcated rights. The tribunal reduced differently valenced legal settlements into a rights formula. One rights formula—diffuse—overdetermined both the 1995/1998 couplet contract and the 2011 Lago Agrio ruling, rendering the two legal processes formally isomeric. The function of this isomeric relation was to make a poorly executed contract negate a seven-year litigation *and* to make a purportedly wrongfully lodged and adjudicated litigation transmuted into a breach of treaty. The tribunal "found" that all four Ecuadorian judgments "rest[ed] upon finding Chevron liable for diffuse claims in noncompliance with" the Republic's "obligations to release Chevron . . . from such liability under the 1995 Settlement Agreement."[53] As a consequence, through the "acts of its judicial branch," the Republic of Ecuador violated its "obligations under Article II(3)(c) of the Treaty, thereby committing international wrongs towards each of Chevron and TexPet."[54]

Conclusion

CONJURING DIFFUSE RIGHTS

The idea that a contract between the corporation and a state ministry could negate virtually fifteen years of litigation (nearly eight in trial and over seven in appeals) seemed absurd to many in Ecuador. As the LAP lawyers saw it, the Lago Agrio lawsuit in no way "deprived" (as Chevron claimed) the 1995/1998 couplet contract "of any practical meaning." Regardless of what one thought, the couplet contract was enforceable under Ecuadorian law for its named parties. The language of the contract precisely named those parties and what was being released by whom: "the Government and Petroecuador . . . release[d], acquit[ed], and forever discharge[d] Texpet . . . Texaco, Inc. . . . of all the Government's and Petroecuador's claims against the Releases."[55] Moreover, the memorandum of understanding (MOU) signed in anticipation of the 1995 agreement explicitly excluded and protected the rights of third parties.[56] The LAP lawyers underscored that this was in line with all prior agreements. When, over the course of Texaco's operations, the ministry sanctioned the company for a spill or rupture, the state explicitly expressed in writing that the rights of third parties were not compromised by those resolutions. A number of documents in the Lago Agrio case file demonstrated this. Despite the corporation's pleas, the republic refused to include local inhabitants in the 1995/1998 couplet contract. The contract could not and did not bind the Lago Agrio plaintiffs.

Sitting in the legal team's Lago Agrio office surrounded on three sides by floor-to-ceiling bookshelves crammed with the litigation case file, Pablo Fajardo explained it like this:

> Our lawsuit was filed in late 1993. The 1995 contract was signed eighteen months later. And why was Chevron so fervent about sealing that deal? Because it wanted to use it as leverage and get the New York judge to dismiss our case. Chevron didn't return [to the Amazon] to remediate because it was nice, because it wanted to do the right thing. No, it came back because it wanted to terminate our legal action and it made this argument before every court presiding over it: "With all due respect, your Honor, you must nullify this case because I've already negotiated and settled the issue." Think about it. I file a lawsuit against you, Suzana, because you perpetrated a harm against me. Tell me, can you go and negotiate with my government my right to file a legal claim? Take our case. If the lawsuit is filed on behalf of Amazonian peoples, how is it that Chevron can negotiate with the state over a lawsuit in which the state is not party? Like I said, the state has no capacity to negotiate away the rights of third parties. In the United States, can Mr. Donald Trump bargain away your rights? No! Never would he be allowed that. Why is it that here the rights of people are negotiable?[57]

The "substantive law" under which Chevron was tried was tort law set out in the Republic's Civil Code of 1861. "If you want to fall for the trumped-up game of qualifying rights as individual or diffuse," Fajardo continued, "the Civil Code long predates any theorizing along those lines. Think about it—in the 1800s the concept of diffuse rights did not exist."[58] Indeed, the debate over what a diffuse right might constitute was quite recent in Ecuador—it appeared in the state's legal lexicon with the 2008 Constitution. It followed then that this notion was outside the realm of legal discourse not only in the mid-nineteenth century but also at the end of the twentieth century. The term *diffuse* appeared nowhere in the 1995/1998 couplet contract, nor, according to the Republic of Ecuador's lawyers, could the notion of "diffuse rights" have been a concept in the minds of the lawyers negotiating it at the time.[59]

Scratch the surface only slightly and the tribunal's assertion that Article 19.2 of the 1979 Constitution represents "diffuse rights" becomes problematic. To begin, the term "diffuse rights" appears nowhere in the 1979 Constitution. To wit, Article 19.2 does concern "the right to live in an environment free of contamination." And, yes, it declares that it is "the State's duty to ensure that this right not be violated and to safeguard the preservation of nature."[60] But two elements trouble the tribunal's claim that a diffuse right is at play here. In the 1979 Constitution, Article 19.2 appears in "Section I: On the Rights of Persons" under "Title II: On Rights, Duties, and Guarantees." The article before it (19.1) concerns "the inviolability of life and personal integrity." Importantly, the right to live in a clean environment directly follows the right to life. And although these rights are safeguarded by the state—in fact, the preamble to Article 19 declares that "the State guarantees" every enumerated right under it—the rights elaborated here are individual rights. They are among "the Rights of Persons." Moreover, after noting the state's duty, Article 19.2 goes on to clarify that "the law will establish restrictions on the exercise of determined rights and liberties in order to protect the environment." That is, specific statutes in Ecuadorian law detail the measures needed to ensure a clean environment. Infringement of those statutes is a violation of "the right to live in an environment free of contamination" and thus would become the basis for how that right is violated. The Lago Agrio lawsuit against Chevron detailed precisely that.

Moreover, as the Republic of Ecuador's lawyers had made clear, it was long-standing tort provisions of Ecuador's Civil Code, not Article 19.2 of the Constitution, that grounded the Lago Agrio complaint and 2011 judgment. As noted, a particularly important provision was Article 2236, giving persons the right to file an action to remedy or prevent an imminent or contingent harm endangering their person and that of others. By definition, a favorable ruling under the "acción popular" of Article 2236 would address harm to the persons filing the legal action *and* then extend beyond them to others. This does not make the right to such mitigation a diffuse right. It does not turn the specific actions that plaintiffs press into a diffuse claim. And it does not turn the remedial action awarded by a court into a diffuse judgment. As a lawyer for the Republic of Ecuador noted, "the legal community began developing the notion of diffuse rights more than a century and a half" after the Civil Code was written. This makes it impossible to claim "that Article 2236 was . . . enacted as a mechanism to assert only diffuse claims."[61]

The Lago Agrio legal complaint and the 2011 judgment did not seek simply to address a generic requisite to live in an environment free of contamination. Rather, the 2003 claim and 2011 judgment attended to specific threatening wrongs resulting from Texaco's negligent and reckless conduct: formation waters siphoned into specific streams and swamps, hydrocarbons seeping into specific soil systems, crude oil filtering into specific adjacent waters, gas flares torching the skies of specific regions. These claims are claims of discrete contamination that exposed local peoples to crude oil and noxious fumes in waterways, in soils, and in the air. The 1995/1998 couplet contract was never intended to cover claims by third parties of the type pressed in the New York federal court in 1993 or refiled in the Provincial Court of Justice of Sucumbíos in 2003.

ARBITRAL ALCHEMY

The debate over dichotomizing rights as singularly individual or collectively diffuse was clearly more than a game of semantics. It rubbed hard against the legal logic that sent the lawsuit to Ecuador in the first place back in 2001. Recall that after the *Aguinda* lawsuit was first filed in the New York federal court in November 1993, it spent nearly ten years ricocheting between the New York district court and the US Court of Appeals for the Second Circuit (Sawyer 2001, 2002). During that time Texaco argued vehemently that the case should be litigated in Ecuador, plying the court with numerous submissions pronouncing Ecuador as a more effective and appropriate forum. The corporation submitted fourteen affidavits by Ecuadorian lawyers and legal scholars who assured the US Court of Appeals for the Second Circuit that Ecuadorian courts were fair and that plaintiffs would find no impediments to refiling their claims in Ecuador. As such, in remanding the case to Ecuador, the US federal court needed to assure that the Ecuadorian courts were competent to hear plaintiffs' claims and sought-after relief.[62]

Adjusted to abide by Ecuadorian law, the lawsuit filed in Lago Agrio in 2003 mirrored the one filed in New York in 1993. The claims made and the relief sought by the plaintiffs in the New York suit were repeated in the Lago Agrio suit in substantially identical terms—both sought relief in three areas: remediation of contamination in soils and streams, implementation of a medical monitoring system, and compensation for destroyed Indigenous lifeways. In fact, the very same court that had remanded the case in 2002 had reiterated in 2011 (in the case in which the Republic of Ecuador sought to stay the BIT arbitration at the American Arbitration Association): "Chevron's contention that the Lago Agrio litigation is not the refiled Aguinda action is without merit. The Lago Agrio plaintiffs are substantially the same as those who brought the suit in the Southern District of New York, and the claims now being asserted in Lago Agrio are the Ecuadorian equivalent of those dismissed on forum non conveniens grounds."[63]

Chevron's argument, however, before the international tribunal hinged on those claims not being the same: the corporation maintained that, in the United States, individual rights were what plaintiffs sought to vindicate, whereas, in Ecuador, their claim concerned diffuse rights. Once coded as diffuse rights, the Lago Agrio claim was barred by the 1995/1998 couplet contract. Remarkably, the PCA tribunal concurred, determining that the plaintiffs' 1993 claims—viable in New York then twenty-five years earlier—were barred in Ecuador. The arbiters declared that the Lago Agrio lawsuit comprised diffuse rights already and forever settled and released by a corporate-state contract executed six years prior to Texaco convincing the New York federal courts in 2001 that Ecuador was a more convenient forum in which to adjudicate plaintiffs' claims.

In the eyes of the LAP lawyers, *this* was the "fraud"—fraud on the US federal court, fraud on Indigenous rights, fraud on fundamental notions of legal justice. As the legal team for the Republic of Ecuador made clear before the tribunal, "As a matter of U.S. law, the Second Circuit in New York would not and could not have granted [Chevron's] requested dismissal on *forum non conveniens* if the same claims could not be heard and adjudicated in Ecuador."[64] As Chevron well knew, dismissal on the grounds of *forum non conveniens* necessitated securing an adequate alternate legal forum; if such a forum could not be secured, the case would have to be litigated in the New York district court.

Marginalized and humble forest peoples would watch their 1993 New York claim be sent to Ecuador in 2001, and would participate in that claim's litigation for nearly eight years (2003–11) and prevail, only to have an international arbitration, in which they had no part, tell them in 2018 that they no longer had a claim and that the 2011 judgment award was null and void. The cruel imperial irony reverberated deeply. The LAP insisted that not one of the claims presented in the Lago Agrio lawsuit—of contaminated lands and waterways, of ravaged health, of destroying Indigenous lifeways—was settled by the 1995/1998 agreement. Even Judge Sands of the US District Court for the Southern District of New York, upon hearing the same breach-of-contract claim, determined that it would have been "highly unlikely that a settlement entered into while *Aguinda* was pending [in the New York federal court] would have neglected to mention the third-party claims contemporaneously made in Aguinda, if it had been intended to release those claims."[65] The tribunal's actions deprived Indigenous and non-Indigenous people of their ability to have their claims heard by a court of law.

Established in 1899, the PCA was originally designated as a forum for resolving disputes between states. Over the past fifty years, the court transformed to become a key institution shaping public and private international commercial law. When in 1976 the UN Commission on International Trade Law (UNCITRAL) adopted its arbitration rules (later revised in 2010), UN member states appointed the PCA the role of overseeing arbitration proceedings between parties representing different legal, economic and social systems from around the world.[66] As the PCA increasingly implemented UNICTRAL rules, the court increasingly arbitrated trade disputes involving various combinations of state entities, intergovernmental organizations, and private parties.

As one among other arbitral bodies around the world, the PCA partakes in a modestly shrouded "gentlemanly" system of dispute resolution where an elite cadre of private jurists—who have called themselves "the Club"—circulate as judge and jury in arbitration cases unfolding around the globe (Dezalay and Garth 1996). The amount earned by arbiters and spent on legal fees easily rises into the multi-multi-millions when large corporations are concerned. This generates perverse incentives in the shape and duration of arbitral litigation. The secrecy, monetary incentives, and unbridled jurisdictional authority empowering arbitral proceedings have been roundly critiqued (Cutler 2016; Cutler and Lark 2017; Gélinas 2017; Goldhaber 2013; Van Harten 2008, 2013). The near decade-long proceedings in the Chevron / Republic of Ecuador arbitration were conducted behind closed doors, barring the LAP from attending as observers, witnesses, or derivative respondents. Partially redacted transcripts of months of hearings, years of submissions to the tribunal, and the tribunal interim and final judgments were eventually

TABLE 7 **Major International Oil Spills**

Oil Spill, Location, and Year	Magnitude of Spill	Cleanup Costs and Damages (in US $ 2008)	Reference
Prestige oil spill, Coast of Spain, 2002	20 million gallons; cleanup began immediately	$2 to $3 billion in cleanup costs (actual) $1.2 billion in damage claims	IOPCF 2002 New York Times 2003a, 2003b
Exxon Valdez, Valdez, Alaska, 1989	11 million gallons; cleanup began immediately	$2.9 billion in cleanup costs (actual) $4.1 billion in damages claims (settled for $1 billion) $3.6 billion in punitive damages (reduced to $500 million on appeal)	Exxon Valdez Oil Spill Trustee Council 2017 Duffield 1997
Amoco Cadiz, Brittany, France, 1978	186 miles of coastline; cleanup began immediately	$3.4 billion in cleanup costs and damages (actual)	New York Times 1989 Lenntech 2006
Oil spills in Kuwait from Gulf War, 1991	100 square miles; contaminated for several years before cleanup	$2.2 billion in cleanup costs (claim amount granted by UNCC)	UNCC Governing Council 2005

Source: "Track 2 Rejoinder on the Merits of the Republic of Ecuador (Part 1: Response to Factual Predicate to Claimants' Claims)," In the Matter of an Arbitration, Case No. 2009-23, December 16, 2013, p. 92, PCA, The Hague.

made public after the fact, following public pressure, document leaks, and the parties' consent.

The Republic of Ecuador invested heavily in this arbitration. Its prosecutor general's office hired top-notch US corporate lawyers and scientific experts to build and argue their case. In one of the legal team's final submissions to the tribunal they presented information regarding international oil spills (see table 7).

The $5.4 billion that the 2011 Ecuadorian judgment had designated for cleanup costs was hardly out of the ordinary. Contemporaneous corporate and ministry documents showed that, at a minimum, Texaco operations spilled sixteen million gallons of crude oil and twelve to twenty billion gallons of toxic-produced water into the Upper Amazon. Far from manipulating evidence to give presence to contamination where it was not, the LAP legal and scientific team presented to the Lago Agrio court a reality reflective of the degrading effects of decades of oil-industry activities in the rainforest. Extensive analysis conducted by scientific experts on behalf of the Republic of Ecuador thoroughly substantiated this enchainment for the tribunal.

Ultimately, however, the concept of "diffuse rights" proved ingenious in its capacity to sway the tribunal's arbiters—all lawyers schooled in the liberal sanctity of contract law and seasoned in investment arbitrations. Diffuse rights served as the catalyst that performed both a formal isomorphism (making isomers of the Lago Agrio lawsuit and the 1995/1998 couplet contract) and metamorphosis (turning a contamination lawsuit into an investment treaty dispute). In chemistry, a catalyst is a body relationally beckoned to trigger a chemical reaction yet in the reactive process is not consumed—meaning, a catalyst retains its form and continued capacity to excite further reactions. Clearly, the industry anticipated future legal successes by deploying the notion of diffuse rights. In a public presentation to industry representatives, Chevron counsel guaranteed that through the invocation of "diffuse rights" parties could be "assured their settlement is *forever*."[67]

This diffuse/individual rights bifurcation was absurd in the Amazon. As Pablo Fajardo repeatedly clarified (most recently in his UDAPT office in July and August 2019), Ecuadorian tort law, as codified in Ecuador's Civil Code of 1861, determined that individuals had the right to sue an entity on behalf of others for negligent actions that caused present or future threat to person, property, and their attachments. The law clearly understood that any threatening harm when remediated would reach far beyond the interest of one individual. Removal of an impending harm would benefit a collectivity. This was a plurivalent composite right, neither singular nor collective, neither individual nor diffuse. It simultaneously enfolded individual rights and far exceeded them. It fell outside a binary legal lexicon. Outside the realm of dichotomized categorical distinctions, this rights-worlding inherent in Ecuadorian law generated spaces for the recognition of speculative dangers to an array of living beings. Indeed, over 150 years ago, Ecuador's Civil Code presaged the need for slowing down, the necessity for precaution, when considering the effects of an unscrupulously wily industry.

The metamorphic journey of the lawsuit against Chevron is, in turn, inspiring, tortured, and distressing. My intention in *The Small Matter of Suing Chevron* has been to invite you to accompany me in exploring how the legal saga's brilliantly emboldened and debilitating changing composition came to be assembled. During the litigation of the original lawsuit in Lago Agrio, the spheres of chemistry, of epidemiology, and of contract law creatively amassed distinctive bonds to constellate opposing arguments, while the elemental emplacement performed through the judicial inspections exposed a distinct compositional uniformity whose prehensive knowing seared through scientific and legal uncertainty and doubt. The 2011 Ecuadorian ruling entrained all these phenomena—enlisting them in the flow of its argument—subsuming them and reckoning their valenced import when filtered through a precautionary logic. Whiplashed into a US judicial forum, the lawsuit transmogrified. A contamination lawsuit over industrial activity, which the 2011 Ecuadorian judgment determined to have debilitated life-forms, then mutated into a fraud-and-racketeering scandal suitable for the mob. Here constitutive elements essential to the case in Ecuador (the materiality of Texaco's operations and their chemical, statistical, contractual, and sensate embodying) morphed into purported machinations of conspiracy, fraud, and corruption. The ease with which a limitlessly funded civil RICO case could be launched and prevail was astounding.

Twenty-five years after the lawsuit was first filed in 1993, a tribunal of the PCA in The Hague performed an ultimate alchemical act. Traversing across time and space—a quarter century, three continents, and three jurisdictions— the tribunal's ruling transformed a contamination lawsuit and judgment into the breach of a BIT in which the Lago Agrio plaintiffs had neither voice nor standing. In the tribunal's "juris diction," misconstrued chemistry and misguided contamination assessments triggered (together with the distortion of LAP slips) an elaborate narrative of connivance, corruption, and fraud that transmogrified the Lago Agrio litigation into a "denial of justice" under treaty terms. What surfaced, smugly untainted, was a corporate-state deal—the 1995/1998 couplet contract. Authored under less than transparent circumstances and inadequately fulfilled, the Texaco-ministry agreement alchemically emerged unscathed as a contract blessed under the US-Ecuador BIT of 1997. Reading the couplet contract narrowly and prejudicially—such that it applied to beings not party to it—the tribunal declared that the Republic of Ecuador breached its treaty obligations and international law by allowing the Lago Agrio litigation to proceed and by upholding its judgment. To transmogrify a "contamination claim" into a "breach of treaty" was to execute the achievement of a piercingly malevolent mixt.

The Small Matter of Suing Chevron has taken as its sphere of inquiry crude's "valence of truths." Leaning on insights from the philosophy of chemistry, it has delved into how competing truth facts at the core in this legal saga were, far from absolutes, emergences of collective composition: the often-arduous, agitated, and viscously transformative combining effects of, with, and through crude oil. The world of chemistry is one of compositional entities. Chemistry offers a grammar for understanding as collective, for capturing the different modalities that constitute relational beingness, and for knowing that complex entities are never only the sum of their parts. To speak of valence means to speak of a relationally constitutive reality in which entities are never singular or fixed but rather are always vibrant achievements of merged becomings. It is to hold the world and worlding as composed of entities with "relations integrally implied" (de la Cadena 2015: 32). Thinking with valence—that combining power (with its tendencies and affordances) of chemical agencies—as a method device guided me into the complexity of the amalgam of law, science, and crude. Similarly, probing crude's valence of truths has triggered reflection on how method makes manifests its object.

Increasingly, conflicts over the effects of industrial activity, extractive activity in particular, are morphing into environmental legal disputes in jurisdictions across the globe. Given that ever-multiplying events of indeterminate harm are seen to result from this corporate activity, it matters how we engage the valence of truths surfacing in these controversies. Thinking with philosophers of chemistry provided an alternative modality for considering a legal trilogy haunted by a righteous urgency that is palpable from all sides— and it extends similar possibilities for considering other socioecological legal conflicts. The anticipated narrative here is that of the rapacious corporation— the corporation that, after decades of imperious extraction, is finally called to account, only to then turn around and shaft its victims. This story is not wrong. Indeed, it resonates with the all-too-predictable events defining lived experience for many in the Amazon. But it does so both too tightly (narrowly) and too loosely (vaguely). Thinking with chemical philosophy permitted my thought to abide in rich relational paradox—to think with a science whose fundamental elements evince simultaneously as abstract absolutes, material fabrications, *and* wily uncertainties of immanence.

Rendered in very different (but related) litigations over very different (but related) claims in very different legal forums, the 2011 Ecuadorian, 2014 US, and 2018 international rulings reflect divergent modalities for making legal truths. As *Small Matter* has detailed, in the sphere of the law, the making of legal truths about harm (whether of alleged contamination or alleged fraud) was materialized, and dematerialized, molecularly, scientifically, technically, legally, and vernacularly. Corporate action clearly was ensconced in those achievements. But so too were other compositions, associations, and assemblages. Consequently, as I have argued here, the tractable and intractable conditions of harm wrought and subsequently disavowed by petrocapital cannot merely be denounced as the workings of corporate power. Although crucial, denunciation is also insufficient. Denunciation offers "the seductive clarity" (Redfield 2005: 349), but it also elides the dilemma that we in the petro-techno zones of privilege are profoundly complicit in the very industry we condemn. Complicity invites discomfort and asks more of us—a tact, a discernment, a sensibility that eschews comforting binaries, hierarchies, and transcendence. And it encourages us to inspect (not to portray as separate) the bonds we sustain with and through crude oil. Doing so interrupts the idea that what corporations do is simply lie. They very well may. But what makes the oil industry, and Chevron in this case, so powerful is not that they lie about and falsify the real. It is that they generate entire worlds, and those worlds compose and recompose a plethora of entities, processes, and beings in coalescing "truths." Key to this process is law and liberal legality more broadly. This orientation differs from that of many Chevron critics— where accusations of judicial prejudice, bias, and unfair treatment constellate the core critique against the corporation and its RICO trial. It also sustains a mode of analysis that may prove more difficult for corporate proponents to dismiss.

Not uncommonly, our language infuses law with a moral imperative, making it seem like the law's ultimate work is to allocate justice. But justice and the law are two different concepts. The law sets the terms and parameters for what might *constitute* justice. It has no morality of its own. Perhaps more accurately, following a number of legal scholars (Beck 1992; Jain 2006; Jasanoff 1995, 2012), law functions as a system that calibrates, in specific contexts and within specific technical constraints, what compositional relations might acceptably hold among persons and things, among bodily health and economic vitality, among corporeal cohesion and corporate profit, among contractual leverage and constitutional obligations. *The Small Matter of Suing Chevron* has attended to the valence whereby crude in association with law, understood as such, partakes in configuring how risk and reward, deprivation and disavowal, suffering and surfeit become legally distributed all too often unequally across the globe.

Inspired by Lexie Groper's work with Amisacho Restauración (https://amisacho.com)

FIGURE 56 *Pleurotus djamor,* or the pink oyster mushroom, as vibrant and inspired as Amisacho. Source: http://amisacho.com/cultivo-de-hongos/. Photo by Lexie Groper. Used with kind permission.

In a region of verdant tropical diversity, crude slips along waters, envelops surfaces, embeds into texture, wafts through airs, seeps through interstices, depletes life. It transudes and alters earthly forms. The Moderns—those equivocal souls living under the delusion that this earth is indifferent to their fantasies of endless growth—turned subterranean ecologies into assets as lived ecologies gave way to bottom lines. Black gold's glimmering promise blinds. The abundance of once-thriving forests is of no consequence, and its loss less so.

And yet. Pause for a moment, if you can, amid the compromised undergrowth of contaminated lands. Pay heed to the flourishing below fallen trees. Or beneath the organic debris that holds hydrocarbon residue from the waste pit a kilometer away. Sporing potentialities germinate and spread resilient roots within the forest floor. Communicating signals, transferring energy, sharing nutrients, absorbing knowings, composing new worlds in reactive combination with other worlds around. Animated by mycelial networks, these sporing beings impart secrets for recuperating forests and tending terrestrials. The two-legged, the four-legged, the winged, the squigglers, the rooted, and the airborne earthbounds.

Moving and moved, sporing faculties spawn orbital portals for collective and collaborative concrescence amid disrupted and impaired ecologies. Fungi and bacteria, nutrients and elements, leaf litter and scrappy trees, critters and humans—young, aging, women, men, Indigenous, mestizo—compose pluralities compelled to intensify their valenced connections. They are metamorphizing agents constitutive of the immanent processual ecologies of their Amazonian home.

Weighed by six decades of slowly seeping violence, sporing nonhumans and enseeded humans are through their distributed agencies forming compositions that animate dependencies, limitations, transformation. Tweaking the words of Gwendolyn Brooks, they are "each other's magnitude and bond" (Brooks 2005: 228).

"Amisacho" is what A'i Kofan called their Indigenous forest community that once thrived on the site where equivocal foreign souls christened Lago Agrio a petroempire they soon forgot. Amisacho Restauración seeks to create anew its postpetroleum figuration. Located in the epicenter of Texaco's historic oil fields, with a capped Texaco oil well fewer than fifty meters away, Amisacho Restauración is a tendrilled project more than a place. An inspired collective that welcomes, houses, and draws within its pull that which prepetrol Amisacho relished. Commitments for abiding in and with. Bonding magnitudes that generate attachments—attachments that alter participants and constitutive earthly processes.

And, thus, tantalizing possibilities emerge from Amisacho's very form. Once a site of industrial refuse and waste from Pozo-26 and the site of infertile pastures from the colonos who homesteaded, Amisacho allures the airborne, the rooted, the squigglers, the winged, and the multilegged to collaboratively feast and share. It has taken over a decade, but Amisacho is no longer its former degraded pastures and contaminated soils. The collective tending of mycelial soils has yielded for Amazonian plant, animal, and human souls.

The two-leggeds at Amisacho also thrive in alliance with other two-leggeds—those of the Unión de Afectados y Afectadas por las Operaciones Petroleras de Texaco (the entity that peasant and indígena plaintiffs formed upon suing Chevron), La Clinica Ambiental (an innovative project of socioecological reparations and healing), and Alianza Ceibo (an Indigenous alliance that, together with Amazon Frontlines, advocates for Indigenous communities in oil-ravaged lands of the Upper Amazon). Each of these collectives deeply warrants attention. And, surely, the writing of Lindsay Ofrias and Amelia Fiske, among others, will open toward that end.

For now, sink deeper into the project of Amisacho. A venture, an obligation, to collaborate in regenerating ruined land and ruined bodies by guiding mycelial bioremediation and distilling fungi medicines. A colaboring with diverse Amazonian communities—other-than-human and human—affected by oil operations. In dense collaboration, Amisacho thrives to transform the possibilities for being well amid the Upper Amazon's imperial ruin.

Among the array of initiatives spawned from Amisacho's rainforest island, the mycelial trials highly intrigue. Beneath a ceiba tree, a small laboratory experiments with fungal adventures in bioremediation. To remediate is to heal, again. To bioremediate is to catalyze collectivities, to enact interdependencies, where fungi and bacteria, plants and critters co-labor with humans in a healing art of transfiguration. Enlaced in rapport with bacteria and microorganisms, suffusing soils and borne across air, fungi break down entities into their molecular constituents. The Amisacho lab seeks to multiply those reactions.

For years now, knowers have enlisted bacteria to biodegrade light aromatic and aliphatic petroleum hydrocarbons—a process shown to be effective in quelling benzene contamination immediately following a spill. Only more recently, however, have fungi joined remediating initiatives, spawning experimental efforts in mycoremediation. An alchemical craft in which fungi are experts. Fungi have loyally served as the earth's great decomposers long before the leggeds found their balance. Infusing an earthly substrate with legions of mycelial rootlike filaments, fungi exude enzymes from their threaded laby-rinth that transform matter into metabolites and diffuse them toward the my-celium. Imagine. Extracellularly, with an alchemical capacity that extends be-yond their somata, fungi break down some of the largest molecules in formerly living matter. This includes lignin—that complex organic compound worthy of holding trees as they stretch to the skies . . . a molecular structure analogous to that of complex petroleum aromatic hydrocarbons (PAHs, those compounds composed of multiple benzene rings) with high molecular weight.

In rainforest shade, mycelial trials nurture the performance of a fungal menagerie, perfecting its achievements of decomposition. Experiments play with devising alternative adequate conditions for the earth's fungal capac-ities to decompose crude compounds, making them available perhaps to deeper nutrient cycles. Composition and decomposition, the intimacies and proximities, the intensities and temporalities, a worlding of postpetro-leum earthbounds composed of beings with relations constitutively implied. Where being each other's magnitude and bond is not additive but triggers metamorphosis.

FIGURE 57 Mycelia encountering crude oil. Source: Amisacho Restauración web page, http://amisacho.com/. Photo by Lexie Groper. Used with kind permission.

NOTES

Fraud

1 "Chevron/Ecuador: The 'Legal Fraud of the Century' in 3 Minutes," posted March 25, 2015, https://www.chevron.com/ecuador/.
2 This quote is directly from the text of the web page "Ecuador Lawsuit," Chevron website, accessed May 1, 2015, https://www.chevron.com/ecuador/. All other quotes are from the Chevron video cited in the previous note.

Opening

1 *Maria Aguinda Salazar y Otros v. ChevronTexaco Corp.*, Case No. 002-2003-P-CS-JNL (2011-63-1), Provincial Court of Justice of Sucumbíos, Nueva Loja, Ecuador. Ecuadorian plaintiffs filed their complaint on May 7, 2003, in the Superior Court of Justice of Nueva Loja (renamed the Provincial Court of Justice of Sucumbíos). Judge Nicolás Zambrano Lozada rendered his judgment on February 14, 2011. Having merged to form ChevronTexaco in 2001, the corporation changed its name again to Chevron in 2005.
2 *Chevron Corp. v. Donziger*, 974 F. Supp. 2d 362 (S.D.N.Y. 2014), ECF 11 Civ. 0691 (LAK).
3 The following year, the US Supreme Court declined to review that decision.
4 I am indebted to other anthropologists who have intervened in controversies concerning the lethal or less-than-lethal effects of corporate activity: among them are Hannah Appel (2012, 2019), Andrew Barry (2013, 2020), Kim Fortun (2001, 2010, 2012), Stuart Kirsch (2006, 2014, 2018), and Sara Wylie (2018).
5 My analysis in this book is deeply informed by and in conversation with critical studies by anthropologists and kin: Kim Fortun (2001), Aya Hirata Kimura (2016), Lochlann Jain (2006, 2013), Stuart Kirsch (2018), Max Liboiron (2012),

Michelle Murphy (2006), Adriana Petryna (2002), and Sara Wylie (2018). Each of these works explores in distinct spaces and times the tense intersections of modernist production, science, and law. Susanna Rankin Bohme (2014) richly chronicles a notable exception of claims of the marginalized people (in this case from Central America) prevailing against a corporation (Dole) in the US court of law.

6 In November 1993, US-based lawyers first filed the class-action lawsuit against Texaco Inc. (*Aguinda et al. v. Texaco Inc.*, ECF 93 Civ. 7527 [VLB] [S.D.N.Y. complaint filed November 3, 1993]). Three years after its original filing, the case was dismissed from the New York federal court in November 1996. In light of new evidence, the plaintiffs petitioned later that year that the court reconsider its decision. In August 1997, the district court dismissed the case once more. The following year, in October 1998, the US Court of Appeals for the Second Circuit reversed the lower court decision and reinstated the case. Three years later, in May 2001, the New York district court dismissed the case once more. In August 2002, the same court of appeals heard the case again but this time upheld the lower court's decision and ruled the case be sent to Ecuador. In May 2003, the case was accepted in the Superior Court of Justice of Nueva Loja. For ethnographic analysis of the US legal claim, see Sawyer 2001, 2002. See also *Aguinda v. Texaco Inc.*, 142 F. Supp. 2d 534 (S.D.N.Y. 2001); and *Aguinda v. Texaco Inc.*, 303 F.3d 470, 473 (2d Cir. 2002).

7 See the works of Constable (2014a, 2014b) and Cormack (2007) that brilliantly and distinctively explore the matter of truth and the truth of the matter.

8 Etymologically, *verdit* is Middle English from medieval Latin *veredictum* (true saying) from classical Latin *vere* (truly) + *dictum* (a thing said). The term *verdict* is conventionally used in common law to refer to a "jury's findings or conclusions on the factual issues presented by a case," although the term may also refer to a "judge's resolution of issues in a bench trial" (Cornell Law School, Legal Information Institution, accessed June 14, 2019, https://www.law.cornell.edu/wex/verdict). Not all legal dictionaries concur, however, claiming that a "judgment by a judge sitting without a jury is not a verdict" (Gerald and Kathleen Hill, *The People's Law Dictionary*, accessed June 14, 2019, https://dictionary.law.com/Default.aspx?selected=2217).

Despite this equivocation, the United States court, in a "Glossary of Legal Terms," defines verdict as "the decision of a trial jury or a judge that determines the guilt or innocence of a criminal defendant, or that determines the final outcome of a civil case" (Administrative Office of the US Courts on Behalf of the Federal Judiciary, "Glossary of Legal Terms," accessed November 14, 2019, https://www.uscourts.gov/glossary#letter_v).

In Ecuador the term *veredicto*, while less commonly used than *sentencia* or *laudo arbitral*, refers to a ruling of the court.

9 As Strathern notes, "Complexity in this sense denotes systems not just heterogenous in composition but open-ended in extent" (1995: 40). And Law relates that "events and processes are not simply complex in the sense that they are

technically difficult to grasp (though this is certainly often the case). Rather, they are also complex because they *necessarily exceed our capacity to know them"* (2004: 6; emphasis in the original).

10 Strathern's words are "it matters what ideas one uses to think other ideas (with)" (1992: 10). See also Haraway: "It matters what matters we use to think other matters with; it matters what stories we tell to tell other stories with; it matters what knots knot knots, what thoughts think thoughts, what descriptions describe descriptions, what ties tie ties" (2016: 12).

11 Feature-length documentary in production, directed by Lindsay Ofrias, produced by Myles Estey, Leo Cerda, and Jonathan Gray.

12 *Maria Aguinda Salazar y Otros v. ChevronTexaco Corp.*, Case No. 002-2003-P-CS-JNL, filed May 7, 2003; Ley de Gestión Ambiental [Environmental Management Law], Official Registry, Record No. 245, Articles 41–43 (July 31, 1999).

13 Ecuadorian tort law, as codified in the 1861 Civil Code (Article 2260 of the Civil Code, later renumbered as Article 2236 [paragraph V.1(b)]), stipulated that an individual or group of individuals threatened by a future risk could sue the offending party and demand the threat be remediated. Although remediating measures invariably would mitigate risk to others in addition to the suing individual(s), no specific legal procedure existed in Ecuador for collective legal action for environmental harm. This was not seen, however, as an obstacle to its enactment.

See also Ley de Gestión Ambiental, No. 99–37. Article 41 of this law reads as follows: "In order to protect individual or collective environmental rights, public action is granted to natural persons, legal entities or human groups to denounce the violation of environmental norms, without prejudice to the constitutional protection action provided for in the Political Constitution of the Republic." Article 43 of the law reads as follows: "Natural or juridical persons or human groups, linked by a common interest and directly affected by the harmful action or omission, may file actions before the competent judge for damages and losses and for deterioration caused to health or the environment, including the biodiversity with its constituent elements."

14 The judges presiding over the litigation, in order, were Alberto Guerra Bastidas (May 2003–January 2004), Efraín Novillo Guzmán (January 2004–January 2006), German Yánez Ruíz (February 2006–August 2007), Efraín Novillo Guzmán (August 2007–August 2008), Juan Evangelista Núñez Sanabria (August 2008–September 2009), Nicolás Zambrano Lozada (September 2009–February 2010), Leonardo Ordóñez Piña (February 2010–August 2010), and Nicolás Zambrano Lozada (August 2010–March 2011).

15 This was the January 26, 2012, opinion of the US Court of Appeals for the Second Circuit, which rescinded an interim judicial ruling by the district court hearing Chevron's countersuit after, on March 7, 2011, it placed a global injunction on the 2011 Ecuadorian ruling (*Chevron Corp. v. Naranjo*, 667 F.3d 232, 234 (2d Cir. 2012); ECF 11-1150-cv [L], DI 644 and 648; 11-1264-cv (CON)). Judge Gerard E. Lynch wrote the opinion. For Judge Kaplan's March 7, 2011, preliminary

injunction, see *Chevron Corp. v. Donziger*, ECF 11 Civ. 0691 (LAK), DI 181. Also available at 768 F. Supp. 2d 581 (S.D.N.Y. 2011).

16 DeLanda is reflecting on Ian Hacking's book *Representing and Intervening* (1983).

17 As the judicial inspections proceeded during the trial in Ecuador, there was suspicion that Chevron had begun to tamper with the samples it extracted. But alleged tampering occurred four years after the inspections began. By that time, substantial amounts of analytic data had already been generated from the analysis of soil and water samples taken by both parties.

18 In writing about chemistry, I draw from the work of Bensaude-Vincent and Simon (2012); Bensaude-Vincent and Stengers (1996); DeLanda (2015); Duhem (2002); Heilbron (2003); Lavoisier (1789); Wallace and BelBruno (2006); and Woody et al. (2012).

19 See also "The Chemical Elements," Elementymology and Elements Multidict, accessed January 14, 2017, http://www.vanderkrogt.net/elements/element.php?.

20 The "laboratory," Holmes reminds us, was originally "the space in which chemists 'elaborated' chemical and medicinal substances"—the workshop for the practice of craft (Holmes 2003: 145).

21 Key thinkers include Bensaude-Vincent (2008, 2014); Hacking (2002); Latour (1988, 1993, 1999, 2005); Law (2004); Mol (2002); Stengers (2005, 2010); and Shapin and Schaeffer (1985).

22 As Bachelard observed, "The real in chemistry is a realization" (cited in Bensaude-Vincent 2014: 66).

23 FARC-EP, or the Revolutionary Armed Forces of Colombia—People's Army, was the guerrilla movement engaged in armed conflict in Colombia from 1964 to 2017.

24 Via kinship, I have been long imbricated in multinational oil extraction since the early decades of the 1900s. My grandfather, Guy H. Sawyer, began his oil career in Bolivia as a civil engineer for Standard Oil in 1921. Subsequently, he worked for Standard Oil in Argentina and then Venezuela until his death in 1948. My uncle, Herbert Sawyer, worked in Venezuela and Cuba in the 1940s and 1950s. As I was growing up, my father, J. Allan Sawyer, worked as a petroleum geologist in Libya (my birthplace) for nearly fifteen years and then in Peru and Panama. And though we moved to the United States just before my thirteenth birthday, he continued to do exploratory work in Venezuela and Surinam. After I graduated from high school, while I worked as a ballet dancer in Europe and then later when I was in college, I visited my parents over an eight-year period in Colombia, Argentina (my father's birthplace), and Egypt.

25 Over the course of this legal saga, a number of individuals in Ecuador and the United States have found their lives rattled in varying degrees by Chevron. While never of great significance, upon occasion, I, too, have experienced Chevron's overbearing reach. The most recent was via the National Endowment for the Humanities (NEH). In 2018, I had applied for an NEH fellowship. My proposal was reviewed by eight external reviewers (more than the norm) and each reviewer rated my proposal "excellent"—the ranking necessary for

obtaining funding. To my confusion, however, the NEH withheld funding from me. The reasons were as follows: "Despite the strong ratings, concerns were raised at later stages of review about the possibility of additional court appeals in the case that is the centerpiece of your project. While the actual legal case seems to be closed, the repercussions might not be. The project seemed premature for federal funding when the final outcome is not yet settled" (NEH email, March 8, 2019). "Premature"? At that point, this legal saga had been transpiring for nearly twenty-six years.

When I made further inquiries, an NEH senior program coordinator kindly looked into the matter and later responded as follows: "As explained in our guidelines and in the application process information, there are three stages of review. The peer review is the first stage, from which you received your project's evaluations. The next stage of review is conducted by the National Council for the Humanities, a presidentially selected body of twenty-six people confirmed by the senate. The final review falls in the purview of the chairman of the NEH, who by law makes all funding decisions. Despite the high peer review ratings, concerns were raised at the later stages of review regarding the pendency of the litigations" (NEH email, March 25, 2019).

The legal details that followed included different events and court submissions made to the Supreme Court of Canada (where enforcement proceedings were being pursued) and the ruling of the PCA (which had yet to make its determination at the time I submitted my proposal in the spring of 2018). Of particular concern apparently was the fact that Chevron in 2019 had only recently submitted the PCA ruling to the Canadian Supreme Court and that this would "affect the narrative arc of [the] project and render a final call premature. Another concern was raised regarding a balanced representation in [the] narrative concerning Chevron and the U.S. court system" (NEH email, March 25, 2019).

When would a 3,000-word application that was ranked "excellent" by eight academic reviewers be deemed ineligible for "federal funding" because details that had yet to occur were not addressed in the application? Any serious scholar following this legal saga would enfold consequential legal events and allow them to shape her analysis. Apparently, someone—perhaps familiar, perhaps not, with my work and connected to the National Council for the Humanities or the NEH chairman—did not want me to write about the Chevron legal saga. It stung, of course, not to be given the support that reviewers believed I merited; the funds and time would have been warmly welcomed. But it also shocked me (perhaps naively) that the NEH would interfere in academic freedom in this way. I share this episode because it is suggestive of the extent of Chevron's tentacles. In this case, those tentacles stretched to undermine both the integrity of the peer review process and the confidence that academics place in it.

Hearing

1 For a glimpse into the pretrial hearings in the United States between 1993 and 2001, see Sawyer (2001, 2002, 2006).

2 All quotations in this section attributed to Dr. Adolfo Callejas are parts of the text of Chevron's legal response to the lawsuit filed against it on October 21, 2003. "Response of ChevronTexaco Corporation," *Maria Aguinda Salazar y Otros v. ChevronTexaco Corp.*, Case No. 002-2003-P-CSJNL (2011-63-1).

3 The corporate legal logic went as follows: "On October 9, 2001, pursuant to the terms agreed in a document called 'Merger Agreement and Plan,' the meetings of shareholders of Chevron Corporation and Texaco Inc. approved a merger or union of companies, which actually occurred on that date between Texaco Inc. and a company called Keeper Inc., which was a wholly owned subsidiary of Chevron Corporation. The result of that legal transaction was that Texaco Inc. survived the merger, inasmuch as it fully absorbed Keeper Inc. without, therefore, losing its legal personhood and its capacity to acquire rights and contract obligations.

"On that same date Chevron Corporation, which was and continues to be a completely different company from Texaco Inc., amended its corporation by-laws according to which it changed its name to 'ChevronTexaco Corporation'" ("Response of ChevronTexaco Corporation," October 21, 2003, p. 3).

4 When Texaco's Ecuador subsidiary, the Texaco Petroleum Company, was created in 1964, its shortened name was spelled "Texpet." This is the spelling used in Ecuador and used in historical documents. And this is the spelling I use throughout this text, despite a number of US courts, Chevron itself, and a number of press reports referring to this entity as "TexPet."

Chapter One. Chemical Agency

A shorter version of this chapter was published as "Crude Contamination: Law, Science, and Indeterminacy in Ecuador and Beyond," in *Subterranean Estates: Life Worlds of Oil and Gas*, ed. Hannah Appel, Arthur Mason, and Michael Watts (Ithaca, NY: Cornell University Press, 2015), 126–46.

1 Each party in the lawsuit submitted a list of the oil facilities they wanted the judge to inspect. Some locations on their lists overlapped, leaving a total of 122 sites for judicial inspections. As the litigation proceeded, the plaintiffs' lawyers petitioned the court to retract a number of the sites they had initially requested because each inspection demanded a considerable investment of time and expense. Actual site visits necessitated weeks of planning, scheduling, and arranging, not a simple logistical feat in the rainforest. The scientific analysis of extracted samples and the ultimate reports—usually completed months later—required precise expertise. After having completed a couple dozen inspections, the plaintiffs' lawyers felt they had made their case (contamination was rampant) and further inspections were not needed. The court honored the

plaintiffs' request, reducing the total number of inspected sites to fifty-four. In its public relations campaign, Chevron claimed that this court decision was nefarious and circumspect. This was not so, however; either party could retract would-be inspection sites that it proposed. The parties completed judicial inspections for all the sites that Chevron's lawyers had requested be inspected.

2   Lago Agrio plaintiffs (LAP), "Informe del perito de la inspección judicial en el pozo Sacha-57," p. 42, *Maria Aguinda Salazar y Otros v. Chevron Texaco Corp.*, Case No. 002-2003-P-CSJNL (2011-63-1), Ing. Edison Camino Castro, L. P. 05-17-259.

3   Chevron Corporation, "Informe de la inspección judicial del pozo Sacha-57," June 2005, p. v, *Maria Aguinda Salazar y Otros v. Chevron Texaco Corp.*, Case No. 002-2003-P-CSJNL (2011-63-1), Doctor Gino C. Bianchi Mosquera.

4   This is information from Texaco's drilling logs as compiled by DINAPA (Dirección Nacional de Protección Ambiental) in 1993 (Cabrera Vega 2008: appendix F, p. 15, and appendix U3). Texaco extracted 1,312,940,910 barrels of oil; dumped 379,246,100 barrels of wastewater into the environment; and burned 230,464,948 cubic feet of gas.

5   The first twenty-five judicial inspections exhibit this trend. Thereafter, significant anomalies appeared in Chevron's results as they shifted their testing methods.

6   My contention builds on the work of anthropologists and science studies scholars. A basic premise of this scholarship asserts that there is no outside, independent, ahistorical world—usually captured by the term "nature"—waiting patiently to be discovered. Rather, following Latour, what he calls the modernist settlement—the divide between nature and society—is a sixteenth-century legacy that has prevented us from seeing the way the worlds really work through wily and hybrid movements of humans and nonhumans. First exploring this position through historical and contemporary studies of laboratory practices (and beyond), Latour unraveled the tangle of practices, techniques, and technologies whereby humans and nonhumans bring forth action, properties, propensities, and potentialities into being. "Instead of starting with entities that are already components of the world," Latour explored "the complex and controversial nature of what it is for an actor to come into existence. The key is to define the actor by what it does—its performance—under laboratory trials" (1999: 303) or conditioned arrangement. He thus demonstrated the entanglements that allow for truth claims to be pronounced.

7   The description of Texaco's operations comes from my own research beginning in 1992 and an array of overlapping sources: extended conversations (1993–95) with geologists and petroleum engineers who had worked for Texaco in the Ecuadorian Amazon; discussions (2000, 2003, 2007, 2010, 2019) with oil workers who maintained wellheads and flowlines for Texaco; and Judith Kimerling's work (1991a, 1991b, 1993, 1996).

8   During the period of Texaco's Ecuadorian operations, it was standard industrial practice in the United States to reinject formation waters and subterranean sands at least one mile below the surface of the earth and to process

chemical solvents until they were environmentally safe. The lawsuit alleges that, between 1964 and 1992, Texaco made strategic decisions in its New York headquarters to maximize its corporate profits by using substandard technology in its Ecuadorian oil operations. As Steven Donziger observed in 2003 at the onset of the Ecuadorian litigation, "Contaminated water is not the effect of just random spills. It is the result of decisions made by Texaco to install a type of drilling process that would lead to the systematic dumping of toxins. Texaco made a decision to dump these toxins into the Amazon to save money and increase its profits" (personal communication with author, November 2003). The decision not to reinject formation waters back into the subterranean strata from which they emerged allegedly reduced the company's per-barrel production costs by approximately $3 and saved the parent corporation roughly $5 billion over the course of its operations in Ecuador.

9   See also David Russell, court transcripts, October 16, 2013, p. 310, *Chevron Corp. v. Donziger*, ECF 11 Civ. 0691 (LAK), DI 1790.

10 Railroad Commission of Texas, "Open Pit Storage Prohibited," Texas Statewide Order No. 20-804, July 31, 1939.

11 Louisiana Department of Natural Resources, "SONRIS/2000: SRCN4282K Injection Wells by Parish," last report run on February 9, 2019, 5:34 p.m., http://reports.dnr.state.la.us/reports/rwservlet?SRCN4282K_p.

12 Texaco's reinjection technology patents were as follows: United States Patent 3,680,389, Frederick H. Binkley Jr. et al., assignor to Texaco Inc., August 1, 1972; United States Patent 3,817,859, Jack F. Tate, assignor to Texaco Inc., June 18, 1975.

    In 1962, the American Petroleum Institute (API) published *The Primer of Oil and Gas Production* in which, among other things, it voiced its concern about the dangers of formation waters.

13 Understandings of the total number of oil wells that Texaco drilled vary. According to DINAPA (1993), Texaco drilled 356 wells, 20 of which were injection wells. Powers and Quarles (2006) estimate 336 oil wells. Stratus Consulting estimates 336 oil wells and an additional 20 injection wells drilled late on (author had access to Stratus's internal database from 2008 to 2009). Kimerling estimates that "Texaco drilled 339 wells [with]in 442,965 hectares. Some 235 wells are [presently] active, generating over five million gallons of toxic waste every day" (1996: 64). The hectarage that Kimerling notes is incorrect, as will become evident in chapter 3.

14 Chevron asserted that "earthen pits are an integral part of petroleum exploration and production operations and are used worldwide." Claimants' "Memorial of Merit," September 6, 2010, pp. 19–20, *In the Matter of an Arbitration*, Case No. 2009-23, Permanent Court of Arbitration (PCA), The Hague.

15 Method 418.1, in EPA 1983.

16 Method 418.1 consists of solvent extraction followed by treatment in a silica gel column and infrared spectroscopy. A minor fault with this method (which was

of no concern in the lawsuit against Chevron) is that it tends to overestimate the quantity of TPH in soil samples when TPH concentrations are low (approximately 100 to 200 ppm of TPH).

17 This same figure also appears unattributed in an API publication (2001: 35). Both publications have the following description of the graph: "shows the overlap between the carbon number ranges of different hydrocarbon products as well as the overlap in the corresponding TPH analytical methods. For example, this figure demonstrates that a TPH method designed for gasoline range organics (i.e., C_6 to C_{12}) may report some of the hydrocarbons present in diesel fuel (i.e., C_{10} to C_{28}). The same is also true for TPH analytical tests for diesel range organics which will identify some of the hydrocarbons present in gasoline-contaminated soils. Lastly . . . [h]owever, crude oil may contain hydrocarbons with carbon numbers that range from C_3 to $C_{45}+$ and are not fully addressed even with the use of all three TPH methods" (McMillen, Rhodes, et al. 2001: 60).

18 Among other things, the API sets voluntary standards for best practices. In 1921, API established the API gravity scale through which to assess the density and quality of different crude oils. The scale is set in relationship to the density of water. The higher on the API scale, the less dense the crude.

19 The volume also states, "The TPHCWG is made up of industry, government, and academic scientists, working to develop a broad set of guidelines to be used by engineering and public health professionals in decisions on petroleum contaminated media" (TPHCWG 1998: 10–11).

20 As the ATSDR states, "Although chemicals grouped by transport fraction generally have similar toxicological properties, this is not always the case. For example, benzene is a carcinogen, but toluene, ethylbenzene, and xylene are not. However, it is more appropriate to group benzene with compounds that have similar environmental transport properties than to group it with other carcinogens such as benzo(a)pyrene that have different environmental transport properties" (1999: 13).

21 The highest TPH readings were 900,000 ppm (or mg/kg) obtained from a Chevron sample taken from Sacha-14 and obtained from a plaintiffs' sample taken from oil well Shushufindi-4.

22 Chevron, "Informe de la inspección judicial del pozo Sacha-57," Dr. Gino C. Bianchi Mosquera (2005: vii).

23 "The 13 TPH fractions are based on 'equivalent carbon' (EC) numbers rather than 'carbon numbers.' ECs are related to the boiling point of individual compounds in a boiling point GC [gas chromatography] column, normalized to the boiling point of a normal alkane. Thus, for compounds where only a boiling point is known, the EC can be readily calculated. For example, the EC of benzene is 6.5 because its boiling point and GC retention time are approximately halfway between those of n-hexane and n-heptane. Benzene's EC number is greater than that of n-hexane because its ring structure results in a higher boil-

ing point. The TPHCWG chose the concept of EC numbers because these values are more logically related to compound mobility in the environment than carbon numbers" (McMillen, Rhodes, et al. 2001: 62).

24 This quote comes from Dr. Alberto Wray (then chief lawyer for the LAP) in his "Observaciones" of the judicial inspection of oil well Sacha-57, November 22, 2004, 15.

25 Decree 1215—"Environmental Regulations for Hydrocarbon Operations in Ecuador," Official Registry No. 265, February 13, 2001. The Unified Text of Secondary Environmental Legislation by the Ministry of Environment (TULA), Executive Decree 2825, Official Registry No. 623, July 22, 2002, sets standards for all other analytes in crude oil and the oil process.

26 See also ATSDR, "Overview of Total Petroleum Hydrocarbons," accessed September–November 2006, www.atsdr.cdc.gov/toxprofiles/tp123.html, p. 19.

27 The TPHCWG details three assumptions embedded in the TPHCWG's indicator/surrogate model, including "(1) toxicity of the fractions being tested does not change with weathering; (2) that the composition of the fraction will not vary from the surrogate tested; (3) that the interaction of the various fractions is additive" (1997b: 3). The scenario these assumptions paint is quite conservative. Significant research demonstrates that toxicity is not static, that fractional composition may vary overtime, and that the effects or capacities of complex chemicals in collectivity is not additive but nonlinear, stochastic, and exponential. That is, twice the amount of a harmful chemical is not twice the risk but possibly four times or even one hundred times the risk.

28 In chemistry, isomers are molecules whose chemical formula is the same (i.e., C_{25}) but whose chemical architecture differs. This difference in structural and spatial arrangements compels molecules to exhibit different properties and fates.

29 "The term 'hazardous substance' is defined under CERCLA Section 101(14) to include approximately 714 toxic substances listed under four other environmental statutes, including RCRA. Both the definition of hazardous substance and the definition of 'pollutant or contaminant' under Section 104(a) (2) exclude 'petroleum, including crude oil or any fraction thereof,' unless specifically listed under those statutes" (EPA 1987: 2).

Inspection

1 *Maria Aguinda Salazar y Otros v. Chevron Texaco Corp.*, Case No. 002-2003-P-CS-JNL (2011-63-1), p. 97,526.

2 *Maria Aguinda Salazar y Otros v. Chevron Texaco Corp.*, case file p. 97,527.

3 *Maria Aguinda Salazar y Otros v. Chevron Texaco Corp.*, case file p. 97,535.

4 *Maria Aguinda Salazar y Otros v. Chevron Texaco Corp.*, case file p. 97,536.

5 *Maria Aguinda Salazar y Otros v. Chevron Texaco Corp.*, case file p. 97,544.

6 *Maria Aguinda Salazar y Otros v. Chevron Texaco Corp.*, case file p. 97,542.

7 *Maria Aguinda Salazar y Otros v. Chevron Texaco Corp.*, case file p. 97,544.

8 *Maria Aguinda Salazar y Otros v. Chevron Texaco Corp.*, case file pp. 97,570, 97,571.

Chapter Two. Exposure's Orbitals

1 "Declaration" by Miguel San Sebastián before Judge Alberto Guerra Bastidas, then president of the Superior Court of Justice of Nueva Loja, October 27, 2003.

2 These articles do not represent all the epidemiological research submitted to the Superior Court of Justice of Nueva Loja. One additional cluster of research documented popular epidemiological surveys of community-health concerns and rural appraisals, mapping the location of oil infrastructure and human habitation (Acción Ecológica 1993; Garzón 1995; Pallares 1999). A second cluster analyzed water contamination, health clinical surveys, and/or socioeconomic and legal issues (Center for Economic and Social Rights 1994; Kimerling 1991b; San Sebastián, Córdoba, and Apostólico de Aguarico 1999; UPPSAE 1993). In this chapter, I restrict my analysis to the epidemiological reports published in peer-reviewed journals.

3 San Sebastián and colleagues also published other studies. One found that women living near oil fields were two and a half times more likely to have their pregnancies end in spontaneous abortions than did those living distant from oil fields (San Sebastián, Armstrong, and Stephens 2002). Another found that in one community, surrounded by a separation station and numerous oil wells, local streams contained TPH levels up to 288 times higher than the permissible level in the European Community, and the area was marked by an excess of cancer deaths among men (San Sebastián, Armstrong, and Stephens 2001). And another study indicated that cancer rates among Indigenous people are higher than those of non-Indigenous people living in the Oriente region (San Sebastián and Hurtig 2004a).

4 In order to establish a rate of cancer incidence or mortality for a designated group—in this case, a canton—research teams for each study needed to estimate the total annual population per canton during the period they were examining. Since official annual population data did not already exist, researchers estimated the population size from Ecuadorian national census data taken once every ten years. For instance, investigating cancer incidence between 1985 and 1998, the San Sebastián teams used projections generated by the Ecuadorian National Institute of Statistics and Census of the estimated population per canton in 1992—the midpoint of their study period. They then used this midpoint population size as constant throughout the period of their study.

By contrast, the Kelsh team estimated population differently. Investigating cancer mortality between 1990 and 2005, these researchers used the 1990 and the 2001 national census survey data to calculate the total population of each canton for each year between 1990 and 2000; and for the years 2002 to 2005, the Kelsh team projected the total annual population per canton from the 2001 census, assuming a constant rate of change. According to Chevron experts (Arana and Arellano 2007; Kelsh, Morimoto, and Lau 2009), San Sebastián and colleagues' methodology was flawed, and they miscalculated population size; the Chevron experts maintained that the San Sebastián teams "likely underestimated the population of the exposed regions" (Kelsh, Morimoto, and Lau 2009: 392). Kelsh maintained that underestimating "the population at risk in exposed regions would artificially inflate cancer rates in these areas and may explain their observation of elevated cancer risks" (392).

5 Kelsh and colleagues note only that these data were obtained from IHS Energy (http://energy.ihs.com), without giving more details.

6 Cuyabeno—formerly the eastern half of Lago Agrio—was newly defined as a separate canton in 1998 and consequently was already included (as part of Lago Agrio) in Hurtig and San Sebastián's research.

7 See the *American Association of Petroleum Geologists Bulletin (AAPGB)* publications from 1972 to 1982.

8 In 1976, OKC relinquished its concession and left the country.

9 See the *AAPGB* publications from 1981 to 1986.

10 By the end of 1972, Texaco had "so far discovered 11 oil fields and has 99 productive wells producing 220,000 barrels daily" (Sawyer 1975: 1149).

11 First, possible exposure in Cascales, Putumayo, or Cuyabeno would have been ten to twenty years after possible exposure in the Lago Agrio, Sachas, Shushufindi, or Orellana cantons where oil production began in 1969 for the local refinery and in 1972 for export. At the earliest, exposure occurred in 1986 in Cascales, when the Berjemo oil field first produced, and in Putumayo, when the Cuyabeno, Sansahuari, and Tetete fields started to produce; in 1980 in Cuyabeno when the Fanny, Joan, and Mariam wells first produced; and as late as 1990 for other wells. Second, given the time lapse that the onset of cancer as a disease necessitates, let alone death by cancer for fatal cancers, death records would not be registering mortality by cancer until years after the onset of operations. For example, those wells drilled in Cascales in 1967 (that were not dry) would in actuality never have emitted potential exposure until 1986 and most likely would not lead to measurable increases in cancer mortality rates until Kelsh's study was virtually over.

12 Brown and Williamson, "Smoking and Health Proposal," document no. 68056778–1786 (1969), available at http://legacy.library.ucsf.edu/tid/nvs40foo, last accessed March 13, 2014, cited in Michaels (2008: 11). See also Proctor (2011: 289).

13 At the time that Michael Kelsh and colleagues conducted and published this research they worked for Exponent—"a leading engineering and scientific consulting firm providing solutions to complex technical problems" (http://www.exponent.com/history)—where Kelsh was the principal scientist for Epidemiology, Biostatistics, and Computational Biology. Subsequently, Chevron commissioned Kelsh to write three more expert opinions (2009, 2010, 2011) that the corporation submitted to the court in its defense. Each of these opinions attacks the logic and science of epidemiological claims supporting an association between oil extraction and cancer.

14 Michaels continues: "For most chemicals, . . . cohort mortality studies that examine workers whose exposures began less than twenty years earlier will not show an effect" (2008: 63). These studies deserve critical review, given that the "negative results" obtained can "intentionally misinform those not trained in the subtleties of epidemiology" (63).

15 See *Industrial Union Department v. American Petroleum Institute*, 448 US 607 (1980).

16 *Industrial Union Department v. American Petroleum Institute*, pp. 664–65.

17 *Industrial Union Department v. American Petroleum Institute*, p. 634.

18 Not infrequently, this research is funded by the API and the IPIECA (the global oil and gas industry association for environmental and social issues).

19 A meta-analysis combines the results of several similar studies in order, theoretically, to obtain statistically powerful results. As Michaels writes, they can be a "recipe for countering the results of a well-conducted study: just mix the good study with several weak studies or badly designed ones, and you will get a 'no findings' conclusion . . . [thereby] trump[ing] the results of that one pesky study" (2008: 68).

20 Michaels writes, "For purposes of litigation protection, ChemRisk [consulting firm] continues to churn out assessments of benzene exposure in different industries. . . . By itself, the for-hire provenance gives away the game with such studies. . . . Litigation . . . studies of no value whatsoever in the regulatory arena can be quite valuable for corporate defendants in the courtroom" (2008: 75).

21 Wong and Raabe together also published more articles, which are cited in Michaels (2008).

22 Chevron detailed these assessments on a corporate website set up to address the ongoing litigation in Ecuador. That website no longer exists and was removed sometime in 2015. The URL was http://www.texaco.com/sitelets /ecuador, last accessed March 13, 2014.

23 This statement is from Chevron's former website dedicated to the lawsuit, now removed from the internet: http://www.texaco.com/sitelets/ecuador /en/responsetoclaims/default.aspx, last accessed March 13, 2014.

24 This statement is from Chevron's former website dedicated to the lawsuit, now removed from the internet: http://www.texaco.com/sitelets/ecuador/en /responsetoclaims/default.aspx, last accessed March 13, 2014.

25 This statement is from Chevron's former website dedicated to the lawsuit, now removed from the internet: http://www.texaco.com/sitelets/ecuador/en /responsetoclaims/default.aspx, last accessed March 13, 2014. The report that is referenced is UNICEF 1992.

26 This statement is from Chevron's former websites dedicated to the lawsuit, now removed from the internet: http://www.texaco.com/sitelets/ecuador/en /releases/2005-02-02.aspx, and http://www.texaco.com/sitelets/ecuador/en /PlaintiffsMyths.aspx, both last accessed April 1, 2014.

27 This statement is from Chevron's former websites dedicated to the lawsuit, now removed from the internet: http://www.texaco.com/sitelets/ecuador/en /releases/2005-02-02.aspx, and http://www.texaco.com/sitelets/ecuador/en /PlaintiffsMyths.aspx, both last accessed April 1, 2014.

28 Hurtig and San Sebastián (2005) similarly engage in this distinction.

29 "Declaration" by Miguel San Sebastián before Judge Alberto Guerra Bastidas, then president of the Superior Court of Justice of Nueva Loja, October 27, 2003.

Death

1 Frente de Defensa de la Amazonía, "Legendary Ecuadorian Nurse Who Hosted Celebrities and Battled Chevron over Pollution Tragically Dies of Cancer," January 4, 2017, https://chevroninecuador.org/news-and-multimedia /2017/0104-legendary-ecuadorian-nurse-tragically-dies-of-cancer.

Catch

1 The video is posted at https://theamazonpost.com/chevron-to-ecuador-keep -your-promise-clean-up-the-amazon, accessed July 23, 2021. It is also available on the Chevron Corporation's website, www.chevron.com/ecuador. The video first existed on Texaco sites (http://www.texaco.com and http://www.texaco .com/sitelets/ecuador) created specifically for the lawsuit—and then was moved to Chevron's home page.

Chapter Three. Alchemical Deals

1 "Acta of the Sacha-57 Judicial Inspection" (SA57), from *Maria Aguinda Salazar y Otros v. ChevronTexaco Corp.*, Case No. 002-2003-P-CSJNL (2011-63-1), case file p. 11,914, Provincial Court of Justice of Sucumbíos, Nueva Loja, Ecuador.

2 SA57, case file p. 11,916.

3 SA57, case file p. 11,922.

4 SA57, case file p. 11,923.

5 SA57, case file p. 11,940.

6 SA57, case file p. 11,941.

7 SA57, case file p. 11,940.

8 SA57, case file p. 11,927.

9 SA57, case file pp. 11,942–43.

10 The "settlement agreement"—titled "Contract for Implementing of Environmental Remedial Work and the Release from Obligations, Liability, and Claims"—was signed in Quito on May 4, 1995, by Ricardo Reis Veiga, then vice president of Texaco and now vice president of Chevron; Dr. Rodrigo Pérez Pallares, then and later the legal agent for Texaco (now Chevron) in Ecuador;

Dr. Galo Abril Ojeda, the Ecuadorian minister of energy and mines; and Dr. Frederico Vintimilla Salcedo, the president of Petroecuador, the state oil company.

11 Supreme Decree No. 205-A, February 5, 1964, Official Registry No. 186, February 21, 1964. Formerly archived in the library of the Ecuadorian National Congress and presently archived in the library of the Ecuadorian Constitutional Court in Quito.

12 "Texaco-Gulf Napo Agreements and Joint Operating Agreement" (plaintiff's exhibit), January 1, 1965, *Jota et al. v. Texaco Inc.*, ECF 97-9102 (2d Cir. 1998, decided), DI A2794–A2859.

13 "In 1937, the Anglo-Saxon Petroleum Co., Ltd. ('Shell'), received a concession from Ecuador for the entire Oriente region, but it abandoned its effort in 1950 after drilling six wells with unsatisfactory results." Exhibit C-410, Donald G. Sawyer, "Report: Response to Evidentiary Request No. 29," July 1, 2010, pp. 1–2, in "Claimants Memorandum of Merits," September 6, 2010, p. 12n52, *In the Matter of an Arbitration*, Case No. 2009-23, Permanent Court of Arbitration (PCA), The Hague.

14 Acuerdo No. 844 of December 20, 1965 (Official Registry No. 655, December 27, 1965), authorizing the transfer of 650,000 hectares of hydrocarbon land from Minas y Petróleos del Ecuador S.A. to las Cías. Petrolera Pastaza C.A. y Petrolera Aguarico S.A. Formerly archived in the library of the Ecuadorian National Congress and presently archived in the library of the National Assembly and Constitutional Court in Quito.

15 *Norsul Oil and Mining Co., Ltd. v. Texaco Inc.*, 703 F. Supp. 1520 (S.D. Fla. 1988), "Memorandum Opinion," p. 11; Hydrocarbons Law, Decree No. 1459, September 27, 1971, Official Registry No. 322, October 1, 1971. Specifically, the new Hydrocarbons Law included limitations on the maximum concession and exploitation areas, increased annual surface taxes, and increased government royalties. In 1989, CEPE was renamed Empresa Estatal Petróleos del Ecuador (Petroecuador).

16 "Memorandum Opinion," p. 12, *Norsul Oil and Mining Co., Ltd. v. Texaco Inc.*, 703 F. Supp. 1520 (S.D. Fla. 1988).

17 Supreme Decree No. 925, August 4, 1973, Official Registry No. 370, August 16, 1973; agreement between the Government of Ecuador, Ecuadorian Gulf Oil Company, and Texaco Petroleum Company, August 6, 1973. Formerly archived in the library of the Ecuadorian National Congress and presently archived in the library of the National Assembly and Constitutional Court in Quito.

18 The agreement among the Republic of Ecuador, CEPE, and Ecuadorian Gulf Oil Company was signed on May 27, 1977, but made effective as of the last day of 1976. See "Claimants' Memorial on the Merit," September 6, 2010, p. 14n65, *In the Matter of an Arbitration*, Case No. 2009-23, PCA, The Hague.

19 According to DINAPA (1993), this is the number of wells that Texaco reported at the time of transfer of the concession to Petroecuador.

20 The first environmental audit was conducted by HBT AGRA Limited (1993). The second environmental audit was conducted by Fugro-McClelland Inc. (1992). HBT AGRA later changed its name to AGRA, Earth and Environmental, Ltd.

21 In 1995, Texpet developed a "scope of work (SOW) of environmental reparation and a remediation action plan (RAP)" to direct and guide the recovery program (Texaco and Ministry of Mines and Energy 1995; Woodward-Clyde International 2000). Woodward-Clyde International (2000) carried out the remediation work between 1995 and 1998.

22 After reviewing and accepting Texpet's work, the government of Ecuador signed the "final release of claims," or acta final, in September 1998. On September 30, 1998, the Ministry of Energy and Mines issued a certificate of completion to accept the finalization of the process. Chevron produced a video about its remediation work and the Texaco-ministry contracts, available on YouTube: see Amazon Post, "Crude Reality: Texpet's Successful Remediation," https://www.youtube.com/watch?v=XZJnqxWiLiY, posted September 11, 2013.

23 He read Chevron's written response to the legal complaint during the Audiencia de Conciliación. *Maria Aguinda Salazar y Otros v. ChevronTexaco Corp.*, case file p. 253.

24 Ecuadorian Civil Code, Art. 2348; "Claimants' Memorial on the Merits," September 6, 2010, p. 192, *In the Matter of an Arbitration*, Case No. 2009-23, PCA, The Hague.

25 Ecuadorian Civil Code, Art. 2362; "Claimants' Memorial on the Merits," September 6, 2010, p. 192, *In the Matter of an Arbitration*, Case No. 2009-23, PCA, The Hague.

26 "Claimants' Memorial on the Merits," September 6, 2010, pp. 196–99, *In the Matter of an Arbitration*, Case No. 2009-23, PCA, The Hague.

27 "Claimants' Memorial on the Merits," September 6, 2010, p. 204, *In the Matter of an Arbitration*, Case No. 2009-23, PCA, The Hague.

28 1979 Political Constitution of the Republic of Ecuador, Art. 19(2).

29 "Claimants' Memorial on the Merits," September 6, 2010, p. 203, *In the Matter of an Arbitration*, Case No. 2009-23, PCA, The Hague.

30 "Track I Hearing on Interim Measures," London, pp. 80–90, February 11, 2012, *In the Matter of an Arbitration*, Case No. 2007-02/AA277, PCA, The Hague.

31 Judge Nicolás Zambrano, judgment, February 14, 2011, p. 30, *Maria Aguinda Salazar y Otros v. ChevronTexaco Corp.*, Case No. 002-2003-P-CSJNL (2011-63-1), Provincial Court of Justice of Sucumbíos, Nueva Loja, Ecuador.

32 Zambrano, judgment, February 14, 2011, p. 31.

33 Texaco and Ministry of Energy and Mines 1995, Art. 5.1, p. 9; Zambrano, judgment, February 14, 2011, p. 32 (emphasis added).

34 Zambrano, judgment, February 14, 2011, p. 32.

35 Zambrano, judgment, February 14, 2011, p. 32.

36 "Claimants' Memorial on the Merits," September 6, 2010, p. 44, *In the Matter of an Arbitration*, Case No. 2009-23, PCA, The Hague.

37 Lago Agrio plaintiffs (LAP), final plea, January 24, 2011, p. 61, *Maria Aguinda Salazar y Otros v. ChevronTexaco Corp.*

38 For oily wastes like those found in TexPet's former concession, the EPA developed another analytical method (Extraction Procedure for Oily Wastes— Method 1330) that measures hazardous components not only in aqueous leachate, but also in oil embedded in soil.

39 LAP, final plea, January 24, 2011, p. 57.

40 Maest, Quarles, and Powers 2006: 6n8; LAP, final plea, January 24, 2011, p. 57.

41 LAP, final plea, January 24, 2011, p. 57.

42 LAP, final plea, January 24, 2011, p. 56.

43 Contraloría General del Estado 2002: 13, citing Memo 301-PAB-96, May 7, 1996; comptroller general's report, "Special Analysis on the Agreement for Performance of Environmental Remediation Works," DA3-25-2002, April 9, 2003. Archived in the Office of the State General Comptroller in Quito. See also Republic of Ecuador Exhibit R-78, "Comptroller General's Report No. DA3-25-2002," *In the Matter of an Arbitration*, Case No. 2009-23, PCA, The Hague.

44 Contraloría General del Estado 2002: 16–18, citing Memo 494-SPA-96, September 23, 1996.

45 A similar process took place in the office of the district prosecutor of Pichincha, the province in which Quito, the country's capital, is located.

46 Prosecutor General Alfredo Alvear Enríquez, prosecutorial opinion, 09-2008 DRR/PVC/ASC, April 29, 2010, pp. 118–19 (Ecuador). See also https://the amazonpost.com/post-trial-brief-pdfs/brief/10PX0272.pdf.

47 Personal communication, April 16, 2008.

48 Enríquez, prosecutorial opinion, April 29, 2010.

49 The other officials were Patricio Ribadeneira García, ex-minister of energy and mines; Ramiro Gordillo García, former executive president of Petroecuador; Luis Albán Granizo, former general manager of Petroproducción; Martha Susana Romero de la Cadena; Jorge Rene Dután Erraez; Alix Paquito Suárez Luna; and Marcos Fernando Trejo Ordóñez.

50 Enríquez, prosecutorial opinion, April 29, 2010, p. 117. The legal concept was developed by Professor Franz Eduard Ritter von Liszt, a nineteenth-century German jurist and criminologist; Professor Eduard Mezgel, an early twentieth-century German criminologist; and Professor Hans Mayer, a twentieth-century jurist and literary scholar.

51 Enríquez, prosecutorial opinion, April 29, 2010, p. 117.

52 Enríquez, prosecutorial opinion, April 29, 2010, p. 118.

53 Enríquez, prosecutorial opinion, April 29, 2010, pp. 4 and 120.

54 See "Second Partial Award on Track II," August 30, 2010, part IV, p. 36, *In the Matter of an Arbitration*, Case No. 2009-23, PCA, The Hague.

55 Enríquez, prosecutorial opinion, April 29, 2010, p. 131.

56 Decision of the First Criminal Chamber of the National Court of Justice, Case No. 150–209WO, June 1, 2011 (Ecuador). See also "Appendix Exhibits"

to Republic of Ecuador's "Amicus Brief" at *Chevron Corp. v. Donziger*, 833 F.3d 74 (2d Cir. 2016) [ECF 14-0826 (L), DI 256-7, 312852, Exhibit 13, pp. 1–72].

57 "Second Partial Award on Track II," August 30, 2018, part IV, p. 38.

58 "Second Partial Award on Track II," August 30, 2018, part IV, p. 38.

59 Track II hearing, April 21, 2015, pp. 3031–32, *In the Matter of an Arbitration*, Case No. 2009-23, PCA, The Hague.

60 I address concerns surrounding an ex-judge named Alberto Guerra in a separate publication (Sawyer and Ofrias, forthcoming).

61 "Amended Complaint," *Republic of Ecuador v. ChevronTexaco Corp.*, ECF 04 Civ. 8378 (LBS), DI 26 (S.D.N.Y.; filed December 8, 2004).

62 Cited in Honorable Leonard B. Sand, "Opinion and Order," p. 17, ECF 04 Civ. 8378 (LBS), DI 67 (S.D.N.Y.; filed June 27, 2005). See also *Republic of Ecuador v. ChevronTexaco Corp.*, 376 F. Supp. 2d 334, 374 (S.D.N.Y. 2005), p. 342.

63 *Republic of Ecuador v. ChevronTexaco Corp.*, 499 F. Supp. 2d 452 (S.D.N.Y. 2007).

64 *Republic of Ecuador v. ChevronTexaco Corp.*, 296 Fed. Appx. 124 (2d Cir. 2008).

65 *ChevronTexaco Corp. v. Republic of Ecuador*, 557 US 936 (2009) [ECF 08-1123].

66 The videos have since been removed from Chevron's main website but are stored, in various forms, on the corporation's site dedicated to the lawsuit, the *Amazon Post*. "Judge Nuñez Misconduct Overview—Chevron Ecuador Lawsuit," https://www.youtube.com/watch?v=var67Gg9rKs, posted August 31, 2009, last accessed July 23, 2021.

67 Clifford Krauss, "Revelation Undermines Chevron Case in Ecuador," *New York Times*, October 29, 2009, https://www.nytimes.com/2009/10/30/world/americas/30ecuador.html.

68 Grant W. Fine, esq., report of investigation from Fine and Associates, Inc., April 5, 2010, available at https://chevroninecuador.org/assets/media/borja/borja-report.pdf.

69 Audio transcript of Skype conversation between Diego Borja and Santiago Escobar, October 1, 2009—11:47:55, p. 6 of the transcript, posted on the Chevron in Ecuador website as "Borja Audio Files and Chat Record," https://chevroninecuador.org/news-and-multimedia/borja-report/audio-and-chat-files, last accessed July 24, 2021.

70 Twenty-seven separate audio recordings were made public. Nineteen of those occurred on October 1, 2009. See Chevron in Ecuador, "Borja Audio Files and Chat Record," https://chevroninecuador.org/news-and-multimedia/borja-report/audio-and-chat-files, last accessed July 24, 2021.

71 See "R-325, Summary of Chevron Payments to or on Behalf of Diego Borja," *In the Matter of an Arbitration*, Case No. 2009-23, PCA, The Hague.

72 *The Republic of Ecuador and Dr. Diego García Carrión, Applicants, for the Issuance of a Subpoena for the Taking of a Deposition and the Production of Documents in a Foreign Proceeding Pursuant to 28 U.S.C. § 1782*, Case No. 10-mc-80225 (CRB), DI 115, 116 (under seal), 184 (N.D. Cal). These documents were subsequently used in the RICO trial *Chevron Corp. v. Donziger*, ECF 11 Civ 0691 (LAK), DI 174–176, 187 (under seal), exhibits 0000343–0000411.

73 Within the US District Court of the Northern District of California, there were eight additional cases related to Borja: ECF C-10-mc-80324 (CRB), ECF 3:11-mc-80087 (CRB), ECF 3:11-mc-80110 (CRB; decided 2012), ECF 10-mc-80087 CRB (NC), ECF 3:11-mc-80171 (CRB), ECF 3:11-mc-80172 (CRB), ECF 3:11-mc-80217 (CRB), ECF 3:11-mc-80219, ECF 3:11-mc-80237 (CRB).

74 Chevron hired the firm Arguedas, Cassman, and Headley to defend Borja. Cristina Arguedas is a nationally prominent criminal defense lawyer who in 2011 had defended Barry Bonds, and "in 1995 she was invited to join the 'Dream Team' defending O.J. Simpson on double-murder charges. Her job was to put Simpson through a grueling mock cross-examination to help the defense team determine whether he should testify in court. (He didn't.)"—as noted in her law firm biosketch, Cristina C. Arguedas, Arguedas, Cassman, Headley, and Goldman LLP, accessed October 28, 2021, https://achlaw.com/cristina-c-arguedas/.

75 "Claimant's Notice of Arbitration," September 23, 2009, *In the Matter of an Arbitration*, Case No. 2009-23, PCA, The Hague.

Clandestine

1 A subset of these tapes can be found at https://amazonwatch.org/news/2015/0408-the-chevron-tapes-video-shows-oil-giant-allegedly-covering-up-amazon-contamination.

2 The transcript below is of footage beginning at minute 4:30 of the video.

3 The transcript below is of footage beginning at minute 57:42 of the video.

Chapter Four. Radical Inspections

1 Originally, 122 inspections were slated. After conducting about half of the inspections they proposed to examine, the plaintiffs' lawyers asked to retract the remaining sites on their list; they felt they had reached their burden of proof (i.e., proven their case) and that more inspections would incur an unnecessary expense. The court granted the plaintiffs' request. All sites that Chevron requested to inspect were examined during the judicial inspections.

2 If being is a process, and a process that is continually and infinitely engaged, then connection, interaction, touch, responsivity, being affected or moved or shifted or influenced or co-becoming through the interactive process is what is constitutive of reality. Enveloping that emanates from enfolding that multiplies, that overflows capacity to cognitively contain. Each iteration allowed for transformation, for a novel configuration.

3 For a riveting analysis of a very different geologic narrative emerging from the Soviet Union and post-Soviet Russia, see Rogers (2018, n.d.).

4 See *Oxford English Dictionary Online*, s.v. "palaeo," accessed June 14, 2017, https://www.oed.com/view/Entry/136163#eid32498764; and *Oxford English Dictionary Online*, s.v. "zoe," accessed June 14, 2017, https://www.oed.com/view/Entry/232964?redirectedFrom=zoe#eid.

5 It was not until massive geological forces pushed up the Andes range during the mid-Cenozoic Era forty-five million years ago that expansive inlet seas retreated and former seabeds formed the Upper Amazon.

6 Johannes van der Waals was an influential Dutch theoretical physicist whose foundational work in the late nineteenth and early twentieth centuries affirming the existence of molecules and the capacity to assess their size, shape, attractive strength, and intermolecular action set the tone of the molecular science of the twentieth century.

7 When immersed with water, distinct hydrocarbon components of crude oil disaggregate.

8 José Segundo Córdova Encalada, testimony, *Maria Aguinda Salazar y Otros v. Chevron Texaco Corp.*, Case No. 002-2003-P-CSJNL (2011-63-1), case file p. 97,539; Judge Nicolás Zambrano, judgment, February 14, 2011, p. 141, *Maria Aguinda Salazar y Otros v. Chevron Texaco Corp.*, Case No. 002-2003-P-CSJNL (2011-63-1).

9 In line with Ecuadorian legal procedure, each party could present individuals to the judge, and those individuals would then answer questions posed by the judge (and those by the opposing team) and testify to their experiences. The judge, thus, partook in devising this testimonial space during the trial's five-year evidentiary phase.

10 Hugo Ureña, testimony, *Maria Aguinda Salazar y Otros v. Chevron Texaco Corp.*, case file p. 97,537.

11 Gerardo Plutarco Gaibor, testimony, *Maria Aguinda Salazar y Otros v. Chevron Texaco Corp.*, case file pp. 82,612–13; Zambrano, judgment, February 14, 2011, p. 142.

12 Guamán Romero, testimony, *Maria Aguinda Salazar y Otros v. Chevron Texaco Corp.*, case file p. 141,008-141,009; Zambrano, judgment, February 14, 2011, p. 148.

13 Lilia Perpetua Mora Verdesoto, testimony, *Maria Aguinda Salazar y Otros v. Chevron Texaco Corp.*, case file p. 56,738; Zambrano, judgment, February 14, 2011, p. 141.

14 Gustavo Ledesma Riera, testimony, *Maria Aguinda Salazar y Otros v. Chevron Texaco Corp.*, case file p. 74,885; Zambrano, judgment, February 14, 2011, p. 140.

15 José Holger García Vargas, testimony, *Maria Aguinda Salazar y Otros v. Chevron Texaco Corp.*, case file pp. 97,542–43; Zambrano, judgment, February 14, 2011, p. 142.

16 Aura Fanny Melo Melo, testimony, *Maria Aguinda Salazar y Otros v. Chevron Texaco Corp.*, case file p. 74,987; Zambrano, judgment, February 14, 2011, p. 143.

17 Gaibor, testimony, case file p. 82,614; Zambrano, judgment, February 14, 2011, p. 142.

18 Miguel Zumba, testimony, *Maria Aguinda Salazar y Otros v. Chevron Texaco Corp.*, case file p. 11,722; Zambrano, judgment, February 14, 2011, p. 151.

19 Amada Francisca Armijos Ajila, testimony, *Maria Aguinda Salazar y Otros v. Chevron Texaco Corp.*, case file p. 123,098; Zambrano, judgment, February 14, 2011, p. 140.

20 Máximo Celso, testimony, *Maria Aguinda Salazar y Otros v. ChevronTexaco Corp.*, case file p. 41,659; Zambrano, judgment, February 14, 2011, p. 168.

21 Celso, testimony, case file p. 41,659; Zambrano, judgment, February 14, 2011, p. 150.

22 Simón José Robles, testimony, *Maria Aguinda Salazar y Otros v. ChevronTexaco Corp.*, case file p. 41,659; Zambrano, judgment, February 14, 2011, p. 168.

23 Carlos Quevedo Quevedo, testimony, *Maria Aguinda Salazar y Otros v. Chevron-Texaco Corp.*, case file p. 56,742; Zambrano, judgment, February 14, 2011, p. 149.

24 Emergildo Criollo, testimony, *Maria Aguinda Salazar y Otros v. ChevronTexaco Corp.*, case file p. 155,978.

25 Gaibor, testimony, case file p. 82,613; Zambrano, judgment, February 14, 2011, p. 150.

26 Celso, testimony, case file p. 41,659; Zambrano, judgment, February 14, 2011, p. 168.

27 Carlos Cruz Calderón, testimony, *Maria Aguinda Salazar y Otros v. Chevron-Texaco Corp.*, case file p. 122,533; Zambrano, judgment, February 14, 2011, p. 151.

28 Luis Vicente Albán, testimony, *Maria Aguinda Salazar y Otros v. ChevronTexaco Corp.*, case file p. 122,533; Zambrano, judgment, February 14, 2011, p. 151.

29 Zumba, testimony, case file p. 11,722; Zambrano, judgment, February 14, 2011, pp. 151–52.

30 Ureña, testimony, case file p. 97,537; Zambrano, judgment, February 14, 2011, p. 149.

31 Melo Melo, testimony, case file p. 74,987; Zambrano, judgment, February 14, 2011, p. 151.

KuanKuan

1 Huangana is a singular and plural noun.

Chapter Five. Plurivalent Rendering

1 Judge Nicolás Zambrano Lozada, judgment, February 14, 2011, *Maria Aguinda Salazar y Otros v. ChevronTexaco Corp.*, Case No. 002-2003-P-CSJNL (2011-63-1).

2 Appellate rapporteur Judge Milton Toral Zavellos, judgment, January 3, 2012, p. 12, *Maria Aguinda y Otros v. ChevronTexaco Corp.*, Case No. 2011-0106-P-CPJS (2012-78-6), Appeal Division of the Sole Chamber, Provincial Court of Justice of Sucumbíos, Nueva Loja, Ecuador.

3 Zavellos, judgment, January 3, 2012, p. 12.

4 Zavellos, judgment, January 3, 2012, p. 12, citing Case No. 126-2005, Exp. 308.06.

5 *Chevron Corp. v. Donziger*, 833 F.3d 74, 84–85 (2d Cir. 2016). Also available through the ECF 14-0826 (L), DI 484, pp. 84–85.

6 The company brought one thousand motions to the provincial court, increased the case file by twenty thousand pages when it appealed to the Lago Agrio appellate court, and submitted another ten thousand pages of documentation after appealing to the national supreme court.

7 Callon, Lascoumes, and Barthe (2009) would call this a "hybrid forum." They use the term throughout their book.

8 All in-text parenthetical citations are from Judge Zambrano's 2011 ruling unless stated otherwise.

9 For example, in the 1974 Hydrocarbons Law, Official Registry No. 616, dated August 14, 1974; and in the 1978 Hydrocarbons Law, Official Registry No. 711, dated November 15, 1978.

10 Zambrano noted that the 1937 Petroleum Law was in effect until 1971, which meant that Texpet had already drilled several wells while the law was still in effect. As such, Texpet "should have fulfilled its provisions, which ordered that the performance of work had to be in accordance with the Technical Regulations for Petroleum Work [Reglamento Técnico de Trabajos Petrolíferos]." As far as the court could ascertain, however, the 1937 regulations were never issued, "which means that, from the start of the Texpet operations until 1971, that is, the initial period in which a considerable part of the consortium's facilities were built, there was no law in effect that established specific technical obligations that the operator should fulfill from the perspective of hydrocarbon technology" (Zambrano, judgment, February 14, 2011, p. 61).

11 This sentiment was voiced many times by Zambrano.

12 Zambrano in fact detailed the multiple warnings and penalties noted in the record that Ecuadorian regulatory agencies had leveled at Texpet "for failing to adopt the measures necessary to avoid the contamination of waters" and the environment (citing the General Manager of Hydrocarbons, pp. 71–72).

13 This primer is reproduced in the case files of *Maria Aguinda Salazar y Otros v. ChevronTexaco Corp.* as pp. 140,620–698.

14 See the case files of *Maria Aguinda Salazar y Otros v. ChevronTexaco Corp.*, pp. 153,722–725.

15 US Patent Office 3.817.859, June 18, 1974; *Maria Aguinda Salazar y Otros v. ChevronTexaco Corp.*, p. 104,363; partially quoted in Zambrano, judgment, February 14, 2011, p. 162.

16 See the case files of *Maria Aguinda Salazar y Otros v. ChevronTexaco Corp.*, p. 3,892; also cited in Zambrano, judgment, February 14, 2011, p. 161.

17 See the case files of *Maria Aguinda Salazar y Otros v. ChevronTexaco Corp.*, pp. 3,892–93; cited in Zambrano, judgment, February 14, 2011, p. 162.

18 See the case files of *Maria Aguinda Salazar y Otros v. ChevronTexaco Corp.*, pp. 3,893.

19 See the case files of *Maria Aguinda Salazar y Otros v. ChevronTexaco Corp.*, p. 155,522; cited in Zambrano, judgment, February 14, 2011, p. 80.

20 See the case files of *Maria Aguinda Salazar y Otros v. ChevronTexaco Corp.*, p. 140,601; cited in Zambrano, judgment, February 14, 2011, p. 113.

21 CESR (1994) and Kimerling (1991a, 1993) suggest that 4.3 million gallons of formation waters were released every day for twenty years, which would be 29 billion gallons in total.

22 Zambrano also noted that "the understanding that this court has acquired on the thematic constituting this lawsuit allows it to observe that the eventual use of the technology described in the patent would have replaced the gooseneck [pipes], not the pits, thus the argument related to the pits in other countries is irrelevant as regards this evidence." Zambrano, judgment, February 14, 2011, p. 164.

23 Ernesto Baca, testimony, case files of *Maria Aguinda Salazar y Otros v. Chevron Texaco Corp.*, p. 101,133; cited in Zambrano, judgment, February 14, 2011, p. 118.

24 Memo from Engineer Granja, technical assistant manager, to Michael Martínez, manager of Texaco Petroleum Company, November 24, 1976, case files of *Maria Aguinda Salazar y Otros v. ChevronTexaco Corp.*, p. 101,106; cited in Zambrano, judgment, February 14, 2011, p. 118.

25 The quotation here is from expert Jorge Bermeo, case files of *Maria Aguinda Salazar y Otros v. ChevronTexaco Corp.*, pp. 159,373–376; cited in Zambrano, judgment, February 14, 2011, pp. 130–31.

26 Although not noted by Zambrano, "Yana Curi" represents a portion of Miguel San Sebastián's doctoral thesis for the London School of Medicine and Tropical Hygiene. He conducted this work in conjunction with the department of health for the vicarage of Aguarico. Because it is not peer reviewed, I did not discuss it in chapter 3.

27 Simply to note, Zambrano incorrectly titles this study "Cancer in the Ecuadorian Amazon."

28 Roberto Bejarano and Monserrat Bejarano, "Study to Ascertain the Scope of the Effects of Contamination at the Oil Wells and Areas Drilled before 1990: The Lago Agrio, Dureno, Atacapi, Guanta, Shushufindi, Sacha, Yuca, Auca, and Cononaco Fields," October 22, 2003, vol. 7, p. 614.

29 Ecuadorian Code of Civil Procedure, Title XXXIII, fourth book of Articles 2241 through 2261; cited in Zambrano, judgment, February 14, 2011, p. 74.

30 The case is *Delfina Torres Vda. de Concha v. Petroecuador,* Case No. 229 R.O. 43 Supreme Court of Justice, First Chamber of Civil and Commercial Claims, Quito, Ecuador (decided October 29, 2002, and published in Official Registry No. 43, dated March 19, 2003).

31 The Supreme Court ruling in *Delfina* incorrectly read the maxim as "ubi emolumentum ibi llus." Case No. 229 R.O. 43, October 29, 2002, p. 28.

32 *Derecho Civil, Volumen III, De las obligaciones*, 8th ed. (Bogotá: Temis, 1990), pp. 230–31, cited in Zambrano, judgment, February 14, 2011, p. 84. The cited articles are equivalent to Articles 2247 and 2249 of the Ecuadorian Civil Code.

33 Darío Preciado Agudelo, *Indemnización de perjuicios. Responsabilidad civil contractual, extracontractual y delictual, Volumen II* (Bogotá: Ediciones Librería del Profesional, 1988), pp. 805 and 806, cited in Zambrano, judgment, February 14, 2011, p. 86.

34 Judge Nicolás Zambrano Lozada, clarification order, issued on March 4, 2011, addressing Chevron's motion for clarification and expansion of the Ecuadorian court's February 14, 2011, judgment.

35 Zambrano, clarification order, March 4, 2011, p. 12.

36 Zambrano, clarification order, March 4, 2011, 12.

37 Zambrano, clarification order, March 4, 2011, 13.

38 Zambrano, clarification order, March 4, 2011, 13.

39 Zambrano, clarification order, March 4, 2011, 14.

40 Zambrano, clarification order, March 4, 2011, 14.

41 LAP's "Legal Compliant" filed May 3, 2003, *Maria Aguinda Salazar y Otros v. ChevronTexaco Corp.*, pp. 1–16.

42 In 2007, Europe launched REACH (the Registration, Evaluation, Authorisation, and Restriction of Chemicals program). The European Commission's website about the enforcement of REACH reads as follows: "Manufacturers and importers are required to gather information on the properties of their chemical substances, which will allow their safe handling, and to register the information in a central database run by the European Chemicals Agency in Helsinki." See European Commission, "REACH Enforcement," accessed September 23, 2021, http://ec.europa.eu/environment/chemicals/reach/enforcement _en.htm. See also the European Commission Communication on the Precautionary Principle, explained for a broad audience here: European Commission 2017.

43 "Art. 2236.—Por regla general se concede acción popular en todos los casos de daño contingente que por imprudencia o negligencia de alguno amenace a personas indeterminadas. Pero si el daño amenazare solamente a personas determinadas, sólo alguna de éstas podrá intentar la acción." [Art. 2236—As a general rule, an acción popular are all cases of contingent harm in which an act of recklessness or negligence threatens an indeterminate number of people. If the harm, however, threatens a determinate number of people, only one of that group may file a claim.] Code of Civil Procedure, Book 4, "On General Obligations and Contracts," Section xxxiii, "On Crimes and Quasi-Crimes."

44 *Chevron Corp. v. Donziger*, 833 F.3d 74, 84 (2d Cir. 2016). Also available through ECF 14-0826 (L), DI 484, pp. 84.

Chapter Six. Bonding Veredictum

1 *Chevron Corp. v. Donziger*, 974 F. Supp. 2d 362 (S.D.N.Y. 2014). The case file can be found in the US court electronic filing system at *Chevron Corp. v. Donziger*, ECF 11 Civ. 0691 (LAK), in United States District Court, Southern District of New York. The case was filed on February 1, 2011. Steven Donziger was the US-based advisory lawyer for the Lago Agrio plaintiffs (LAP) and had been part of the legal team since the lawsuit against Texaco was first filed in 1993. He gar-

nered crucial funding and public attention for the litigation in Ecuador, and he assisted in developing the LAP legal strategy. In the United States, discussion of this legal saga positions him as the key character. Donziger was clearly important to the contamination case. But he is *not* the case, despite Chevron's relentless attempts to make him precisely that (see Sawyer and Ofrias forthcoming).

2   The Racketeer Influenced and Corrupt Organizations Act (RICO, 18 USC §§ 1961–68) statute of the Organized Crime Control Act of 1970 is a legal framework designed to litigate racketeering in the United States. RICO provides for extended criminal penalties and a civil cause of action for acts performed as part of an ongoing criminal organization. Chevron claimed to be the victim of racketeering, fraud, and corruption perpetrated by fifty-six named individuals and organizations (the Ecuadorians' legal team—Steven Donziger being the first on the list—two scientific experts, and forty-eight Indigenous peoples and rainforest peasants, most of whom have never left the Ecuadorian Amazon) and eighteen nonparty coconspirators (ten individuals, four law firms, one environmental assessment firm, one financial firm, and two nongovernmental environmental and Indigenous rights organizations).

The countersuit was crafted to delegitimize the anticipated adverse judgment in Ecuador. Chevron's original claim alleged, first, that the Ecuadorian legal team (everyone else is a coconspirator) bribed Zambrano and conspired to have his judgment ghostwritten by an ex-judge and, second, that the Ecuadorian judicial system is so systematically inadequate that any judgment emanating from it is unworthy of recognition.

3   Comity ensures that a sovereign nation adopts or enforces the law of another out of deference, mutuality, and respect. Although not inscribed in international law, this long-standing principle of states accepting each other's laws, political systems, and customs embodies international goodwill among nations.

4   "Stipulated and Ordered," June 27, 2001, *Aguinda et al. v. Texaco Inc.*, 93 Civ. 7527 (JSR), DI 159; *Jota et al. v. Texaco Inc.*, 94 Civ. 9266 (JSR).

5   The next entry in *Black's Law Dictionary* recites a maxim in Roman law: "*Veredictum, quasi dictum veritatis; ut judicium quasi juris dictum.* . . . The verdict is, as it were, the dictum [say-ing] of truth; as the judgment is the dictum of law" (Black [1891] 1968: 1732).

6   In this chapter, I do not delve into Chevron's complicated relationship with its chief witness, Alberto Guerra. Much has been written in the press about this audaciously compromised relationship. I engage with the figure of Guerra in another piece (Sawyer and Ofrias forthcoming).

7   A version of the 2011 judgment was found on Guerra's computer. However, it was clearly added onto his computer weeks after the February 14 judgment was submitted and rendered through Ecuador's electronic court filing system.

8 At the time, the team included Cristóbal Bonifaz, Alberto Wray, Monica Pareja, and Steven Donziger.

9 Declaration of David L. Russell, October 3, 2013, p. 4, *Chevron Corp. v. Donziger*, ECF 11 Civ. 0691 (LAK) DI 1561-1 (filed October 16, 2013).

10 Russell, declaration, October 3, 2013, pp. 4 and 12.

11 Russell, declaration, October 3, 2013, p. 5.

12 Courtroom transcript, October 17, 2013, p. 339, *Chevron Corp. v. Donziger*, ECF 11 Civ. 0691 (LAK) DI 1792.

13 Courtroom transcript, October 17, 2013, p. 385.

14 Courtroom transcript, October 17, 2013, p. 385.

15 Courtroom transcript, October 17, 2013, p. 386.

16 Courtroom transcript, October 17, 2013, pp. 365–86.

17 Courtroom transcript, October 17, 2013, pp. 399–400.

18 Courtroom transcript, October 17, 2013, p. 394.

19 Courtroom transcript, October 17, 2013, pp. 394–95.

20 Courtroom transcript, October 17, 2013, p. 406.

21 Courtroom transcript, October 17, 2013, p. 406.

22 Courtroom transcript, October 17, 2013, p. 407.

23 Courtroom transcript, October 17, 2013, p. 407.

24 Courtroom transcript, October 17, 2013, p. 408.

25 Judge Lewis A. Kaplan, "Opinion," March 4, 2014, p. 43, *Chevron Corp. v. Donziger*, 974 F. Supp. 2d 362 (S.D.N.Y. 2014), ECF 11 Civ. 0691 (LAK).

26 Kaplan, "Opinion," March 4, 2014, p. 43.

27 Courtroom transcript, October 17, 2013, p. 408, cited in Kaplan, "Opinion," March 4, 2014, p. 59.

28 Courtroom transcript, October 17, 2013, pp. 408:22–409:2.

29 Courtroom transcript, October 17, 2013, pp. 408:10–14, cited in Kaplan, "Opinion," March 4, 2014, p. 59.

30 Tribunal hearing, April 21, 2015 (Intervention of Republic of Ecuador lawyer Gregory Ewing), *In the Matter of an Arbitration*, Case No. 2009-23, p. 278, Permanent Court of Arbitration (PCA), The Hague.

31 Kaplan, "Opinion," March 4, 2014, p. 59.

32 Kaplan, "Opinion," March 4, 2014, p. 59n242.

33 This description came from the website of Gibson, Dunn & Crutcher, accessed September 2, 2021, https://www.gibsondunn.com/lawyer/mastro-randy-m/.

34 Andrew J. Hawkins, "Randy Mastro: The Maestro in Chief," *Crain's New York Business*, March 5, 2013, http://www.crainsnewyork.com/article/20130505/POLITICS/305059982/randy-mastro-the-maestro-of-mischief#.

35 This description of the RICO case, one of the "major litigation matters handled by Randy Mastro," comes from the website of Gibson, Dunn & Crutcher, accessed September 2, 2021, http://www.gibsondunn.com/lawyers/rmastro-randy-m/.

36 The bio was on the website of Gibson, Dunn & Crutcher, accessed September 2, 2021, http://www.gibsondunn.com/lawyers/rmastro-randy-m/.

37 Courtroom transcript, November 5, 2013, pp. 1607–8, *Chevron Corp. v. Donziger*, ECF 11 Civ. 0691 (LAK).

38 Courtroom transcript, November 5, 2013, pp. 1608–11.

39 Courtroom transcript, November 5, 2013, pp. 1611–13.

40 Courtroom transcript, November 5, 2013, pp. 1613–14.

41 Courtroom transcript, November 5, 2013, pp. 1614–15.

42 Courtroom transcript, November 5, 2013, pp. 1711–12.

43 Kaplan, "Opinion," March 4, 2014, p. 182.

44 Judge Nicolás Zambrano Lozada, judgment, February 14, 2011, p. 107, *Maria Aguinda Salazar y Otros v. ChevronTexaco Corp.*, ECF 002-2003-P-CSJNL (2011-63-1).

45 Zambrano, judgment, February 14, 2011, p. 88 (emphasis mine). The English-language translation in the text here was provided by Chevron during the RICO trial.

46 Zambrano, judgment, February 14, 2011, p. 88.

47 Zambrano, judgment, February 14, 2011, pp. 80 and 90.

48 Zambrano, judgment, February 14, 2011, p. 88.

49 Zambrano, judgment, February 14, 2011, pp. 20 and 21.

50 Track 2 hearing, May 8, 2015, and April 21, 2015, pp. 2852–54, 328–30 (intervention of lawyers for Winston and Strawn, representing the Republic of Ecuador), *In the Matter of an Arbitration*, Case No. 2009-23, PCA, The Hague.

51 Track 2 hearing, April 21, 2015, p. 329.

52 Errors occurred. For example, when, in October 2010, Chevron filed its thirty-nine separate motions challenging a single Lago Agrio court order, in response, the court addressed all thirty-nine concerns, even though only thirty-five of Chevron's motions actually appear in the official record. Track 2 hearing, April 21, 2015, p. 325.

53 Kaplan, "Opinion," March 4, 2014, pp. 182, 182–83, 184–87 (internal citations omitted for ease of reading).

54 Track 2 hearing, May 8, 2015, pp. 2796, 2809–10.

55 These include legal proceedings under 28 USC § 1782 in New York (the residence of Joe Berlinger, the producer and director of the documentary film *Crude*), Colorado (the location of Stratus Consulting), New Mexico (the base of E-Tech), California (the base of William Powers, an environmental consultant with E-Tech), Texas (the location of 3TM Consulting), North Carolina (the residence of Charles Champ, an environmental consultant), and Tennessee (the residence of Mark Quarles, an environmental consultant with E-Tech).

56 Zambrano, judgment, February 14, 2011, p. 51.

57 Judge Zambrano, clarification order, March 4, 2011, p. 8, *Maria Aguinda Salazar y Otros v. ChevronTexaco Corp.*, ECF 002-2003-P-CSJNL (2011-63-1).

58 Zambrano, judgment, February 14, 2011, pp. 124 and 125.

59 James I. Ebert, witness statement, October 31, 2013, p. 7, *Chevron Corp. v. Donziger*, ECF 11 Civ. 0691 (LAK) DI 1651-1.

60 Spencer Lynch, witness statement, October 7, 2013, p. 43, *Chevron Corp. v. Donziger*, ECF 11 Civ. 0691 (LAK) DI 1584-1.

61 Courtroom transcript, October 21, 2013, pp. 639–40, *Chevron Corp. v. Donziger*, ECF 11 Civ. 0691 (LAK).

62 In its "Track 2 Supplemental Counter-Memorial on the Merits" (November 7, 2014, pp. 77–82), the Republic of Ecuador complements and extends the analysis I provide. *In the Matter of an Arbitration*, Case No. 2009-23, PCA, The Hague.

63 Judge Lewis A. Kaplan, "Appendices to Opinion," March 4, 2014, pp. 47–48, *Chevron Corp. v. Donziger, ECF 11 Civ. 0691 (LAK) DI 1874-1.

64 John Connor, "2013 Expert Report," table C.1.A (49 remediated pits, 15 NFA pits, 1 COC pit, and 83 non-RAP pits), p. 89, in "Track 2 Rejoinder on the Merits of the Republic of Ecuador (Part I: Response to Factual Predicate to Claimants' Claims)," December 16, 2013, *In the Matter of an Arbitration*, Case No. 2009-23, PCA, The Hague.

65 Zambrano, judgment, February 14, 2011, p. 181.

66 Zambrano, judgment, February 14, 2011, p. 181.

67 Circuit Judges Amalya L. Kearse, Barrington D. Parker, and Richard C. Wesley, "Opinion," p. 51 (brackets in the original), *Chevron Corp. v. Donziger*, 833 F.3d 74 (2d Cir. 2016), ECF 14-0826 (L), DI 484.

68 Kaplan, "Opinion," March 4, 2014, p. 46.

Metamorphic Reprise

1 The three arbiters for the tribunal were Dr. Horacio A. Grigera Naón, Professor Vaughan Lowe QC, and V. V. Veeder QC (president).

2 "Second Partial Award on Track II," tribunal decision, August 30, 2018, pp. X:2–3, *In the Matter of an Arbitration*, Case No. 2009-23, Permanent Court of Arbitration (PCA), The Hague.

3 "Claimants' Notice of Arbitration," September 23, 2009, *In the Matter of an Arbitration*, Case No. 2009-23, PCA, The Hague.

4 US-Ecuador BIT, Art. II (7), p. 13, signed August 27, 1993, and entered into enforcement May 11, 1997. These claims are based on infringements to two clauses (3 and 7) within Article II of the BIT. Clause 3 affirms that "(a) Investments shall at all times be accorded fair and equitable treatment," that "(b) Neither Party shall in any way [be] impair[ed] by arbitrary or discriminatory measures," and that "(c) Each Party shall observe any obligation it may have entered with regard to investment." Clause 7 states that "[e]ach Party shall provide effective means of asserting claims and enforcing rights with respect to investment, investment agreements, and investment authorizations."

5 "Claimants' Memorial on the Merits," September 6, 2010, pp. 275–76, *In the Matter of an Arbitration*, Case No. 2009-23, PCA, The Hague.

6 "Claimants' Memorial on the Merits," September 6, 2010, pp. 276–77.

7 "Claimants' Memorial on the Merits," September 6, 2010, pp. 196–99.

8 "Claimants' Memorial on the Merits," September 6, 2010, p. 204.

9 1979 Political Constitution of the Republic of Ecuador, Art. 19(2).

10 "Claimants' Memorial on the Merits," September 6, 2010, p. 203.

11 "Track I Hearing on Interim Measures," London, February 11, 2012, pp. 80–90, *In the Matter of an Arbitration*, Case No. 2007-02/AA277, PCA, The Hague.

   This tactic traces processes that Kim Fortun (2001) eloquently analyzes for the aftermath of the devastating chemical explosion in Bhopal, India. The republic argued that under well-settled principles of Ecuadorian law there is no *parens patrie* doctrine.

12 "Claimants' Memorial on the Merits," September 6, 2010, p. 199, quoting Professor Enrique Barros, first expert report, *In the Matter of an Arbitration*, Case No. 2009-23, PCA, The Hague.

13 "Claimants' Memorial on the Merits," September 6, 2010, p. 199, quoting Professor A. Oquendo, expert report, *In the Matter of an Arbitration*, Case No. 2009-23, PCA, The Hague.

14 "Tribunal Track II Hearing," April 21, 2015, transcript p. 261, *In the Matter of an Arbitration*, Case No. 2009-23, PCA, The Hague.

15 "Art. 2236.—Por regla general se concede acción popular en todos los casos de daño contingente que por imprudencia o negligencia de alguno amenace a personas indeterminadas. Pero si el daño amenazare solamente a personas determinadas, sólo alguna de éstas podrá intentar la acción." Civil Code, book 4, "De las Obligaciones en General y de los Contratos," Titulo xxxiii, "De los Delitos y Cuasidelitos."

16 Ecuador's National Court of Justice and Constitutional Court independently affirmed this.

17 "Tribunal Track II Hearing," April 21, 2015, transcript p. 261.

18 "Tribunal Track II Hearing," April 21, 2015, transcript p. 262.

19 "Tribunal Track II Hearing," April 21, 2015, transcript p. 262.

20 "Tribunal Track II Hearing," April 21, 2015, transcript p. 262.

21 "Tribunal Track II Hearing," April 21, 2015, transcript p. 262.

22 "Tribunal Track II Hearing," April 21, 2015, transcript pp. 262–63.

23 "Tribunal Track II Hearing," April 21, 2015, transcript p. 263.

24 "Tribunal Track II Hearing," April 21, 2015, transcript p. 288.

25 "Tribunal Track II Hearing," May 8, 2015, transcript p. 2905, *In the Matter of an Arbitration*, Case No. 2009-23, PCA, The Hague.

26 "Tribunal Track II Hearing," May 8, 2015, transcript pp. 2905–6.

27 "Tribunal Track II Hearing," May 8, 2015, transcript p. 2927.

28 "Tribunal Track II Hearing," May 8, 2015, transcript p. 2927. Ecuador Civil Code Article 2349 states that a contract can "only compromise a person possessing the right to dispose of the objects included in the transaction," thus foreclosing the possibility of one party settling rights that do not belong to it. Article 2354 mandates that a "settlement over the rights of others or over non-existent rights is invalid," while Article 2363 dictates that a "settlement or transaction shall only take effect between the transacting Parties."

29 "Tribunal Track II Hearing," May 8, 2015, transcript p. 2927.

30 "Tribunal Track II Hearing," April 21, 2015, transcript p. 265.

31 "Tribunal Track II Hearing," April 21, 2015, transcript p. 265.

32 "Tribunal Track II Hearing," April 21, 2015, transcript p. 265, and May 8, 2015, transcript p. 2929.

33 "Track 2 Supplemental Rejoinder on the Merits of the Republic of Ecuador," March 17, 2015, p. 28, *In the Matter of an Arbitration*, Case No. 2009-23, PCA, The Hague.

34 "Track 2 Supplemental Rejoinder on the Merits of the Republic of Ecuador," March 17, 2015, p. 28.

35 "Track 2 Supplemental Rejoinder on the Merits of the Republic of Ecuador," March 17, 2015, p. 28.

36 "First Partial Award on Track I," September 17, 2013, "Part E: The Operative Part," p. 43, *In the Matter of an Arbitration*, Case No. 2009-23, PCA, The Hague.

37 "First Partial Award on Track I," September 17, 2013, "Part E: The Operative Part," p. 43.

38 "First Partial Award on Track I," September 17, 2013, "Part E: The Operative Part," p. 43.

39 "First Partial Award on Track I," September 17, 2013, "Part E: The Operative Part," p. 43.

40 "First Partial Award on Track I," September 17, 2013, "Part E: The Operative Part," p. 43.

41 "First Partial Award on Track I," September 17, 2013, "Part E: The Operative Part," p. 43.

42 "First Partial Award on Track I," September 17, 2013, "Part E: The Operative Part," p. 45.

43 "First Partial Award on Track I," September 17, 2013, "Part E: The Operative Part," p. 45.

44 "Decision on Track 1B," March 12, 2015, *In the Matter of an Arbitration*, Case No. 2009-23, PCA, The Hague.

45 "Decision on Track 1B," March 12, 2015, p. 47.

46 "Decision on Track 1B," March 12, 2015, p. 47.

47 "Decision on Track 1B," March 12, 2015, p. 48.

48 "Second Partial Award on Track II," August 30, 2018, "Part V: Judgments," p. 75, *In the Matter of an Arbitration*, Case No. 2009-23, PCA, The Hague.

49 "Second Partial Award on Track II," August 30, 2018, "Part V: Judgments," p. 75.

50 "Decision on Track 1B," March 12, 2015, p. 43.

51 "Second Partial Award on Track II," August 30, 2018, "Part V: Judgments," p. 74.

52 "Second Partial Award on Track II," August 30, 2018, "Part V: Judgments," p. 75.

53 "Second Partial Award on Track II," August 30, 2018, "Part VIII: Merits," p. 2, *In the Matter of an Arbitration*, Case No. 2009-23, PCA, The Hague. See also "Part X: The Operative Part," p. 2.

54 "Second Partial Award on Track II," August 30, 2018, "Part VIII: Merits," p. 2.

55 Texaco and Ministry of Energy and Mines 1995, p. 9. The complete sentence reads as follows: "On the execution date of this contract . . . the Government and Petroecuador shall hereby release, acquit, and forever discharge Texpet . . . Texaco, Inc. . . . of all the Government's and Petroecuador's claims against the Releases for Environmental Impact arising from the Operations of the Consortium, except for those related to the obligations conducted hereunder for the performance by Texpet of the Scope of Work."

Significantly, Chevron's claim before the international tribunal misrepresents the scope of the agreement. In its complaint Chevron wrote "the Government and Petroecuador shall hereby release, acquit, and forever discharge Texpet [and its affiliate] for Environmental Impact arising from the Operations of the Consortium, except for those related to . . . the Scope of Work." This wording omitted the fact that the corporation was being released from "all the Government's and Petroecuador's claims against the Releases," not those of the entire citizenry.

56 The memorandum of understanding signed by the parties stated that "the provisions of this [MOU] shall apply without prejudice to the rights possibly held by third parties for the impact caused as a consequence of the operations of the former Petroecuador-Texaco consortium." Texaco and Ministry of Energy and Mines 1994.

57 Personal communication, July 29, 2019. Offices of the Unión de los Afectors por Texaco (UDAPT), Lago Agrio, Ecuador.

58 Personal communication, July 29, 2019. UDAPT offices, Lago Agrio, Ecuador.

59 "Tribunal Track II Hearing," May 8, 2015, transcript p. 2923.

60 1979 Political Constitution of the Republic of Ecuador, Art. 19(2).

61 "Tribunal Track II Hearing," April 21, 2015, transcript p. 260.

62 As the court declared, the New York *Aguinda* lawsuit "sought extensive equitable relief to redress contamination of the water supplies and environment" that entailed "environmental cleanup" and the "creation of an environmental monitoring fund . . . [and] medical monitoring fund." *Aguinda v. Texaco Inc.*, 303 F.3d 470, 472 (2d Cir. 2002), ECF 01-7756 (L), 01-7758 (C).

63 *Republic of Ecuador v. Chevron Corp.*, 638 F.3d 374, 374n5 (2d Cir. 2011), Docket ECF 10-1020-cv (L).

64 "Tribunal Track II Hearing," May 8, 2015, transcript p. 2925.

65 Judge Sands, US District Court for the Southern District of New York, June 17, 2005, citing *Republic of Ecuador v. ChevronTexaco Corp.*, 376 F. Supp. 2d 334, 374 (S.D.N.Y. 2005). Judge Sands ruled in favor of the Republic of Ecuador and had Chevron's claim dismissed from the American Arbitration Association (AAA).

66 The United Nations Commission on International Trade Law (UNCITRAL), established by the UN General Assembly in 1966, was charged with reconciling legal obstacles between states and national laws governing international trade and has since emerged as the UN's core legal body in the area of international trade law.

67 King and Spalding's presentation, "The Emergence of Diffuse Rights and Public Participation in Environmental Cases," 22nd International Petroleum Environmental Consortium Conference, Denver, Colorado, November 17–19, 2015, available at https://cese.utulsa.edu/wp-content/uploads/2017/06 /IPEC-2015-THE-EMERGENCE-OF-DIFFUSE-RIGHTS-AND-PUBLIC -PARTICIPATION-IN-ENVIRONMENTAL-CASES.pdf.

REFERENCES

*

Select Court Cases, Documents, and Agreements

The court citations listed below provide both the *Bluebook* legal citations and, in brackets, the Case Management/Electronic Case Files (CM/ECF) number available through the Public Access to Court Electronic Records (PACER).

Aguinda et al. v. Texaco Inc., 93 Civ. 7527 (VLB) DI 1 (S.D.N.Y. complaint filed November 3, 1993).

Aguinda et al. v. Texaco Inc., 850 F. Supp. 282 (S.D.N.Y. 1994) [ECF 93 Civ 7527 (VLB)].

Aguinda et al. v. Texaco Inc., 175 F.R.D. 50 (S.D.N.Y. 1997) [ECF 93 Civ. 7527 (JSR)].

Aguinda v. Texaco Inc., 142 F. Supp. 2d 534 (S.D.N.Y. 2001) [ECF 93 Civ. 7527, 94 Civ. 9266].

Aguinda v. Texaco Inc., 303 F.3d 470 (2d Cir. 2002) [ECF 01-7756 (L); 01-7758 (C)].

Chevron Corp. v. Donziger, ECF 11 Civ. 0691 (LAK), DI 1 (S.D.N.Y.; filed February 1, 2011).

Chevron Corp. v. Donziger, 768 F. Supp. 2d 581 (S.D.N.Y. 2011) [ECF 11 Civ. 0691 (LAK), DI 181].

Chevron Corp. v. Donziger, 974 F. Supp. 2d 362 (S.D.N.Y. 2014) [ECF 11 Civ. 0691 (LAK), DI 1874].

Chevron Corp. v. Donziger, 833 F.3d 74 (2d Cir. 2016) [ECF 14-0826 (L), DI 484].

Chevron Corp. v. Naranjo, 667 F.3d 232 (2d Cir. 2012) [ECF 11-1150-cv (L); 11-1264-cv (CON)].

ChevronTexaco Corp. v. Republic of Ecuador, 557 US 936 (2009) [ECF 08-1123].

Contraloría General del Estado. 2003. No. DA3-25-2002, "Special Evaluation of the Contract for Implementing of Environmental Remedial Works and Release from Obligations, Liabilities and Claims, Executed on May 4, 1995, by and between the Minister of Energy and Mines on Behalf of the Ecuadorian Government, the Executive President of Petroecuador, and the Vice-President of the Texaco Petroleum Company TexPet." April 9, 2003.

Delfina Torres Vda. de Concha v. Petroecuador, Case No. 229 R.O. 43, Supreme Court of Justice, First Chamber of Civil and Commercial Claims, Quito (decided October 29, 2002).

Fiscalía General del Estado. 2008. Indictment order against Ricardo Reis Vega, Dr. Rodrigo Peréz Pallares, and seven others. August 26.

In the Matter of an Arbitration under the Rules of the United Nations Commission on International Trade Law, Chevron Corporation and Texaco Petroleum Company, Claimants, v. The Republic of Ecuador, Respondent (I). Case No. 2007-02/AA277, Permanent Court of Arbitration (PCA), The Hague.

In the Matter of an Arbitration under the Rules of the United Nations Commission on International Trade Law, Chevron Corporation and Texaco Petroleum Company, Claimants, v. The Republic of Ecuador, Respondent. Case No. 2009-23, Permanent Court of Arbitration (PCA), The Hague.

Jota et al. v. Texaco Inc., 157 F.3d 153 (2d Cir. 1998). [ECF 97-9102 (L), 97-9104 (CON), 97-9108 (CON)].

Maria Aguinda Salazar y Otros v. ChevronTexaco Corp., Case No. 002-2003-P-CSJNL (2011-63-1), Provincial Court of Justice of Sucumbíos, Nueva Loja, Ecuador.

Maria Aguinda Salazar y Otros v. ChevronTexaco Corp., Case No. 2011-0106-P-CPJS (2012-78-6), Appeal Division of the Sole Chamber, Provincial Court of Justice of Sucumbíos, Nueva Loja, Ecuador.

Norsul Oil and Mining, Ltd. v. Texaco Inc., 641 F. Supp. 1502 (S.D. Fla. 1986).

Norsul Oil and Mining, Ltd. v. Texaco Inc., 703 F. Supp. 1520 (S.D. Fla. 1988).

Republic of Ecuador v. ChevronTexaco Corp., 376 F. Supp. 2d 334 (S.D.N.Y. 2005) [ECF 04 Civ 8378 (LBS)].

Republic of Ecuador v. ChevronTexaco Corp., 499 F. Supp. 2d 452 (S.D.N.Y 2007) [ECF 04 Civ. 8378 (LBS)].

Republic of Ecuador v. ChevronTexaco Corp., 296 Fed. Appx. 124 (2d Cir. 2008) [ECF 07-2868-cv].

Republic of Ecuador v. Chevron Corp., 638 F.3d 384 (2d Cir. 2011) [ECF 10-1020-cv (L)].

Texaco and Ministry of Energy and Mines. 1994. "Memorandum of Understanding between the Government of Ecuador, Petroecuador, and Texaco Petroleum Company." December 14.

Texaco and Ministry of Energy and Mines. 1995. "Contract for Implementing of Environmental Remedial Work and Release from Obligations, Liability and Claims." May 4.

Texaco and Ministry of Energy and Mines. 1998. Final release (acta final) between the government of Ecuador, Petroecuador, and Texaco Petroleum Company. September 30.

Treaty between the United States of America and the Republic of Ecuador Concerning the Encouragement and Reciprocal Protection of Investment, with Protocol and Related Exchange of Letters. 103D Congress, 1st Session, Senate

Treaty Doc. 103-15. Signed August 27, 1993, and entered into enforcement May 11, 1997.

Select Reports from Court Experts

Alvarez, Pedro J., Douglas M. Mackay, and Robert E. Hinchee. 2006. "Evaluation of Chevron's Sampling and Analysis Methods." Expert report commissioned by Chevron. August 28, 2006.

Bjorkman, B. 2006. "Site Production Station Sacha Sur." Expert report commissioned by Chevron. July 2006.

Cabrera Vega, Richard Stalin. 2008. "Technical Summary Report: Expert Opinion" (also known as the Cabrera report). March 24, 2008. *Maria Aguinda Salazar y Otros v. ChevronTexaco Corporation*. Case No. 002-2003. Superior Court of Justice of Nueva Loja.

Chevron Corporation. 2005. "Informe de la inspección judicial del pozo Sacha-57." *Maria Aguinda Salazar y Otros v. ChevronTexaco Corp*. Case No. 002-2003. Superior Court of Justice of Nueva Loja. Perito: Doctor Gino C. Bianchi Mosquera. June 2005.

Chevron Corporation. 2007. "Affirming the Truth: About Texaco's Past Operations and Questions of Health." Corporate publication.

Christopher, John P. 2010a. "Evaluation of the Doctoral Thesis of Plaintiffs' Expert Dr. Miguel San Sebastián." Expert report commissioned by Chevron. September 29, 2010.

Christopher, John P. 2010b. "Evaluation of the Scientific Value of the Published Work of Thesis of Plaintiffs' Experts, Dr. Miguel San Sebastián and Colleagues." Expert report commissioned by Chevron. September 6, 2010.

Fugro-McClelland. 1992. *Final Environmental Field Audit for Practices 1964–1990, Petroecuador-Texaco Consortium, Oriente, Ecuador*. October 1992.

Green, Laura C. 2005. "Analysis of Various Studies of Communities Near the Former TexPet Oil Operations in Ecuador." Expert report commissioned by Chevron. January 26, 2005.

Hewitt, David J. 2005. "Comment Regarding Causal Association between Ecuador Oil Exploration and Health Claims." Expert report commissioned by Chevron. January 24, 2005.

Kelsh, Michael A. 2006. "Review of Epidemiologic Studies of Cancer, Reproductive Outcomes, and Health Symptoms among Populations in the Amazon Region of Ecuador." Expert report commissioned by Chevron. March 7, 2006.

Kelsh, Michael A. 2009. "Rebuttal to Mr. Cabrera's Responses to Health-Related Questions." Expert report commissioned by Chevron. January 23, 2009.

Kelsh, Michael A. 2010. "Expert Rebuttal: Response to Reports of Dr. Daniel Rourke and Dr. Carlos Picone." Expert report commissioned by Chevron. October 5, 2010.

Kelsh, Michael A. 2011. "Lack of a Scientific Basis for Health Claims and

Health-Related Awards in the *Sentencia*." Expert report commissioned by Chevron. June 20, 2011.

Lago Agrio plaintiffs (LAP). 2005. "Observaciones por Alberto Wray y Pablo Fajardo Mendoza." *Maria Aguinda y Otros v. ChevronTexaco Corp.* Case No. 002-2003. Superior Court of Justice of Nueva Loja.

Lago Agrio plaintiffs (LAP). 2005. "Informe del perito de la inspección judicial en el pozo Sacha-57." *Maria Aguinda y Otros v. ChevronTexaco Corp.* Case No. 002-2003. Superior Court of Justice of Nueva Loja. Expert: Ing. Edison Camino Castro, L.P. 05-17-259.

Lynch, Spencer, Chevron Expert "Direct Testimony," provided to United States District Court for the Southern District of New York, ECF 11 Civ. 0691 (LAK), DI 1584-1 (written October 7, 2013, submitted to the court October 21, 2013).

Maest, Ann, Mark Quarles, and William Powers. 2006. "How Chevron's Sampling and Analysis Methods Minimizes Evidence of Contamination." Expert report commissioned by LAP. March 8, 2006.

McHugh, T. Thomas E. 2008. "Response to the Allegations of Mr. Cabrera Regarding the Potential Human Health Risk Associated with Hydrocarbons and Metals in the Petroecuador-Texaco Concession Area." Expert report commissioned by Chevron. September 8, 2008.

McHugh, Thomas E. 2011. "Lack of Evidence of Health Risks Associated with Hydrocarbons and Metals in the Former Concession Area." Expert report commissioned by Chevron. June 10, 2011.

Powers, Bill, and Mark Quarles. 2006. "Critical Analysis of Chevron's Science: Submission 2. Texaco's Waste Management Practices in Ecuador Were Illegal and Violated Industry Standards." Expert report commissioned by the LAP. April 5, 2006.

Rothman, Kenneth J., and Félix M. Arellano. 2005. Expert opinion commissioned by Chevron. February 1, 2005.

Strauss, Haley. 2013. "Regarding Human Health-Related Aspects of the Environmental Contamination from Texpet's E&P Activities in the Former Napo Concession, Oriente Region, Ecuador." Expert report commissioned by the Republic of Ecuador, *In the Matter of an Arbitration*, PCA Case No. 2009-23. February 18, 2013.

Woodward-Clyde International. 2000. *Remedial Action Project, Oriente Region, Ecuador. Final Report.* Vol. 1. May. Prepared for Texaco Petroleum Company, White Plains, NY. May 2000.

Literature

Acción Ecológica. 1993. *Investigación sobre cuatro estudios de caso de poblaciones afectadas por Texaco.* Quito: Acción Ecológica.

Agency for Toxic Substances and Disease Registry (ATSDR). 1995. *Toxicology Profile for Polycyclic Aromatic Hydrocarbons.* Atlanta, GA: US Department of Health and Human Services.

Agency for Toxic Substances and Disease Registry (ATSDR). 1999. *Toxicology Profile for Total Petroleum Hydrocarbons (TPH)*. Atlanta, GA: US Department of Health and Human Services.

Agency for Toxic Substances and Disease Registry (ATSDR). 2007. *Toxicological Profile for Benzene*. Atlanta, GA: US Department of Health and Human Services.

Alexander, Martin. 1995. "How Toxic Are Toxic Chemicals in Soil?" *Environmental Science and Technology* 29 (11): 2713–17.

Alford, Roger P. (2009) 2012. "Ancillary Discovery to Prove Denial of Justice." *Virginia Journal of International Law* 53 (1): 127–56.

Aman, Alfred C., Jr., and Carol J. Greenhouse. 2017. *Transnational Law: Cases and Problems in an Interconnected World*. Durham, NC: Carolina Academic Press.

Amat, A., T. Burgeot, M. Castegnaro, and A. Pfohl-Leszkowcz. 2006. "DNA Adducts in Fish Following an Oil Spill Exposure." *Environmental Chemistry Letters* 4: 93–99.

American Association of Petroleum Geologists Bulletin (AAPGB). 1968. "Developments in South America and Caribbean Areas." 52 (8): 1366–438.

American Association of Petroleum Geologists Bulletin (AAPGB). 1971. "Developments in South America and Caribbean Areas." 55 (9): 1418–82.

American Association of Petroleum Geologists Bulletin (AAPGB). 1972. "Developments in South America and Caribbean Areas." 56 (9): 1602–60.

American Association of Petroleum Geologists Bulletin (AAPGB). 1973. "Developments in South America and Caribbean Areas." 57 (10): 1968–33.

American Association of Petroleum Geologists Bulletin (AAPGB). 1974. "Developments in South America and Caribbean Areas." 58 (10): 1902–73.

American Association of Petroleum Geologists Bulletin (AAPGB). 1976. "Developments in South America and Caribbean Areas." 60 (10): 1640–703.

American Association of Petroleum Geologists Bulletin (AAPGB). 1984. "Developments in South America and Caribbean Areas." 68 (10): 1467–92.

American Association of Petroleum Geologists Bulletin (AAPGB). 1987. "Developments in South America and Caribbean Areas." 71 (10B): 337–63.

American Association of Petroleum Geologists Bulletin (AAPGB). 1988. "Developments in South America and Caribbean Areas." 72 (10B): 343–66.

American Petroleum Institute (API). 1948. *API Toxicological Review: Benzene*. New York: API Department of Safety.

American Petroleum Institute (API). 2001. *Risk-Based Methodologies for Evaluating Petroleum Hydrocarbon Impacts at Oil and Natural Gas E&P Sites*. API Publication Number 4709. Washington, DC: API Publishing Services.

Amsterdam, Anthony G., and Jerome Bruner. 2002. *Minding the Law: How Courts Rely on Storytelling, and How Their Stories Change the Ways We Understand the Law—and Ourselves*. Cambridge, MA: Harvard University Press.

Appel, Hannah. 2012. "Offshore Work: Oil, Modularity, and the How of Capitalism in Equatorial Guinea." *American Ethnologist* 39 (4): 692–709.

Appel, Hannah. 2019. *The Licit Life of Capitalism: US Oil in Equatorial Guinea*. Durham, NC: Duke University Press.

Arana, Alejandro, and Felix Arellano. 2007. "Cancer Incidence Near Oilfields in the Amazon Basin of Ecuador Revisited." *Occupational and Environmental Medicine* 64 (7): 490.

Baby, Patrice, Marco Rivadeneira, and Roberto B Arragán. 2004. *La cuenca Oriente: Geologia y petróleo.* Quito: Petroproducción y IRD Institut de Recherché pour le Dévéloppement.

Bachelard, Gaston. 1953. *Le matérialisme rationnel.* Paris: Presses Universitaires de France.

Barrett, Paul. 2014. *Law of the Jungle: The $19 Billion Legal Battle over Oil in the Rain Forest and the Lawyer Who'd Stop at Nothing to Win.* New York: Penguin Random House.

Barry, Andrew. 2005. "Pharmaceutical Matters: The Invention of Informed Materials." *Theory, Culture and Society* 22 (1): 51–69.

Barry, Andrew. 2013. *Material Politics: Disputes along the Pipeline.* London: Wiley-Blackwell.

Barry, Andrew. 2015. "Thermodynamics, Matter, Politics." *Distinktion: Scandinavian Journal of Social Theory* 16 (1): 110–25.

Barry, Andrew. 2017. "Manifesto for a Chemical Geography." Inaugural lecture, Gustave Tuck Lecture Theatre, University College London. January 24, 2017.

Barry, Andrew. 2020. "The Material Politics of Infrastructure." In *TechnoScience-Society: Technological Reconfigurations of Science and Society,* edited by Sabine Maasen, Sascha Dickel, and Christoph Schneider, 91–109. Cham, Switzerland: Springer.

Barry, Andrew, and Evelina Gambino. 2020. "Pipeline Geopolitics: Subaquatic Materials and the Tactical Point." *Geopolitics* 25 (1): 109–42. https://doi.org/10.1080/14650045.2019.1570921.

Beck, Ulrich. (1986) 1992. *Risk Society: Towards a New Modernity.* London: Sage.

Bensaude-Vincent, Bernadette. 1986. "Mendeleev's Periodic System." *British Journal for the History of Science* 19: 3–17.

Bensaude-Vincent, Bernadette. 2008. "Philosophy of Chemistry." In *French Studies in the Philosophy of Science,* edited by Anastasio Brenner and Jean Gayon, 165–85. Dordrecht, Netherlands: Springer.

Bensaude-Vincent, Bernadette. 2014. "Philosophy *of* Chemistry or Philosophy *with* Chemistry?" *Hyle: International Journal for Philosophy of Chemistry* 20 (1): 58–76.

Bensaude-Vincent, Bernadette, and Jonathan Simon. 2012. *Chemistry: The Impure Science.* London: Imperial College Press.

Bensaude-Vincent, Bernadette, and Isabelle Stengers. 1996. *A History of Chemistry.* Cambridge, MA: Harvard University Press.

Berlinger, Joe, dir. and prod. 2009. *Crude: The Documentary.* New York: RadicalMedia.

Bernal, Andrés, and Edgar E. Daza. 2010. "On the Epistemological and Ontological Status of Chemical Relations." *Hyle: International Journal for Philosophy of Chemistry* 16 (2): 80–103.

Black, Henry Campbell. (1891) 1968. *Black's Law Dictionary: Definitions of the Terms and Phrases of American and English Jurisprudence, Ancient and Modern*. 4th ed. St. Paul, MN: West Publishing Co.

Black, J. A., W. A. Birge, A. G. Westerman, and P. C. Francis. 1983. "Comparative Aquatic Toxicology of Aromatic Hydrocarbons." *Fundamentals of Applied Toxicology* 3: 353–58.

Blot, W. J., L. A. Brinton, J. F. Fraumeni, and B. J. Stone 1977. "Cancer Mortality in U.S. Counties with Petroleum Industries." *Science* 198 (4312): 51–53.

Bobra, Alice, Wan Ying Shiu, and Donald Mackay. 1983. "Acute Toxicity of Fresh and Weathered Crude Oils." *Chemosphere* 12 (9–10): 1137–49.

Boffetta, P., N. Jourenkova, and P. Gustavsson. 1997. "Cancer Risk from Occupational and Environmental Exposure to Polycyclic Aromatic Hydrocarbons." *Cancer Causes Control* 8: 444–72.

Bohme, Susanna Rankin. 2014. *Toxic Injustice: A Transnational History of Exposure and Struggle*. Berkeley: University of California Press.

Bohme, Susanna Rankin, John Zorabedian, and David S. Egilman. 2005. "Maximizing Profit and Endangering Health: Corporate Strategies to Avoid Litigation and Regulation." *International Journal of Occupational and Environmental Health* 11: 338–48.

Bonassi, S., F. Merlo, N. Pearce, et al. 1989. "Bladder Cancer and Occupational Exposure to Polycyclic Aromatic Hydrocarbons." *International Journal of Cancer* 44: 648–51.

Brette, Fabien, Ben Machado, Caroline Cros, John P. Incardona, Nathaniel L. Scholz, and Barbara A. Block. 2014. "Crude Oil Impairs Cardiac Excitation-Contraction Coupling in Fish." *Science* 343 (6172): 772–76.

Brooks, Gwendolyn. 2005. "Paul Robeson." In *The Essential Gwendolyn Brooks*, edited by Elizabeth Alexander, 228–29. New York: Library of America.

Bue, B. G., S. Sharr, S. D. Moffitt, and A. K. Craig. 1996. "Effects of the Exxon-Valdez Oil Spill on Pink Salmon Embryos and Preemergent Fry." *American Fisheries Society Symposium* 18: 619–27.

Burczynski, M. E., and T. M. Penning. 2000. "Genotoxic Polycyclic Aromatic Hydrocarbon Ortho-Quinones Generated by Aldo-Keto Reductases Induce CYP1A1 via Nuclear Translocation of the Aryl Hydrocarbon Receptor." *Cancer Research* 60: 908–15.

Callon, Michel, Pierre Lascoumes, and Yannick Barthe. 2009. *Acting in an Uncertain World: An Essay on Technical Democracy*. Translated by Graham Burchell. Cambridge, MA: MIT Press.

Canguilhem, Georges. 1991. *The Normal and the Pathological*. Translated by Carolyn R. Fawcett. New York: Zone Books.

Center for Economic and Social Rights (CESR). 1994. *Rights Violations in the Ecuadorian Amazon: The Human Consequences of Oil Development*. New York: Center for Economic and Social Rights.

Cepek, Michael L. 2012. "The Loss of Oil: Constituting Disaster in Amazonian Ecuador." *Journal of Latin American and Caribbean Anthropology* 17 (3): 393–412.

Cepek, Michael L. 2016. "There Might Be Blood: Oil, Humility, and the Cosmopolitics of a Cofán Petro-being." *American Ethnologist* 43 (4): 623–35.

Cepek, Michael L. 2018. *Life in Oil: Cofán Survival in the Petroleum Fields of Amazonia.* Austin: University of Texas Press.

Claff, Roger. 1999. "Primer for Evaluating Ecological Risk at Petroleum Release Sites." *Drug and Chemical Toxicology* 22 (1): 311–16.

Constable, Marianne. 2010. "Speaking the Language of Law: A Juris-dictional Primer." *English Language Notes* 48 (2): 9–15.

Constable, Marianne. 2014a. "Law as Language." *Critical Analysis of Law* 1 (1): 63–74.

Constable, Marianne. 2014b. *Our Word Is Our Bond: How Legal Speech Acts.* Stanford, CA: Stanford University Press.

Coopmans, Catelijne, Janet Vertesi, Michael E. Lynch, and Steve Woolgar. 2014. *Representation in Scientific Practice Revisited.* Cambridge, MA: MIT Press.

Cormack, Bradin. 2007. *A Power to Do Justice: Jurisdiction, English Literature, and the Rise of Common Law.* Chicago: University of Chicago Press.

Cutler, A. Claire. 2016. "Transformations in Statehood, the Investor-State Regime, and the New Constitutionalism." *Indiana Journal of Global Legal Studies* 23 (1): 95–125.

Cutler, A. Claire, and David Lark. 2017. "Theorizing Private Transnational Governance by Contract in the Investor-State Regime." In *The Politics of Private Transnational Governance by Contract*, edited by A. Claire Cutler and Thomas Dietz, 1–36. New York: Routledge.

de la Cadena, Marisol. 2015. *Earth Beings: Ecologies of Practice across Andean Worlds.* Durham, NC: Duke University Press.

de la Cadena, Marisol. 2017. "Matters of Method; Or, Why Method Matters toward a *Not Only* Colonial Anthropology." *Hau: Journal of Ethnographic Theory* 7 (2): 1–10.

DeLanda, Manuel. 2015. *Philosophical Chemistry: Genealogy of a Scientific Field.* London: Bloomsbury Academic.

Deleuze, Gilles. 1978. "Sur Spinoza." Translated by Timothy S. Murphy. Accessed November 5, 2021. https://www.webdeleuze.com/cours/spinoza.

Deleuze, Gilles. 1981. "Sur Spinoza." Translated by Simon Duffy. Accessed November 5, 2021. https://www.webdeleuze.com/textes/191.

Deleuze, Gilles. 1988. *Spinoza: Practical Philosophy.* Translated by Robert Hurley. San Francisco: City Lights Books.

Deleuze, Gilles. 1995. *Difference and Repetition.* Translated by Paul Patton. New York: Columbia University Press.

Deleuze, Gilles. 2005. *Francis Bacon: The Logic of Sensation.* Translated by Daniel W. Smith. Minneapolis: University of Minnesota Press.

Deleuze, Gilles, and Félix Guattari. 1987. *A Thousand Plateaus: Capitalism and Schizophrenia.* Translated by Brian Massumi. Minneapolis: University of Minnesota Press.

Dezalay, Yves, and Bryant G. Garth. 1996. *Dealing in Virtue: International Commercial*

Arbitration and the Construction of a Transnational Legal Order. Chicago: University of Chicago Press.

Dirección General de Medio Ambiente (DIGEMA). 1989. *Propone necesidad de incorporar plan de contingencias para Sistema del Oleoducto Transecuatoriano (SOTE)*, *Boletin de Prensa* 1, 5. Quito: Ministry of Energy and Mines.

Dirección National de Protección Ambiental (DINAPA). 1993. "Environmental Assessment of the Consortium Petroecuador-Texaco, Summary of Findings from Pre-Assessment." April 29.

Doll, Richard, and Austin Bradford Hill. 1950. "Smoking and Carcinoma of the Lung; Preliminary Report." *British Medical Journal* 30 (2): 739–48.

Doll, Richard, and Austin Bradford Hill. 1956. "Lung Cancer and Other Causes of Death in Relation to Smoking: A Second Report on the Mortality of British Doctors." *British Medical Journal* 2 (5001): 1071–81.

Duhem, Pierre. (1902) 2002. *Mixture and Chemical Combination and Related Essays*. Edited and translated, with an introduction by Paul Needham. Dordrecht, Netherlands: Kluwer Academic.

Dumit, Joseph. 2004. *Picturing Personhood: Brain Scans and Biomedical Identity*. Princeton, NJ: Princeton University Press.

Eades, D. 2008. *Courtroom Talk and Neocolonial Control*. Berlin: Mouton de Gruyter.

Eckardt, R. E. 1973. "Recent Developments in Industrial Carcinogens." *Journal of Occupational Medicine*. November 15 (11): 904–7.

Eckardt, R. E. 1983. "Petroleum and Petroleum Products." In *ILO Encyclopedia of Occupational Health and Safety*, edited by L. Parmeggiani. Geneva, Switzerland: International Labor Organization.

Edwards, Deborah A., Ann Tveit, and Michele Emerson. 2001. "Development of Reference Doses for Heavy Total Petroleum Hydrocarbon (TPH) Fractions." In *Risk-Based Decision-Making for Assessing Petroleum Impacts at Exploration and Production Sites*, edited by Sara J. McMillen, Renae I. Magaw, and Rebecca L. Carovillano, 111–20. Tulsa, OK: Petroleum Environmental Research Forum / United States Department of Energy.

Engelmann, Sasha, and Derek McCormack. 2017. "Elemental Aesthetics: On Artistic Experiments with Solar Energy." *Annals of the American Association of Geographers* 108 (1): 241–59.

Environmental Protection Agency (EPA). 1983. "Method 418.1, Total Recoverable Petroleum Hydrocarbon. (Spectrometric, Infrared)." In *Method for the Chemical Analysis of Water and Waste*, vol. 2, 418 1-1 to 1-3. Washington, DC: US Environmental Protection Agency.

Environmental Protection Agency (EPA). 1987. *Scope of the CERCLA Petroleum Exclusion under Sections 101(14) and 104(a)(2)*. Washington, DC: US Environmental Protection Agency.

Environmental Protection Agency (EPA). 1989. *Glossary of Terms Related to Health, Exposure, and Risk Assessment*. Washington, DC: US Environmental Protection Agency.

Environmental Protection Agency (EPA). 2006. *Guide for Industrial Waste Management: Protecting Land, Ground Water, Surface Water, Air.* Washington, DC: National Service for Environmental Publications.

Environmental Protection Agency (EPA). 2010. "Development of a Relative Potency Factor (RPF) Approach for Polycyclic Aromatic Hydrocarbon (PAH) Mixtures (Draft)." February. Washington, DC: Environmental Protection Agency.

European Commission. 2017. "Future Brief: The Precautionary Principle: Decision-Making under Uncertainty," no. 18, September. http://ec.europa.eu /environment/integration/research/newsalert/pdf/precautionary_principle _decision_making_under_uncertainty_FB18_en.pdf.

Fiske, Amelia. 2017. "Attending to the Senses in Toxic Exposure." *Medical Anthropology Quarterly*, March 29. http://medanthroquarterly.org/?p=401.

Fiske, Amelia. 2018. "Dirty Hands: The Toxic Politics of Denunciation." *Social Studies of Science* 48 (3): 389–413.

Fiske, Amelia. 2020. "Naked in the Face of Contamination: Thinking Models and Metaphors of Toxicity Together." *Catalyst: Feminism, Theory, Technoscience* 6 (1): 1–30.

Fortun, Kim. 2001. *Advocacy after Bhopal: Environmentalism, Disaster, New Global Orders.* Chicago: University of Chicago Press.

Fortun, Kim. 2010. "Essential2life." *Dialectical Anthropology* 34 (1): 77–86.

Fortun, Kim. 2012. "Ethnography in Late Industrialism." *Cultural Anthropology* 27 (3): 446–64.

Fortun, Kim, and Mike Fortun. 2005. "Scientific Imaginaries and Ethical Plateaus in Contemporary U.S. Toxicology." *American Anthropologist* 107 (1): 43–54.

Foucault, Michel. 1980. *The History of Sexuality.* Vol. 1, *An Introduction.* Translated by Robert Hurley. New York: Vintage.

Foucault, Michel. 1983. Preface to *Anti-Oedipus: Capitalism and Schizophrenia* by Gilles Deleuze and Félix Guattari, xi–xiv. Minneapolis: University of Minnesota Press.

Foucault, Michel. 1995. *Discipline and Punish: The Birth of the Prison.* Translated by Alan Sheridan. New York: Vintage.

Fried, Charles. (1981) 2015. *Contract as Promise: A Theory of Contractual Obligation.* New York: Oxford University Press.

Garzón, Paulina. 1995. "Impacto socioambiental de la actividad petrolera: Estudio de caso de las comunidades San Carlos y La Primavera." In *Marea negra en la Amazonía: Conflictos socioambientales vinculados a la actividad petrolera en el Ecuador,* edited by Anamaría Varea, 265–94. Quito: Abya Yala-ILDIS.

Gélinas, Fabien. 2017. "Arbitration as Transnational Governance: Legitimacy beyond Contract." In *The Politics of Private Transnational Governance by Contract,* edited by A. Claire Cutler and Thomas Dietz, 133–50. New York: Routledge.

Goldhaber, Michael D. 2013. "The Rise of Arbitral Power over Domestic Courts." *Stanford Journal of Complex Litigation* 1 (2): 373–416.

Goldhaber, Michael D. 2014a. *Crude Awakening: Chevron in Ecuador.* New York: Rosetta Books.

Goldhaber, Michael D. 2014b. "Inside Gibson Dunn's Winning Chevron Strategy." *American Lawyer*. August 20.

Gomez, Manuel A. 2013. "The Global Case: Seeking the Recognition and Enforcement of the Lago Agrio Judgment outside of Ecuador." *Stanford Journal of Complex Litigation* 2: 429–66.

Gomez, Manuel A. 2015. "A Sour Battle in Lago Agrio and Beyond: The Metamorphosis of Transnational Litigation and the Protection of Collective Rights in Ecuador." *University of Miami Inter-American Law Review* 46 (2): 153–77.

Griffin, Lynn F., and John A. Calder. 1977. "Toxic Effect of Water-Soluble Fractions of Crude, Refined, and Weathered Oils on the Growth of a Marine Bacterium." *Applied and Environmental Microbiology* 33 (5): 1092–96.

Guamán, Adoración. 2019. "Derechos humanos y empresas transnacionales: Las debilidades del tercer pilar derivadas de las normas de promoción de inversiones." *Cuadernos electrónicos de filosofía del derecho* 39: 113–35.

Guterman, Lila. 2005. "Scientists Denounce Tactics of Texaco and Its Academic Consultants in Ecuadorean Oil Dispute."*Chronicle of Higher Education*. April 6, 2005.

Guzmán-Gallegos, María Antonieta. 2012. "The Governing of Extraction, Oil Enclaves, and Indigenous Responses in the Ecuadorian Amazon." In *New Political Spaces in Latin American Natural Resource Governance*, edited by H. Haarstad, 155–76. London: Palgrave Macmillan.

Hacking, Ian. 1983. *Representing and Intervening: Introductory Topics in the Philosophy of Natural Science*. Cambridge: Cambridge University Press.

Hacking, Ian. 2004. *Historical Ontology*. Cambridge, MA: Harvard University Press.

Hamilton, Wayne, H. James Sewell, and George Deeley. 2001. "Technical Basis for Current Soil Management Levels of Total Petroleum Hydrocarbons." In *Risk-Based Decision-Making for Assessing Petroleum Impacts at Exploration and Production Sites*, edited by Sara J. McMillen, Renae I. Magaw, and Rebecca L. Carovillano, 36–45. Tulsa, OK: Petroleum Environmental Research Forum / United States Department of Energy.

Haraway, Donna J. 2015. "Anthropocene, Capitalocene, Plantationocene, Chthulucene: Making Kin." *Environmental Humanities* 6: 159–65.

Haraway, Donna J. 2016. *Staying with the Trouble: Making Kin in the Chthulucene*. Durham, NC: Duke University Press.

HBT AGRA. 1993. *Environmental Assessment of the Petroecuador-Texaco Consortium Oilfields, Volume I: Environmental Audit Report*. October. Calgary, Alberta: HBT AGRA Environment and Engineering Services.

Heath, Jenifer, Kristin Koblis, and Shawn Sager. 1993. "Review of Chemical, Physical, and Toxicological Properties of Components of Total Petroleum Hydrocarbons." *Journal of Soil Contamination* 2 (1): 1–25.

Heilbron, John L., ed. 2003. *The Oxford Companion to the History of Modern Science*. Oxford: Oxford University Press.

Heintz, Ron, Jeffrey Short, and Stanley Rice. 1999. "Sensitivity of Fish Embryos to Weathered Crude Oil, Part II. Increased Mortality of Pink Salmon

(*Oncorhynchus gorbuscha*) Embryos Incubating Downstream from Weathered *Exxon Valdez* Crude Oil." *Environmental Toxicology and Chemistry* 18 (3): 494–503.

Hepler-Smith, Evan. 2019. "Molecular Bureaucracy: Toxicological Information and Environmental Protection." *Environmental History* 24 (3): 534–60.

Hepler-Smith, Evan. 2020. "The Etymology of Chemical Names: Tradition and Convenience vs. Rationality in Chemical Nomenclature." *Ambix* 67 (2): 202–3.

Hill, Austin Bradford. 1965. "The Environment and Disease: Association or Causation?" *Proceedings of the Royal Society of Medicine* 58: 295–300.

Holford, Theodore. 2002. *Multivariate Method in Epidemiology*. Oxford: Oxford University Press.

Holmes, Frederic Lawrence. 2003. "Chemistry." In *The Oxford Companion to the History of Modern Science*, edited by J. L. Heilbron, 145–51. Oxford: Oxford University Press.

Hurtig, Anna-Karin, and Miguel San Sebastián. 2002a. *Cancer en la Amazonia del Ecuador (1985–1998)*. Coca, Ecuador: Instituto de Epidemiologia y Salud Communitaria "Manuel Amunarriz."

Hurtig, Anna-Karin, and Miguel San Sebastián. 2002b. "Geographical Differences of Cancer Incidence in the Amazon Basin of Ecuador in Relation to Residency Near Oil Fields." *International Journal of Epidemiology* 31 (5): 1021–27.

Hurtig, Anna-Karin, and Miguel San Sebastián. 2002c. "Gynecologic and Breast Malignancies in the Amazon Basin of Ecuador, 1985–1998." *International Journal of Gynecology and Obstetrics* 76: 199–201.

Hurtig, Anna-Karin, and Miguel San Sebastián. 2004. "Incidence of Childhood Leukemia and Oil Exploitation in the Amazon Basin of Ecuador." *International Journal of Occupational and Environmental Health* 10: 245–50.

Hurtig, Anna-Karin, and Miguel San Sebastián. 2005. "Epidemiology vs. Epidemiology: The Case of Oil Exploitation in the Amazon Basin of Ecuador." *International Journal of Epidemiology* 34 (5): 1170–72. https://doi.org/10.1093/ije/dyi151.

Incardona, John P., Mark G. Carls, Hiroki Teraoka, Catherine A. Sloan, Tracy K. Collier, and Nathaniel L. Scholz. 2005. "Aryl Hydrocarbon Receptor-Independent Toxicity of Weathered Crude Oil during Fish Development." *Environmental Health Perspectives* 113 (12): 1755–62.

Incardona, John P., Luke D. Gardner, Tiffany L. Linbo, Tanya L. Brown, Andrew J. Esbaugh, Edward M. Mager, John D. Stieglitz, et al. 2014. "Deepwater Horizon Crude Oil Impacts the Developing Hearts of Large Predatory Pelagic Fish." *Proceedings of the National Academy of Sciences* 111 (15): E1510–18.

Incardona, John P., Carol A. Vines, Bernadita F. Anulacion, David H. Baldwin, Heather L. Day, Barbara L. French, Jana S. Labenia, et al. 2012. "Unexpectedly High Mortality in Pacific Herring Embryos Exposed to the 2007 *Cosco Busan* Oil Spill in San Francisco Bay." *Proceedings of the National Academy of Sciences* 109 (2): E51–8.

Infante, P. F., R. A. Risnsky, and J. K. Wagoner, et al. 1977. "Leukaemia in Benzene Workers." *Lancet* 2 (8028): 22516–29.

International Agency for Research on Cancer (IARC). 1983. *Monographs on the Evaluation of the Carcinogenic Risk of Chemicals to Man: Polynuclear Aromatic Hydrocarbons.* Vol. 34. Lyon, France: IARC.

International Agency for Research on Cancer (IARC). 2009. *Chemical Agents and Related Occupations, Volume 100F: A Review of Human Carcinogens.* Lyon, France: IARC Working Group on the Evaluation of Carcinogenic Risks to Humans.

International Agency for Research on Cancer (IARC). 2012. *Chemical Agents and Related Occupations, Volume 100F: A Review of Human Carcinogens.* Lyon, France: IARC Monographs on the Evaluation of Carcinogenic Risks to Humans.

International Journal of Occupational and Environmental Health. 2005. "Texaco and Its Consultants." 11 (2): 217–20.

Jacob, Jürgen. 2008. "The Significance of Polycyclic Aromatic Hydrocarbons as Environmental Carcinogens: 35 Years Research on PAH—A Retrospective." *Polycyclic Aromatic Compounds* 28: 242–72.

Jacquemet, M. 2009. *Credibility in Court. Communicative Practices in the Camorra Trials.* New York: Cambridge University Press.

Jain, Lochlann. 2006. *Injury: The Politics of Product Design and Safety Law in the United States.* Princeton, NJ: Princeton University Press.

Jain, Lochlann. 2013. *Malignant: How Cancer Becomes Us.* Berkeley: University of California Press.

Jarrín Ampudia, Gustavo. 1977. "El petróleo en la vida nacional." In *Ecuador: De la colonia a los problemas actuales.* Guayaquil, Ecuador: Universidad de Guayaquil.

Jasanoff, Sheila. 1995. *Science at the Bar: Law, Science, and Technology in America.* Cambridge, MA: Harvard University Press.

Jasanoff, Sheila. 2002. "Science and the Statistical Victim: Modernizing Knowledge in Breast Implant Litigation." *Social Studies of Science* 32 (1): 37–69.

Jasanoff, Sheila. 2005. "Law's Knowledge: Science for Justice in Legal Settings." *American Journal of Public Health* 95 (S1): 49–58.

Jasanoff, Sheila. 2006. "Just Evidence: The Limits of Science in the Legal Process." *Journal of Law, Medicine and Ethics* 34 (2): 328–41.

Jasanoff, Sheila. 2007. "Making Order: Law and Science in Action." In *Handbook of Science and Technology Studies,* edited by E. Hackett et al., 761–86. Cambridge, MA: MIT Press.

Jasanoff, Sheila. 2012. *Science and Public Reason.* London: Routledge.

Jensen, Casper Bruun. 2012. "The Task of Anthropology Is to Invent Relations." *Critique of Anthropology* 32: 47–53.

Jensen, Casper Bruun. 2018. "Gilles Deleuze in Social Science: Some Introductory Themes." *Annual Review of Critical Psychology* 14: 31–49.

Keefe, Patrick Radden. 2012. "Reversal of Fortune." *New Yorker,* January 9.

Kelsh, Michael A., Libby Morimoto, and Edmund Lau. 2009. "Cancer Mortality and Oil Production in the Amazon Region of Ecuador, 1990–2005." *International Archives of Occupational and Environmental Health* 82 (3): 381–95.

Khatam, Damira. 2017. "Chevron and Ecuador Proceedings: A Primer on Transnational Litigation Strategies." *Stanford Journal of International Law* 53 (2): 249–90.

Kimerling, Judith. 1991a. *Amazon Crude*. New York: Natural Resources Defense Council.

Kimerling, Judith. 1991b. "Disregarding the Environmental Law: Petroleum Development in Protected Natural Areas and Indigenous Homelands in the Ecuadorian Amazon." *Hastings International and Comparative Law Review* 14 (4): 849–903.

Kimerling, Judith. 1993. *Crudo Amazónico*. Quito: Abya Yala-ILDIS.

Kimerling, Judith. 1996. "Oil, Lawlessness, and Indigenous Struggles in Ecuador's Oriente." In *Green Guerrillas: Environmental Conflicts and Initiatives in Latin America and the Caribbean*, edited by Helen Collinson, 61–73. New York: Monthly Review Press.

Kimerling, Judith. 2006. "Indigenous Peoples and the Oil Frontier in Amazonia: The Case of Ecuador, ChevronTexaco, and *Aguinda v. Texaco*." *New York University Journal of International Law and Politics* 38 (3): 413–664.

Kimerling, Judith. 2013a. "Lessons from the Chevron Ecuador Litigation: The Proposed Intervenor's Perspective." *Stanford Journal of Complex Litigation* 2: 242–94.

Kimerling, Judith. 2013b. "Oil, Contact, Conservation in the Amazon: Indigenous Huaorani, Chevron, and Yasuni." *Colorado Journal of International Environmental Law and Policy* 24 (1): 43–115.

Kimura, Aya Hirata. 2016. *Radiation Brain Moms and Citizen Scientists: The Gender Politics of Food Contamination after Fukushima*. Durham, NC: Duke University Press.

Kirsch, Stuart. 2006. *Reverse Anthropology: Indigenous Analysis of Social and Environmental Relations in New Guinea*. Stanford, CA: Stanford University Press.

Kirsch, Stuart. 2014. *Mining Capitalism: The Relationship between Corporations and Their Critics*. Berkeley: University of California Press.

Kirsch, Stuart. 2018. *Engaged Anthropology: Politics beyond the Text*. Berkeley: University of California Press.

Krøijer, Stine. 2003. "The Company and the Trickster: A Study of Secoya Storytelling as a Mode of Government." Master's thesis, University of Copenhagen.

Krøijer, Stine. 2017. "'Being Flexible': Reflection on How an Anthropological Theory Spills into the Contemporary Political Lives of an Amazonian People." *Tipití: Journal of the Society for Lowland South America* 15 (1): 46–61.

Krøijer, Stine. 2019. "In the Spirit of Oil: Unintended Flows and Leaky Lives in Northeastern Ecuador." In *Indigenous Life Projects and Extractivism: Ethnographies from South America*, edited by Cecilie Vindal Ødegaard and Juan Javier Rivera Andía, 95–118. Cham, Switzerland: Palgrave Macmillan.

Langewiesche, William. 2007. "Jungle Law." *Vanity Fair*, May, 1–18.

Latour, Bruno. 1988. *The Pasteurization of France*. Translated by Alan Sheridan and John Law. Cambridge, MA: Harvard University Press.

Latour, Bruno. 1993. *We Have Never Been Modern*. Translated by Catherine Porter. Cambridge, MA: Harvard University Press.

Latour, Bruno. 1999. *Pandora's Hope: Essays on the Reality of Science Studies*. Cambridge, MA: Harvard University Press.

Latour, Bruno. 2004. *Politics of Nature: How to Bring the Sciences into Democracy*. Translated by Catherine Porter. Cambridge, MA: Harvard University Press.

Latour, Bruno. 2005. *Reassembling the Social: An Introduction to Actor-Network-Theory*. Oxford: Oxford University Press.

Latour, Bruno. 2010. *The Making of Law: An Ethnography of the Conseil d'État*. Translated by Marina Brilman and Alain Pottage. Cambridge: Polity.

Latour, Bruno. 2014. "War and Peace in an Age of Ecological Conflicts." *Revue Juridique de l'Environnement* 39 (1): 51–63.

Latour, Bruno, and Steve Woolgar. 1987. *Laboratory Life: The Construction of Scientific Facts*. Edited by Jonas Salk. Princeton, NJ: Princeton University Press.

Lavoisier, Antoine. 1787. *Méthode de nomenclature chimique, proposée par MM. de Morveau, Lavoisier, Bertholet, & de Fourcroy. On y a joint un nouveau système de Caractères chimiques, adaptés à cette nomenclature par MM. Hassanfratz & Adet*. Paris: Cuchet, Libraire.

Lavoisier, Antoine. 1789. *Traité élémentaire de chimie, présenté dans un ordre nouveau, et d'après des découvertes modernes*. Paris: Chez Cuchet, Libraire.

Lavoisier, Antoine. 1790. *Elements of Chemistry, in a New Systematic Order, Containing All the Modern Discoveries. Illustrated with Thirteen Copperplates*. Edinburgh: Printed for William Creech.

Law, John. 2004. *After Method: Mess in Social Science Research*. New York: Routledge.

Law, John, and Annemarie Mol, eds. 2002. *Complexities: Social Studies of Knowledge Practices*. Durham, NC: Duke University Press.

Lewis, Fraser I., and Michael P Ward. 2013. "Improving Epidemiologic Data Analyses through Multivariate Regression Modelling." *Emerging Themes in Epidemiology* 10 (4): 1–10.

Liboiron, Max. 2012. "Redefining Pollution: Plastics in the Wild." PhD diss., New York University.

Lienhardt, Godfrey. 1953. "Modes of Thought in Primitive Societies." *New Blackfriar* 34 (399): 269–77.

Loehr, Raymond. 2001. Preface to *Risk-Based Decision-Making for Assessing Petroleum Impacts at Exploration and Production Sites*, edited by Sara J. McMillen, Renae I. Magaw, and Rebecca L. Carovillano, 1–5. Tulsa, OK: Petroleum Environmental Research Forum / United States Department of Energy.

Mackay, Donald, and Clayton D. McAuliffe. 1989. "Fate of Hydrocarbons Discharged at Sea." *Oil and Chemical Pollution* 5 (1): 1–20.

Markowitz, Gerald, and David Rosner. 2002. *Deceit and Denial: The Deadly Politics of Industrial Pollution*. Berkeley: University of California Press.

Marty, G. D., J. W. Short, D. M. Dambach, N. H. Willits, R. A. Heintz, S. D. Rice, J. J. Stegeman, and D. E. Hinton. 1997. "Ascites, Premature Emergence, Increased Gonadal Cell Apoptosis, and Cytochrome P4501a Induction in Pink Salmon Larvae Continuously Exposed to Oil-Contaminated Gravel during Development." *Canadian Journal of Zoology* 75 (6): 989–1007.

Martz, John. 1987. *Politics and Petroleum in Ecuador*. New Brunswick, NJ: Rutgers University Press.

McGarity, Thomas O., and Wendy E. Wagner. 2008. *Bending Science: How Special Interests Corrupt Public Health Research*. Cambridge, MA: Harvard University Press.

McMillen, Sara, Jill K. Kerr, and David Nakles. 2001. "Composition of Crude Oils and Gas Condensates." In *Risk-Based Decision-Making for Assessing Petroleum Impacts at Exploration and Production Sites,* edited by Sara J. McMillen, Renae I. Magaw, and Rebecca L. Carovillano, 46–57. Tulsa, OK: Petroleum Environmental Research Forum / United States Department of Energy.

McMillen, Sara J., Renae I. Magaw, and Rebecca L. Carovillano, eds. 2001. *Risk-Based Decision-Making for Assessing Petroleum Impacts at Exploration and Production Sites.* Tulsa, OK: Petroleum Environmental Research Forum / United States Department of Energy.

McMillen, Sara J., Renae I. Magaw, Jill M. Kerr, and Lynn Spence. 2001. "Developing Total Petroleum Hydrocarbon Risk-Based Screening Levels for Sites Impacted by Crude Oils and Gas Condensates." In *Risk-Based Decision-Making for Assessing Petroleum Impacts at Exploration and Production Sites,* edited by Sara J. McMillen, Renae I. Magaw, and Rebecca L. Carovillano, 121–32. Tulsa, OK: Petroleum Environmental Research Forum / United States Department of Energy.

McMillen, Sara, Ileana Rhodes, David Nakles, and Robert E. Sweeney. 2001. "Application of the Total Petroleum Hydrocarbon Criteria Working Group (TPHCWG) Methodology to Crude Oils and Gas Condensates." In *Risk-Based Decision-Making for Assessing Petroleum Impacts at Exploration and Production Sites,* edited by Sara J. McMillen, Renae I. Magaw, and Rebecca L. Carovillano, 58–76. Tulsa, OK: Petroleum Environmental Research Forum / United States Department of Energy.

Mella, Rodrigo A. 2017. "The Enforcement of Foreign Judgments in the United States: The *Chevron Corp. v. Donziger* Case." *New York University Journal of International Law and Politics* 49 (2): 635–48.

Mertz, Elizabeth. 2007. *The Language of Law School: Learning to "Think like a Lawyer."* Oxford: Oxford University Press.

Michaels, David. 2008. *Doubt Is Their Product: How Industry's Assault on Science Threatens Your Health.* Oxford: Oxford University Press.

Michelsen, Teresa, and Catherine Petito Boyce. 1993. "Cleanup Standards for Petroleum Hydrocarbons. Part 1. Review of Methods and Recent Developments." *Journal of Soil Contamination* 2 (2): 1–16.

Miettinen, Olli S. 1985. *Theoretical Epidemiology: Principles of Occurrence Research in Medicine.* Hoboken, NJ: Wiley.

Ministerio de Energía y Minas, República del Ecuador. 1989. *Producciones del petróleo, agua de formación y gas natural, I.* December.

Mol, Annemarie. 2002. *The Body Multiple: Ontology in Medical Practice.* Durham, NC: Duke University Press.

Moolgavkar, Suresh H., Ellen T. Chang, Heather Watson, and Edmund C. Lau. 2014. "Cancer Mortality and Quantitative Oil Production in the Amazon Region of Ecuador, 1990–2010." *Cancer Causes Control* 25 (1): 59–72.

Murphy, Michelle. 2006. *Sick Building Syndrome and the Problem of Uncertainty: Environmental Politics, Technoscience, and Women Workers.* Durham, NC: Duke University Press.

Murphy, Michelle. 2008. "Chemical Regimes of Living." *Environmental History* 13 (4): 695–703.

Murphy, Michelle. 2017. "Alterlife and Decolonial Chemical Relations." *Cultural Anthropology* 32 (4): 494–503.

Nagendrappa, Gopalpur. 2001. "Benzene and Its Isomers: How Many Structures Can We Draw for C_6H_6?" *Resonance* 6 (5): 74–78.

National Institute of Standards and Technology (NIST). 2004. "Hydrocarbon Spectral Database." http://www.nist.gov/pml/data/msd-hydro/index.cfm.

Needham, Paul. 2002. Introduction to *Mixture and Chemical Combination and Related Essays* by Pierre Duhem, edited and translated by Paul Needham, ix–xxxii. Dordrecht, Netherlands: Kluwer Academic.

Neff, C. H. 1970. "Review of 1969 Petroleum Developments in South America, Central America, and Caribbean Area." *Bulletin of the American Association of Petroleum Geologists* 54 (8): 1342–406.

North, James. 2015. "Ecuador's Battle for Environmental Justice against Chevron." *Nation*, June 2, 2015.

North, James. 2021. "Steven Donziger Wants to Convince 'a Different Jury.'" *Nation*, May 18, 2021.

Ofrias, Lindsay. 2017. "Invisible Harms, Invisible Profits: A Theory of the Incentive to Contaminate." *Culture, Theory and Critique* 58 (4): 435–56.

Ofrias, Lindsay, and Gordon Roecker. 2019. "Organized Criminals, Human Rights Defenders and Oil Companies: Weaponization of the RICO Law across Jurisdictional Borders." *Journal of Global and Historical Anthropology* 85: 37–50.

Oquendo, Angel R. 2009. "Upping the Ante: Collective Litigation in Latin America." *Columbian Journal of Transnational Law* 47: 248–91.

O'Reilly, Kirk T., Renae I. Magaw, and William G. Rixey. 2001a. "Hydrocarbon Transport from Oil and Soil to Groundwater." In *Risk-Based Decision-Making for Assessing Petroleum Impacts at Exploration and Production Sites*, edited by Sara J. McMillen, Renae I. Magaw, and Rebecca L. Carovillano, 132–41. Tulsa, OK: Petroleum Environmental Research Forum / United States Department of Energy.

O'Reilly, Kirk T., Renae I. Magaw, and William G. Rixey. 2001b. "Predicting the Effect of Hydrocarbon and Hydrocarbon-Impacted Water Soil on Groundwater." *American Petroleum Institute Bulletin*, no. 14: 1–14.

O'Reilly, Kirk T., and Waverly Thorsen. 2010. "Impact of Crude Oil Weathering on the Calculated Effective Solubility of Aromatic Compounds: Evaluation of Soils from Ecuadorian Oil Fields." *Soil and Sediment Contamination: An International Journal* 19: 391–404.

Pallares, Manuel. 1999. *Encuesta de las familias en el Nor Oriente del Ecuador*. Quito: Gerencia de Protección, Petroecuador.

Papadopoulos, Dimitris. 2018. *Experimental Practice: Technoscience, Alterontologies, and More-Than-Social Movements*. Durham, NC: Duke University Press.

Papadopoulos, Dimitris, Maria Puig de la Bellacasa, and Natasha Myers, eds. 2022. *Reactivating Elements: Chemistry, Ecology, Practice*. Durham, NC: Duke University Press.

Parker CM (Marathon Oil). 2002. "Memorandum to Manager of Toxicology and Product Safety (Marathon Oil). Subject: International Leveraged Research Proposal, 2000." Accessed March 24, 2014. http://defendingscience.org/sites /default/files/upload/Parker_benzene.pdf.

Pateman, Carole, and Charles Mills. 2007. *The Contract and Domination*. New York: Polity.

Peterson, Charles H., Stanley D. Rice, Jeffrey W. Short, Daniel Esler, James L. Bodkin, Brenda E. Ballachey, and David B. Irons. 2003. "Long-Term Ecosystem Response to the *Exxon Valdez* Oil Spill." *Science* 302 (5653): 2082–86.

Peterson, Luke Eric. 2009. "Chevron Goes All-In against Ecuador; New Claim Reflects Latest BIT Usage." *Kluwer Arbitration Blog*, September 24. http:// kluwerarbitrationblog.com/blog/2009/09/24/chevron-goes-all-in-against -ecuador-new-claim-reflects-latest-bit-usage/.

Petryna, Adriana. 2002. *Life Exposed: Biological Citizens after Chernobyl*. Princeton, NJ: Princeton University Press.

Pickering, Ronald. 1999. "A Toxicological Review of Polycyclic Aromatic Hydrocarbons." *Journal of Toxicology: Cutaneous and Ocular Toxicology* 18 (2): 101–35.

Porter, Theodore. 1995. *Trust in Numbers: The Pursuit of Objectivity in Science and Public Life*. Princeton, NJ: Princeton University Press.

Proctor, Robert. 1995. *Cancer Wars: How Politics Shapes What We Know and Don't Know about Cancer*. New York: Basic Books.

Proctor, Robert. 2011. *Golden Holocaust: Origins of the Cigarette Catastrophe and the Case for Abolition*. Berkeley: University of California Press.

Puig de la Bellacasa, Maria. 2022. "Embracing Breakdown: Soil Ecopoethics and the Ambivalences of Remediation." In *Reactivating Elements: Chemistry, Ecology, Practice*, edited by Dimitris Papadopoulos, Maria Puig de la Bellacasa, and Natasha Myers, 196–230. Durham, NC: Duke University Press.

Rancière, Jacques. 1999. *Disagreement: Politics and Philosophy*. Translated by Julie Rose. Minneapolis: University of Minnesota Press.

Redfield, Peter. 2005. "Doctors, Borders, and Life in Crisis." *Cultural Anthropology* 20 (3): 328–61.

Rees, Tobias. 2018. *After Ethnos*. Durham, NC: Duke University Press.

Rheinberger, Hans-Jörg. 2010. *An Epistemology of the Concrete: Twentieth-Century Histories of Life*. Durham, NC: Duke University Press.

Rice, Stanley D., Robert E. Thomas, Mark G. Carls, Ronald A. Heintz, Alex C. Wertheimer, Michael L. Murphy, Jeffrey W. Short, and Adam Moles. 2001. "Impacts to Pink Salmon Following the *Exxon Valdez* Oil Spill: Persistence, Toxicity, Sensitivity, and Controversy." *Reviews in Fisheries Science* 9 (3): 165–211.

Richland, Justin B. 2013. "Jurisdiction: Grounding Law in Language." *Annual Review of Anthropology* 42: 209–26.

Riles, Annelise. 2005. "A New Agenda for the Cultural Study of Law: Taking on the Technicalities." *Buffalo Law Review* 53: 973–1003.

Riles, Annelise. 2011. *Collateral Knowledge: Legal Reasoning in the Global Financial Markets*. Chicago: University of Chicago Press.

Rivadeneira, Marco. 2004. "Breve reseña histórica de la explotación petrolera de la cuenca Oriente." In *La cuenca Oriente: Geologia y petróleo*, 205–28. Quito: Petroproducción y IRD lnstitut de Recherché pour le Dévéloppement.

Rivadeneira, Marco, and Patrice Baby. 2004. "Características geológicas generales de los principales campos petroleros de petroproducción." In *La cuenca Oriente: Geologia y petróleo*, 229–95. Quito: Petroproducción y IRD lnstitut de Recherché pour le Dévéloppement.

Röder, Werner, and Herbert A. Strauss. 1999. *Politik, Wirtschaft, Öffentliches Leben*. Frankfurt / New York: Institut für Zeitgeschichte / Research Foundation for Jewish Immigration.

Rogers, Doug. 2018. "Oil into Food: Microbiology, Industry, and the Petroprotein Movement." Paper presented on the panel "Conjuring Resource Worlds," Annual Meeting of American Anthropological Association, November 2018, San Jose, CA.

Rogers, Doug. n.d. "Abiotic Theories of Hydrocarbon Genesis." Author's personal archive.

Ross, Benjamin, and Steven Amter. 2012. *The Polluters: The Making of Our Chemically Altered Environment*. Oxford: Oxford University Press.

Rothman, Kenneth. 1988. *Causal Inference*. Chestnut Hill, MA: Epidemiology Resources.

Rothman, Kenneth. 2012. *Epidemiology: An Introduction*. 2nd ed. Oxford: Oxford University Press.

Rothman, Kenneth, and Sander Greenland. 2005. "Causation and Causal Inference in Epidemiology." *American Journal of Public Health* 95 (51): S144–50.

Rothman, Kenneth, Sander Greenland, and Timothy Lash. 2012. *Modern Epidemiology*. 3rd ed. Philadelphia: Lippincott Williams and Wilkins.

Salm, Melissa. 2014. "Meet Me in the Meta-Temporalities of Evidence in Evidence-Based Public Health." Preliminary examination paper, Department of Anthropology, University of California, Davis.

Sanders, Lane C., and Stefan A. Wise. 2011. *Polycyclic Aromatic Hydrocarbons Structure Index*. NIST Special Publication no. 922. Gaithersburg, MD: National Institute of Standards and Technology.

San Sebastián, Miguel, Ben Armstrong, Juan Antonio Córdoba, and Carolyn Stephens. 2001. "Environmental Exposure and Cancer Incidence near Oil Fields in the Amazon Basin of Ecuador." *Occupational and Environmental Medicine* 58 (8): 517–22.

San Sebastián, Miguel, Ben Armstrong, and Carolyn Stephens. 2001. "La salud de mujeres que viven cerca de pozos y estaciones de petróleo en la Amazonia ecuatoriana." *Pan American Journal of Public Health* 9 (6): 375–84.

San Sebastián, Miguel, Ben Armstrong, and Carolyn Stephens. 2002. "Outcome of Pregnancy among Women Living in the Proximity of Oil Fields in the Amazon Basin of Ecuador." *International Journal of Occupational and Environmental Health* 8: 312–19.

San Sebastián, Miguel, Juan Antonio Córdoba, and Vicariato Apostólico de Aguarico. 1999. *Yanacuri Report: The Impact of Oil Development on the Health of the People of the Ecuadorian Amazon.* Coca, Ecuador: CICAME and Department of Tropical Medicine and Hygiene of the University of London.

San Sebastián, Miguel, and Anna-Karin Hurtig. 2004a. "Cancer among Indigenous People in the Amazon Basin of Ecuador, 1985–2000." *Revista Panamericana de Salud Pública* 16 (5): 328–33.

San Sebastián, Miguel, and Anna-Karin Hurtig. 2004b. "Oil Exploitation in the Amazon Basin of Ecuador: A Public Health Emergency." *Revista Panamericana de Salud Pública* 15 (3): 205–11.

San Sebastián, Miguel, and Anna Karin Hurtig. 2005. "Oil Development and Health in the Amazon Basin of Ecuador: The Popular Epidemiology Process." *Social Science and Medicine* 60: 799–807.

Sass, Louis C., and C. H. Neff. 1964. "Review of 1963 Petroleum Developments in South America and the Caribbean Area." *American Association of Petroleum Geologists Bulletin* 48 (8): 1242–98.

Sass, Louis C., and C. H. Neff. 1966. "Review of 1965 Petroleum Developments in South America and the Caribbean Area." *American Association of Petroleum Geologists Bulletin* 50 (8): 1564–624.

Sawyer, Herbert. 1975. "Latin America, after 1920." In *Trek of the Oil Finders: A History of Exploration for Petroleum,* 960–1252. Tulsa, OK: American Association of Petroleum Geologists.

Sawyer, Suzana. 1997. "Marching to Nation across Ethnic Terrain: The Cultural Politics of Nation, Territory, and Resource-Use in the Ecuadorian Upper Amazon." PhD diss., Stanford University.

Sawyer, Suzana. 2001. "Frictions of Sovereignty: Of Prosthetic Petro-Capitalism, Neoliberal States and the Phantom-like Citizens in Ecuador." *Journal of Latin American Anthropology* 6: 156–97.

Sawyer, Suzana. 2002. "Bobbittizing Texaco: Dis-membering Corporate Capital and Re-membering the Nation in Ecuador." *Cultural Anthropology* 17 (2): 150–80.

Sawyer, Suzana. 2003. "Subterranean Techniques: Corporate Environmentalism, Oil Operations, and Social Injustice in the Ecuadorian Rain Forest." In *In Search of the Rain Forest,* edited by Candace Slater, 69–100. Durham, NC: Duke University Press.

Sawyer, Suzana. 2004a. *Crude Chronicles: Indigenous Politics, Multinational Oil, and Neoliberalism in Ecuador.* Durham, NC: Duke University Press.

Sawyer, Suzana. 2004b. "Crude Properties: The Sublime and Slime of Oil Operations in the Ecuadorian Amazon." In *Property in Question: Value Transformation in the Global Economy,* edited by Katherine Verdery and Caroline Humphrey, 85–111. Oxford: Berg.

Sawyer, Suzana. 2006. "Disabling Corporate Sovereignty in a Transnational Lawsuit." *Political and Legal Anthropology Review* 29 (1): 23–43.

Sawyer, Suzana. 2007. "Empire/Multitude-State/Civil Society: Topographies of Power through Transnational Connectivity in Ecuador and Beyond." *Social Analysis* 52 (2): 64–85.

Sawyer, Suzana. 2009. "Suing ChevronTexaco." In *The Ecuador Reader*, edited by Steve Striffler and Carlos de la Torre, 321–28. Durham, NC: Duke University Press.

Sawyer, Suzana. 2015. "Crude Contamination: Law, Science, and Indeterminacy in Ecuador and Beyond." In *Subterranean Estates: Life Worlds of Oil and Gas*, edited by Hannah Appel, Arthur Mason, and Michael Watts, 126–46. Ithaca, NY: Cornell University Press.

Sawyer, Suzana. n.d. "Corruption Contingent on Contract." Unpublished manuscript.

Sawyer, Suzana, and Lindsay Ofrias. Forthcoming. "Caught in a Legal Vortex: Time, Technique, and Corporate Fraud-Worlding." In *Handbook on Oil and International Relations*, edited by Roland Dannreuther and Wojciech Ostrowksi. Cheltenham, UK: Edward Elgar.

Sellers, Christopher. 1997. *Hazards of the Job: From Industrial Disease to Environmental Health Science*. Chapel Hill: University of North Carolina Press.

Shapin, Steve, and Simon Schaeffer. 1985. *Leviathan and the Air Pump: Hobbes, Boyle, and the Experimental Life*. Princeton, NJ: Princeton University Press.

Shapiro, Nicholas. 2015. "Attuning to the Chemosphere: Domestic Formaldehyde, Bodily Reasoning, and the Chemical Sublime." *Cultural Anthropology* 30 (3): 368–93.

Siemiatycki, Jack. 2002. "Commentary: Epidemiology on the Side of the Angels." *International Journal of Epidemiology* 31 (5): 1027–29.

Sørensen, Lisbet, Elin Sørhus, Trond Nordtug, John P. Incardona, Tiffany L. Linbo, Laura Giovanetti, Ørjan Karlsen, and Sonnich Meier. 2017. "Oil Droplet Fouling and Differential Toxicokinetics of Polycyclic Aromatic Hydrocarbons in Embryos of Atlantic Haddock and Cod." *PLoS One* 12 (7): 1–26.

Spence, Lynn R., Kirk T. O'Reilly, Renae I. Magaw, and William G. Rixey. 2001. "Predicting the Fate and Transport of Hydrocarbons in Soil and Groundwater." In *Risk-Based Decision-Making for Assessing Petroleum Impacts at Exploration and Production Sites*, edited by Sara J. McMillen, Renae I. Magaw, and Rebecca L. Carovillano, 89–110. Tulsa, OK: Petroleum Environmental Research Forum / United States Department of Energy.

Staats, Dee Ann, David Mattie, and Jeffrey Fisher. 1997. "Petroleum-Related Risk Factors and Soil Screening Levels." *Human and Ecological Risk Assessment* 3 (4): 659–81.

Stengers, Isabelle. 1997. *Power and Invention: Situating Science*. Translated by Paul Bains. Minneapolis: University of Minnesota Press.

Stengers, Isabelle. 2005. "Introductory Notes on an Ecology of Practices." *Cultural Studies Review* 11 (1): 183–96.

Stengers, Isabelle. 2010. *Cosmopolitics I*. Translated by Robert Bononno. Minneapolis: University of Minnesota Press.

Stengers, Isabelle. 2011. *Thinking with Whitehead: A Free and Wild Creation of Concepts*. Translated by Michael Chase. Cambridge, MA: Harvard University Press.

Stengers, Isabelle. 2012. "Reclaiming Animism." *e-flux journal*, no. 36: 1–10. https://www.e-flux.com/journal/36/61245/reclaiming-animism/.

Strathern, Marilyn. 1988. *The Gender of the Gift*. Berkeley: University of California Press.

Strathern, Marilyn. 1992. *Reproducing the Future: Anthropology, Kinship, and the New Reproductive Technologies*. Manchester, UK: University of Manchester Press.

Strathern, Marilyn. 1995. *The Relation: Issues in Complexity and Scale*. Cambridge: Prickly Pear Press.

Strathern, Marilyn. 1999a. *Property, Substance and Effect. Anthropological Essays on Persons and Things*. London: Athlone Press.

Strathern, Marilyn. 1999b. "What Is Intellectual Property After?" In *Actor Network Theory and After*, edited by J. Law and J. Hassard, 156–80. London: Blackwell.

Strathern, Marilyn. 2005. *Partial Connections*. Lanham, MD: AltaMira Press.

Sullivan, John B., and Gary R. Krieger. 1992. *Hazardous Materials Toxicology: Clinical Principles of Environmental Health*. Baltimore, MD: Williams and Wilkins.

Susser, Mervyn, and Zena Stein. 2009. *Eras in Epidemiology: The Evolution of Ideas*. Oxford: Oxford University Press.

Tomlinson, Priscilla, and Michael V. Ruby. 2016. "State and Federal Cleanup Levels for Petroleum Hydrocarbons in Soil: State of the States and Implications for the Future." *Human and Ecological Risk Assessment: An International Journal* 22 (4): 911–26.

Total Petroleum Hydrocarbon Criteria Working Group (TPHCWG). 1997a. *Selection of Representative TPH Fractions Based on Fate and Transport Considerations*. Vol. 3. Amherst, MA: Amherst Scientific Publishers.

Total Petroleum Hydrocarbon Criteria Working Group (TPHCWG). 1997b. *Development of Fraction Specific Reference Doses (RfD's) and Reference Concentration (RfC's) for Total Petroleum Hydrocarbons*. Vol. 4. Amherst, MA: Amherst Scientific Publishers.

Total Petroleum Hydrocarbon Criteria Working Group (TPHCWG). 1998. *Analysis of Petroleum Hydrocarbons in Environmental Media*. Vol. 1. Amherst, MA: Amherst Scientific Publishers.

Total Petroleum Hydrocarbon Criteria Working Group (TPHCWG). 1999. *Human Health Risk-Based Evaluation of Petroleum Release Sites: Implementing the Working Group Approach*. Vol. 5. Amherst, MA: Amherst Scientific Publishers.

Trujillo, Jorge. 1992. "La colonización y el desplazamiento de los grupos indígenas." In *Amazonía Presente y . . . ?*, edited by Tierra Viva, 119–48. Quito: Abya Yala-ildis.

Twerdok, Lorainne. 1999. "Development of Toxicity Criteria for Petroleum Hydrocarbon Fractions in the Petroleum Hydrocarbon Criteria Working Group Approach for Risk-Based Management of Total Petroleum Hydrocarbons in Soils." *Drug and Chemical Toxicology* 22 (1): 275–91.

UNICEF. 1992. *Situación de las madres y los niños en la Amazonía Ecuatoriana: Análisis de situación e indicadores de subsistemas demografía, educación, ciudado, y protección en la Amazonia Ecuatoriana*. Quito: UNICEF.

Unión de Promotores Populares de Salud de la Amazonía Ecuatoriana (UPPSAE). 1993. *Culturas bañadas en petroleo: Diagnóstico de salud realiazado por promotores*.

Lago Agrio, Ecuador: Unión de Promotores Populares de Salud de la Amazonía Ecuatoriana.

Uquillas, Jorge. 1985. "Indian Land Rights and National Resource Management in the Ecuadorian Amazon." In *Native Peoples and Economic Development*, edited by T. Macdonald, 26–37. Cambridge, MA: Cultural Survival.

Uquillas, Jorge. 1989. "Social Impacts of Modernization and Public Policy, and Prospects for Indigenous Development in Ecuador's Amazon." In *The Human Ecology of Tropical Land Settlement in Latin America*, edited by D. Schumann and W. Partridge, 407–31. Boulder, CO: Westview Press.

Uquillas, Jorge. 1993. "La tenencia de la tierra en la Amazonía Ecuatoriana." In *Retos de la Amazonía*, edited by Theodoro Bustamante, 61–94. Quito: Abya Yala-ILDIS.

Valdivia, Gabriela. 2007. "The 'Amazon Trial of the Century': Indigenous Identities, Transnational Networks, and Petroleum in Ecuador." *Alternatives* 32 (1): 41–72.

Van Harten, Gus. 2008. *Investment Treaty Arbitration and Public Law*. Oxford: Oxford University Press.

Van Harten, Gus. 2013. *Sovereign Choices and Sovereign Constraints: Judicial Restraint in Investment Treaty Arbitration*. Oxford: Oxford University Press.

Vargas Pazzos, René. 1976. "Petróleo: Desarrollo y seguridad." In *Política petrolera Ecuatoriana, 1972–1976*. Quito: Instituto de Investigaciones Económicas, Universidad Central del Ecuador.

Vertesi, Janet. 2015. *Seeing like a Rover: How Robots, Teams, and Images Craft Knowledge of Mars*. Chicago: University of Chicago Press.

Vickers, William. 1984. "Indian Policy in the Amazonian Ecuador." In *Frontier Expansion in Amazonia*, edited by M. Schmink and C. Wood, 8–32. Gainesville: University of Florida Press.

Vorhees, Donna, and Cheri Butler. 1999. "Calculation of Human Health Risk-Based Screening Levels (RBSLS) for Petroleum." *Drug and Chemical Toxicology* 22 (1): 293–310.

Wallace, Dorothy, and Joseph J. BelBruno. 2006. *The Bell That Rings Light: A Primer in Quantum Mechanics and Chemical Bonding*. Hackensack, NJ: World Scientific Publishing.

Wong, O. 1995. "Risk of Acute Myeloid Leukaemia and Multiple Myeloma in Workers Exposed to Benzene." *Occupational and Environmental Medicine* 52 (6): 380–84.

Wong, Otto, and Gerhard K. Raabe. 1995. "Cell-Type-Specific Leukemia Analyses in a Combined Cohort of More Than 208,000 Petroleum Workers in the United States and the United Kingdom, 1937–1989." *Regulatory Toxicology and Pharmacology* 21 (2): 307–21.

Wong, O., and G. K. Raabe. 1997. "Multiple Myeloma and Benzene Exposure in a Multinational Cohort of More Than 250,000 Petroleum Workers." *Regulatory Toxicology and Pharmacology* 26 (2): 188–99.

Wong, O., and G. K. Raabe. 2000a. "A Critical Review of Cancer Epidemiology in the Petroleum Industry, with a Meta-Analysis of a Combined Database of More Than 350,000 Workers." *Regulatory Toxicology and Pharmacology* 32: 78–98.

Wong, O., and G. K. Raabe. 2000b. "Non-Hodgkin's Lymphoma and Exposure to Benzene in a Multinational Cohort of More Than 308,000 Petroleum Workers, 1937 to 1996." *Journal of Occupational and Environmental Medicine* 42: 554–68.

Woody, Andrea, Robin Findlay Hendry, and Paul Needham, eds. 2012. *Philosophy of Chemistry*. Oxford: New Holland/Elsevier.

Wylie, Sara. 2018. *Fractivism: Corporate Bodies and Chemical Bonds*. Durham, NC: Duke University Press.

Zaitchik, Alexander. 2014. "Sludge Match: Inside Chevron's $9 Billion Legal Battle with Ecuadorean Villagers." *Rolling Stone*, August 28, 2014.

Zevallos, Jose Vincente L. 1989. "Agrarian Reform and Structural Change: Ecuador since 1964." In *Searching for Agrarian Reform in Latin America*, edited by William C. Thiesenhusen, 42–69. Madison: University of Wisconsin Press.

Cabrera, Richard, 284. *See also* Cabrera report

Cabrera report: Chevron's objections, 284-85; RICO countersuit, 283-84, 289-90; Stratus Consulting, 284; waste pit count, 284-89, 307; waste pit volume calculation, 291

cadmium, 239

Callejas Ribadeneira, Adolfo: *Aguinda v. Chevron* opening hearing, 47, 51-52; legal defenses, 153, 348n23; at Sacha-57 inspection, 137-40, 153; at Sacha Sur Production Station inspection, 91-92

Callon, Michel, 203, 246, 256-57, 354n7

cancer: association between incidence and exposure, 112-13, 243; carcinogens, 113-15, 239, 271-72, 275-76; causation, 117-18; developing world challenges, 117-18; IARC research, 113-15; local resident testimonials, 199-200, 203, 205; National Tumor Registry, 243; oil extraction's impact, 98-100; oil industry research funding, 112; smoking, 120. *See also* research used in *Aguinda v. Chevron*, Kelsh's; research used in *Aguinda v. Chevron*, San Sebastián and Hurtig's

carcinogens, 113-15, 239, 271-72, 275-76. *See also* benzene; BTEX hydrocarbons

causation: cancer, 117-18; epidemiology, 114-15, 120-23; Hill criteria, 115, 122-23; oil extraction's health impacts controversy, 97, 114-16, 122, 124; theory of sufficient, 250, 273, 275, 277-78; in Zambrano's ruling, 243, 256, 273, 275, 277-78

chemical philosophy, 20-22, 23-24, 25-27, 29, 327

chemistry: affinity, 21; benzene, 29, 71, 111, 215-17, *218*; bonds, 43-44, 131, 192, 215; catalysts, 325; compounds, 29, 55, 63-64; delocalized stability, 215-17; disciplinary history, 20-22, 26-27; electronic theory of atoms, 44; elements, 20-24, 26-27; hydrocarbons, 70-72, 193-94; isomers, 315; methodology, 11-12, 14; naming system, 20-21; orbitals, 39, 44, 95, 126-27;

organic compounds, 29; practices, 20, 336n20; problematics, 20; propagation, 131; radicals, 131-33, 171; reactants, 307; reactions, 26-27, 29, 131-32; structural atomism, 26; valence, 13, 22, 23-24, 29-30, 44, 55; water, 192-97. *See also* oil's chemistry

Chevron Corporation: American Arbitration Association (AAA) claim, 166; burning and dumping of waste, 54, 339n4; executives indicted in Ecuador, 161; litigation uses, 41-42; name changes, 333n1; power, 336n25; public relations, 1-3, 94, 134-36; shareholder meetings, 220-22; whistleblower recordings, 172-177. *See also* *Chevron v. Republic of Ecuador*; inspections of former oil production sites, Chevron Corporation; *Maria Aguinda Salazar y Otros v. Chevron Texaco Corp.*; Texaco, Inc.; Texpet

Chevron Corporation v. Steven Donziger et al. (New York), 5, 170; adversarial nature, 18; Cabrera report, 283-84, 289-90; Chevron's bribery and ghostwriting complaint, 259, 261, 356n1, 357n2; complexity, 10, 16, 260; contamination evidence exclusion, 294; corruption claims, 5-7, 9, 262-64, 293; Steven Donziger, 269-70; evidence, weakness of, 261; expert testimonies, 262; facts, 33, 261; John Keker's withdrawal, 296-301; Lago Agrio plaintiffs, 261, 264, 279-80; legal argumentation regarding contamination, 260; problems with, 9, 294-95; David Russell's testimony, 262, 264-69, 293-94, 307; waste pit number controversy, 285-91. *See also* *Chevron Corporation v. Steven Donziger et al.*, Chevron's legal strategies; *Chevron Corporation v. Steven Donziger et al.*, Zambrano's cross-examination; *Chevron Corporation v. Steven Donziger et al.* ruling

Chevron Corporation v. Steven Donziger et al., Chevron's legal strategies: creation of reality, 293-94; expert analyses, 218-19, 261, 285-86, 288-90, 293; lawyers hired,

couplet contract and *Aguinda v. Chevron*: applicability, 310–11, 314–15, 318–19, 321–23; Chevron's arguments, 143–44, 153–56; court findings, 155–56

couplet contract and *Chevron v. Republic of Ecuador*: Chevron's claims, 171; Chevron's misrepresentation, 363n55; Chevron's violation of obligation claim, 304, 306, 309–312; diffuse versus individual rights, 311–12, 314–15; equivocation with *Aguinda v. Chevron*, 310–12, 315–19; role in Zambrano's ruling, 308–9; US-Ecuador BIT, 326

Criollo, Emergildo, 209, 211–13

crude oil. *See* oil

Dalton, John, 21

de la Cadena, Marisol, 13, 24, 327

DeLanda, Manuel, 18

Deleuze, Gilles, 12–13, 35–36, 44, 132–33, 182, 185

delocalized stabilities: chemistry, 215–17; Kaplan's ruling, 293–94; Zambrano's ruling, 225, 228–29, 254–57, 263

denunciation, 12, 328

diffuse rights: *Aguinda v. Chevron* ruling, 313–15; *Chevron v. Republic of Ecuador*, 311–18, 314–15, 320, 322, 325; couplet contract, 154–55, 311–12, 320; Ecuadorian law, 320–21; versus individual, 154–55, 310–11, 313–14, 320, 325; oil industry uses, 325

Doll, Richard, 120

Donziger, Steven: *Aguinda v. Chevron*, 265, 356n1, 358n8; Chevron's attacks on, 1, 42, 293; RICO countersuit testimony, 269–70; RICO court's hostility toward, 297–300; Texaco's dumping practices, 339n8

doubt, corporate production of, 37, 80, 110

Duhem, Pierre, 26–27, 29–30

Ebert, James, 285–86, 293

Eckardt, Robert, 111

Ecuador: Code of Civil Procedures, 153–54; Constitution, 155, 311–12, 315–17, 320; contracts with oil companies, 147–49, 347n15; "Environmental Regulations for Hydrocarbon Operations in Ecuador" decree, 81–82, 84–85; Health Code (1971), 231; Hydrocarbons Law (1971), 147–48, 230–31, 347n15; Petroleum Law (1937), 354n10; Supreme Decree 430, 148–49; water cleanliness laws, 232. *See also* Ecuadorian Civil Code; Texaco's oil concession

Ecuadorian Civil Code, 314; diffuse rights, 320–21; Environmental Management Law, 14–15; liability, 247–48; negligence and remediation, 14, 335n13; precautionary principle, 255; rights of individual citizens, 314, 320, 361n28; role of judges, 251–52; sources of obligation, 246; tort law, 14, 313–14, 319–21, 325; unlawful acts, 246

Ecuadorian judge bribery scheme, 168

Engelmann, Sasha, 183

environmental laws for concessions, 230–31, 354n10

Environmental Protection Agency (EPA), 159, 349n38

epidemiology: associative relationships, 115, 123; case-control studies, 101, 120; causal relationships, 114–15, 120–23; certainty, 96; chronic diseases, 120; cohort studies, 100–1, 120; disciplinary, 119–20; ecological studies, 98–101; ethics, 124–25, 127; industry uses, 125; observational versus universal, 117–19, 122, 124; and public health, 120, 122; risk-factor paradigm, 120–22; statistics, 121–25; uncertainty, 110. *See also* oil extraction's health impacts, epidemiological analyses

equivalent carbon (EC) fraction method, 77

Escobar, Santiago, 169–70

evidence. *See* inspections of former oil production sites; oil contamination testimonials

Exxon Valdez disaster, 78, 81

facts, 8–10, 25, 28, 33, 261

Fajardo, Pablo: *Aguinda v. Chevron* opening hearing, 46–51; career, 48; couplet contract, 319; diffuse versus individual

lead, 239

legal truths, 7–8; and facts, 25; finding versus making, 8; making of, 17–18, 33, 327–28; oil extraction's health impacts controversy, 94–95; versus scientific, 210; verdicts, 334n8

Leonard, Tomás, 313–14

liability: couplet contract release, 152–53, 156, 167, 304; favoring of victims, 248; theory of objective, 247–48; use of risky things, 246–49

Lienhardt, Godfrey, 31

Lynch, Gerard, 15–16

Lynch, Spencer, 285–86, 288–89, 293

Maria Aguinda Salazar y Otros v. Chevron-Texaco Corp. (Ecuador), 4, 325–26, 333n1; *Aguinda* lawsuit similarities, 322; case file size, 226, 229, 249, 354n6; central controversies, 16–17; complexity, 10–11, 16, 33–34; corruption optics, 5–6; couplet contract's relevancy, 310; court record errors, 280, 359n52; diffuse rights claims, 315; export report contradictions, 237; facts, 10; industry best practices, 233; inquisitorial nature, 17–18; judges presiding, 15, 335n14; legal procedures, 8–10, 14–15; opening hearing, 46–52; ruling, 5–6, 33–34, 218, 223; significance, 6–7; tort law, 320–21; transmogrifications, 326; truth-making, 10–11; unknowns, 34. *See also* couplet contract, Texpet-Republic of Ecuador; couplet contract and *Aguinda v. Chevron*; inspections of former oil production sites; Lago Agrio plaintiffs (LAP); *Maria Aguinda Salazar y Otros v. ChevronTexaco Corp.*, Chevron Corporation; *Maria Aguinda Salazar y Otros v. ChevronTexaco Corp.* ruling; oil extraction's health impacts controversy

Maria Aguinda Salazar y Otros v. ChevronTexaco Corp., Chevron Corporation: appeal, 223–24; complaint response, 51–52, 338n3; corruption narrative created, 39, 257, 264, 289, 305; countersuits against plaintiffs' experts, 228; criticisms of the court, 228; defenses, 229–30, 280; external counsel, 258; legal strategies, 34; public relations campaigns, 116; residents' poor health defense, 242. *See also* Callejas Ribadeneira, Adolfo; couplet contract and *Aguinda v. Chevron*

Maria Aguinda Salazar y Otros v. Chevron-Texaco Corp. ruling (Zambrano), 5, 218, 223, 253, 326; agreements and laws violated by Texaco/Texpet, 230–32, 236, 238, 240–41, 354n10, 354n12; appeal upholdings, 305; association versus causality, 243; authorship, 275, 280–83, 292–93; basis for amount awarded, 282–83, 290–92; basis of obligation, 246–50; benzene, 275–76; Cabrera report, 284–85, 308; carcinogens, 271–72, 275–76; *Chevron v. Republic of Ecuador*'s invalidation of, 317–18; citations of Supreme Court rulings, 246–50, 254–55, 278; clarification order, 250–51; complexity, 228, 282–83; delocalized stability, 225, 228–29, 254–57, 263; diffuse versus individual rights, 313–14; environmental presence and effects of chemicals, 237–41; epidemiological evidence of health impacts, 241–46; evidence of contamination's adverse effects, 252–53; expert reports considered, 237–38, 242, 250–52, 272–73, 275, 277, 285; fine assessed, 5, 223; ineligibility in US courts, 225; injury assessment, 237–46, 250, 252–53; inspection site samples considered, 238–40; laboratory analyses, 237; legal versus scientific causation, 256; legitimization of non-experts, 253; liability, 246–49; local resident testimonies, 201, 244–46, 250; method of interpretation, 224–26, 278; precautionary principle, 254–57; procedural truth, 245; redress for harm order, 257; Remediation Action Plan's invalidity, 288; research cited, 242–44; in RICO countersuit, 224–25, 256, 259, 271–82, 292–94; right to health, 241; risk, 247–49, 252–54, 278; role of science, 210; sample assessments, 276; scientific certainty, 251; significance, 256–57;

remediation of oil production sites: assessment methods, 159–61, 165, 349n38; bacteria and fungi use, 331–32; cost calculations, 277, 292; criminal fraud investigation, 161–66, 349n49; Ecuadorian Civil Code's applicability, 14, 335n13; inspection results, 157–59, 163, 208, 306. *See also* 1995 settlement agreement; 1998 final release document; Remediation Action Plan (RAP)

research used in *Aguinda v. Chevron*, Kelsh's (Chevron-submitted): cancer incidence versus mortality, 99–100; circumstances and contexts, 119; citations of faulty studies, 114; conclusions, 100, 109; doubt and uncertainty, 110; exposure classification, 101, 103–4, 108–9; mathematical methods versus lived experience, 108; mathematical rigor, 103, 108–9; problems with, 103–9, 126, 344n11; public criticisms of, 116; situatedness in prior work, 113; well density, 103, 106–8; well presence versus operations, 105–7

research used in *Aguinda v. Chevron*, San Sebastián and Hurtig's (LAP-submitted), 97–98, 343nn2–3; associational relationships, 93–94, 118; benzene, 113; cancer and crude oil exposure, 117, 242–43; cancer and hydrocarbons research, 112–13, 242; cancer and oil extraction connections, 98–101, 343–44n4; causal relationships, 114; circumstances and contexts, 118–19; responses to criticism, 118; scholarly criticisms, 125–26; situatedness in prior work, 112–13; strengths, 127; support from other scientists, 243–44; weaknesses, 118; Yana Curí study, 242–43, 355n26; in Zambrano's ruling, 242, 273, 275, 277

res judicata, 140, 143–44, 153–56

RICO countersuit. See *Chevron Corporation v. Steven Donziger et al.*

rights: to clean environment, 155, 311–12, 315–17, 320; to health, 241. *See also* diffuse rights; individual rights

Riles, Annelise, 142–43, 161, 261

risk theory, 247–48

Rosner, David, 110

Rothman, Kenneth, 117, 122, 126

Russell, David: *Aguinda v. Chevron* insignificance, 267; Ecuadorian cleanup cost estimate, 263–66; and LAP, 263, 265, 267; RICO countersuit testimony, 262, 264–69, 293–94, 307; site inspection knowledge, 267–68

Sacha-57 site inspection, 137–40, 153, *158*

Sacha Sur Production Station inspection, 57, 88–92, 198–99

Sands, Leonard B., 323, 364n65

San Sebastián, Miguel: cancer research, 98–101, 112–13, 117, 343n4; Chevron's attacks on, 94, 243, 343–44n4; epidemiology declaration, 93–94; unknowns regarding oil's health impacts, 125. *See also* research used in *Aguinda v. Chevron*, San Sebastián and Hurtig's

scientific knowledge production, 80, 86, 251

scientific truths: Bruno Latour, 126; versus legal, 210; limitations, 197; observational versus universal, 117; oil extraction's health impacts controversy, 101, 108

Sellers, Christopher, 83

sensorial experiences, 180–85, 190, 210

Siemiatycki, Jack, 125

Silver, Nicole, 314

Simon, Jonathan, 23–24, 26–27

Spinoza, Baruch, 13, 44

statute 1782 (US), 283–84

Stein, Zena, 119–20, 122

Stengers, Isabelle, 32, 182, 208

Strathern, Marilyn, 11, 16, 31, 35, 229, 334n9, 335n10

Stratus Consulting, 284

sufficient causation theory, 250, 273, 275, 277–78

Superfund. *See* Comprehensive Environmental Response, Compensation, and Liability Act

Susser, Mervyn, 119–20, 122

Texaco, Inc., 39, 208, 233–34. *See also* Texaco's oil concession; Texaco's oil production in Ecuador; Texpet

Texaco's oil concession: Bermejo, 107; Cascales, 106–8, 344n11; contracts (Texpet), 145, 147–50, 230–32, 347n18; Cuyabeno, 105–6, 108–9, 344n11; environmental audits, 151, 348n21; environmental impact, 150–51; exploration outside original boundaries, 145–47; Lago Agrio, 108–9, 145–46, 344n11; maps, *104, 146, 147, 148*; ownership divisions, 150–52; production rates, 344n10; Putumayo, 107–8, 344n11; residents, 101–2, 198–208, 210, 352n9; reversion to Petroecuador, 151; size decreases, 147, 149–50; size increases, 145; well density, 103–8; well production variance, 108; wells, 188–89. *See also* inspections of former oil production sites; Texaco's oil production in Ecuador; waste pits

Texaco's oil production in Ecuador: cost-cutting practices, 60; defenses of, 63, 340n14; drilling, 58–59, 188, 340n13; environmental impacts, 40, 58, 60, 187–88; exploration, 58, 187–88; formation waters, 54, 58–59, 61–62, 235–36; hazardous waste, 62–63; infrastructure built, 58; production stations, 59; spills and cleanup, 62, 119; well production tests, 59. *See also* oil spills; Texaco's oil concession

Texpet (Texaco Petroleum Company): in *Aguinda v. Chevron* complaint, 51, 91; oil sites, 139; remediation assessment methods, 161; remediation contract, 140, 144, 151–53, 156–61, 161, 165, 348nn21–22; spills and waste, 119. *See also* couplet contract; Texpet–Republic of Ecuador; Texpet's oil concession

Texpet-Gulf consortium, 145, 149–50, 166–67

Texpet's oil concession: 1968 contract, 230; laws violated, 230–31, 354n10, 354n12; operation, 140, 147, 150–51; ownership share, 150

theory of sufficient causation, 250, 273, 275, 277–78

Total Petroleum Hydrocarbon Criteria Working Group (TPHCWG), 69–70, 72, 75–77, 91, 341n19, 342n27

total petroleum hydrocarbons. *See* TPH

toxicity: additive effects, 342n27; arguments as compounds, 55–56; aromatic hydrocarbons, 72, 78, 268–69; carcinogens, 113–15, 239, 271–72, 275–76; claims of, 86; determination processes, 56–57; as mixt, 57, 85–86; oil spills, 78–79; relationality, 55, 85–86; scientific knowledge production, 86. *See also* hydrocarbon toxicity; oil's toxicity; TPH

toxicity characteristic leaching procedure (TCLP), 159–60, 165

toxicology, 83–84

TPH (total petroleum hydrocarbons), 64, 156–57; Chevron's measurement methods knowledge, 67; classification, 72; cleanup standards, 69–70, 75, 81; contamination assessment use, 81–82, 266–67; inspection site samples, 64–68, 74–75, 82, 239, 269, 341n21; methods for measuring, 64–67, 340–41nn16–17; remediation uses, 157–59; research by Chevron scientists, 67, 76–77, 341–42n23; safe levels, 77; toxicity criteria, 75; toxicity indicator use, 69–70, 80; uncertainties, 82–83; in Zambrano's ruling, 238–39, 273–74, 279. *See also* Total Petroleum Hydrocarbon Criteria Working Group

truths: ethnographic, 16; historical ontology, 57; making of, 9–10, 12, 14; procedural, 245; valence of, 326–27. *See also* facts; legal truths; scientific truths

uncertainty, 80; contamination, 250–51; controversies about, 256–57; epidemiology, 110; Kelsh's research on oil extraction's health impacts, 110; scientific knowledge production, 80, 251; toxicology, 83–84; TPH, 82–83; Zambrano's accounting for, 33–34, 228–29, 253